The Primates of Madagascar

THE
PRIMATES
of
MADAGASCAR

IAN TATTERSALL

New York

COLUMBIA UNIVERSITY PRESS

1982

Library of Congress Cataloging in Publication Data

Tattersall, Ian.
The primates of Madagascar.

Bibliography: p.
Includes index.
1. Lemurs. 2. Primates—Madagascar.
3. Mammals—Madagascar. I. Title.
QL737.P95T383   599.8'1   81-15477
ISBN 0-231-04704-5   AACR2

Columbia University Press
New York   Guildford, Surrey

Copyright © 1982 Columbia University Press
All rights reserved
Printed in the United States of America

Clothbound editions of Columbia University Press books are Smyth-sewn and
printed on permanent and durable acid-free paper.

Contents

Preface

"For those who do not know at first hand of the difficulties which still beset the scientific exploration of this great island, it is unbelievable that we know so little of its natural history." *François P. L. Pollen* (1863:277)[1]

Although those difficulties (climate, disease, topography, transport, and communication) about which Pollen complained had largely ceased by early in this century to be real obstacles to scientific exploration in Madagascar, one might nonetheless still have wondered, as much as a hundred years after he wrote, why the unique flora and fauna of this fascinating island, and particularly the lemurs, its original primate inhabitants, remained so little known.[2]

In the past two decades, however, our knowledge of the biology of the lemurs has expanded rapidly, partly in step with the development of field primatology as a scientific discipline, and partly as a result of a general resurgence of interest in the morphology, systematics, and behavior of the "lower" primates as a whole. This new knowledge is far from being evenly distributed among all the various branches of biology involved, or between all of the many primate species which still survive if only precariously in Madagascar; but a more rounded understanding of the diversity of the Malagasy primates, both phylogenetic and adaptive, is beginning to emerge and to dispel the hoary stereotype of the lemurs as a more or less homogeneous assemblage of vaguely "primitive" primates.

Several recent volumes in English have borne witness to this rekindling of interest in the lemurs (Martin et al. 1974; Tattersall and Sussman 1975; Doyle and Martin 1979). But while each of these works focuses on strepsirhine primates, each is in the form of a collection of specialized articles and none is really adequate as an introduction to the Malagasy primates. This is what I have attempted to

---

[1]"Il reste incompréhensible, pour ceux qui ne connaissent pas les obstacles entravent toujours les explorations scientifiques, qu'on sache si peu de l'histoire naturelle de cette grande terre."

[2]By the vernacular term "lemur," I mean simply "Malagasy strepsirhine."

provide here, and although this volume neither is nor is intended to be an encyclopedia of the lemurs, I hope that within its limited space it contains most of the information currently available on these animals to which primatologists and others are likely to require ready access. Where it fails the works cited above or the volume (in French) by Petter et al. (1977) may prove helpful. Although I cannot claim that this book is truly a synthesis, for our knowledge of the lemurs is still too sparse, I hope also, at a time when political circumstances have tragically diminished prospects of future fieldwork on these fascinating animals, that it will provide a useful perspective on what has already been achieved. Writing it, however, has made me acutely aware of the perceptiveness of Paul Valéry's remark that "un oeuvre n'est jamais achévé . . . il est abandonné."

In preparing this review I have enjoyed the assistance of many people at the American Museum of Natural History, the institution which has afforded me the time to write the manuscript and the opportunity to undertake most of the fieldwork on which my acquaintance with the lemurs is based. In particular, I would like to thank Arthur Singer and his colleagues for their photographic help; the library staff for much assistance; Nicholas Amorosi for preparing all of the linework except for figure 3.1, drafted by Marjorie Shepatin; and Nazarie Romain for typing much of the manuscript. Sydney Anderson, Department of Mammalogy, kindly allowed unlimited access to the collections of lemur material in his care, as did those in charge of many such collections in museums elsewhere, notably the AM (the late Paul Radaody-Ralarosy and Césaire Rabenoro); the BMNH (Prue Napier); The MCZ (Edi Rutzmoser); the MNHN, Paris (Francis Petter and Roger Saban), and Brunoy (Jean-Jacques-Petter); the NMNH (Dick Thorington); the NR (Bengt-Olov Stolt); the RMNH (A. M. Husson and Chris Smeenk), and the VNHM (Heinz Kollmann). Arthur and Gisela Tattersall, University College, London, provided extensive translations from Latin and German texts.

My fieldwork in Madagascar and the Comoro Islands, and visits to museum collections in various parts of the world, have been made possible by the financial generosity of several institutions. For such aid I am deeply indebted to the Arthur Williams Wilkins Fund of Yale University; the Wenner-Gren Foundation for Anthropological Research; the Frederick G. Voss Fund for Archaeological and Anthropological Research of the Department of Anthropology, AMNH, the Boise Fund of Oxford University, and above all, the National Geographic Society. I am also indebted to various officials of the

Service des Eaux et Forêts in Madagascar, and of the agricultural services in the Comores and Mayotte, for facilitating my fieldwork in those places.

Quite obviously, I could not have written this review without drawing upon the work and knowledge of many people, far too many to thank individually, to whom my debt is only very inadequately reflected by the bibliography at the book's end. I would, however, like to express my particular appreciation to the following colleagues and friends, not all of whom will agree with or approve of everything contained here, for contributing in many different ways: Lon Alterman; John and the late Vina Buettner-Janusch; Eric Delson; Niles Eldredge; Alison Jolly; Bob Martin; Todd Olson; Jon Pollock; Guy Ramanantsoa; Georges Randrianasolo; Alison Richard; Jeff Schwartz; Elwyn Simons; Mike Stuart; Bob Sussman; John Van Couvering, and Priscilla Ward. And finally, my affection and gratitude to Andrea Dunaif, without whom this book would have been finished much sooner.

ABBREVIATIONS

For the sake of brevity I have, above and in the text which follows, abbreviated the names of various institutions housing collections of lemur material. These are as follows:

AM: Académie Malgache, Tananarive.
AMNH: American Museum of Natural History, New York City.
BMNH: British Museum (Natural History), London.
MCZ: Museum of Comparative Zoology, Cambridge, Massachusetts.
MNHN: Muséum National d'Histoire Naturelle, Paris.
NMNH: National Museum of Natural History, Washington, D. C.
NR: Naturhistoriska Riksmuseet, Stockholm.
RMNH: Rijksmuseum van Natuurlijke Historie, Leiden.
VNHM: Naturhistorisches Museum, Vienna.

Note added in proof

Since much of this book was originally written, many Malagasy place-names have reverted to earlier forms, or their official orthography has been changed. Some of the more important changes are as follows:

Old	*New*
Diego-Suarez	Antseranana
Fénérive	Fenoarivo Atsinanana
Fort-Dauphin	Taolanaro
Ile Ste-Marie	Nosy Borah
Majunga	Mahajanga
Périnet	Andasibé
Tamatave	Toamasina
Tananarive	Antananarivo
Tuléar	Toliara
Vohémar	Vohimarina

The Primates of Madagascar

CHAPTER ONE

The Historical Background

Let the Sieur Etienne de Flacourt, as is no less than his due, have the first word on the lemurs of Madagascar:

There are [in Madagascar] several kinds of monkey, there are big ones, which are white, and have black patches on their ribs, and on their heads, they have a long muzzle like a fox, at Mangabey they call them *varicossy*: they are as fierce as tigers, they make so much noise in the trees, that if there are two of them it seems that there are a hundred. I had two of them brought on our ship; but they threw themselves in the water, they are very difficult to tame unless you have them from when they are young.

There is another species of gray monkey which is smaller and has a very blunt nose, which at Mangabey they call by a name other than vary, which is not difficult to tame, I had one which fell into the sea and drowned while we were passing the Isle Sainte-Marie.

There is another more common species which is called *vary*, which is easy to tame and gets up to much mischief, you must get them when they are young, otherwise they let themselves die of hunger, they are gray, with a long muzzle, and a big velvety tail, like the tail of a fox, which all the other species have as well.

There is yet another species of white monkey, with a tan cap, and which most often goes on its hind feet, it has a white tail and two tan patches on its flanks; it is bigger than the *vari*: but smaller than the *varicossy*, this species is called *sifac*, it lives on beans and there are many of them around Andrioure Damboulombe and Ranofotsy.

In Ampatra and Mahafaly country, there are many of another white monkey which is called *vary,* whose tail is striped in black and white, they go around in groups of 30, 40, and 50, they look like the *varicossi* of Mangabey.

There is another species of gray monkey which would be very beautiful, but is impossible to tame, whose eyes glow like fire, it has short hair, it is always as if enraged, it lets itself die of hunger.

There is a species of gray squirrel which they call *Tsitsihi,* which usually hides in holes trees and is neither beautiful nor good to tame. (Flacourt 1658; 1661 edition pp. 153–154; see appendix at end of this chapter for original.)

Flacourt, obviously a man of remarkable awareness and intellectual curiosity, spent the years from 1648 to 1655 at Fort-Dauphin, on the southeastern tip of Madagascar, as Directeur de la Compagnie Françoise de l'Orient & Commandant pour sa Majesté dans [Madagascar] & és Isles adiacentes. Despite the many difficulties (some, it must be admitted, self-inflicted) and hardships involved in maintaining and defending his tiny settlement unsupplied from France (as his lieutenant the Sieur Angeleaume wrote in a letter to France on February 28, 1654; "nous sommes réduits à aller nuds comme les Negres, iusqu'à Monsieur de Flacourt qui n'a pas vne chemise" [quoted in Flacourt, 1661:405]), Flacourt managed during his stay to compile a work on Madagascar which remained unequaled in its scope for over two centuries.

The passage reproduced provides fully recognizable portraits of *Varecia variegata* (the "varicossy"), of *Hapalemur griseus* (the "gray monkey"), of *Propithecus verreauxi* (the "sifac"), of *Lemur catta* (the "vary . . . whose tail is striped"), of *Microcebus murinus* (the "tsitsihi"), and of *Lemur fulvus* (the "vari . . . which gets up to much mischief"; fig. 1.1A). The identity of the "gray monkey which would be very beautiful . . . whose eyes glow like fire" is less certain, but it may well have been *Avahi laniger,* which occurs in the humid forests to the north of Fort-Dauphin, and which Flacourt might also have had the opportunity to see during his journey from Fort-Dauphin to the Ile Sainte-Marie in 1651.

Although Flacourt's is the best-known and most elaborate of the early descriptions of the Malagasy primates, it is by no means the first. It may seem on the face of it unlikely that during the century following the discovery of Madagascar by the Portuguese in 1500, no written record was made of lemurs by some Portuguese, Dutch, or other navigator, but perhaps it is significant in this connection that in 1609 the hardworking Hieronymus Megiser, compiler of the earliest general account of Madagascar, described nothing which might convincingly be viewed as a lemur. Megiser's *Beschriebung der*

A   B

Figure 1.1 The two earliest representations of lemurs. A: The "Vari ou Singe" of Flacourt, 1658, facing p. 151; B: Mundy's lively sketch of August 1655 (1936, p. 45). In the original the sketch also bears the words: "A Bugee, A pretti Animal."

Insul Madagascar is a valiant attempt to provide an unembroidered account of the island, but although Megiser chided Marco Polo for having mixed Malagasy zoological fact and fiction, he himself fell prey, albeit understandably, to the same failing. Megiser's quite elaborate account of the animals of Madagascar contains descriptions not only of several readily recognizable Malagasy vertebrates, but also, for instance, of elephants, giraffes, and oryx. His only possible description of a lemur translates roughly as follows: "In the southwest side of this island one finds cats which live off tamarinds and indeed spend their time in them. They have sharp snouts, short feet, and long, speckled tails" (Leipzig edition 1623:32). This animal might conceivably be *Lemur catta*, but its description as a "cat" with "short feet" perhaps makes it more likely the viverrid *Galidia*. In any event, the account is too vague to permit a reliable judgment of the animal's identity.

Shortly before Megiser wrote, however, the English East India Company had dispatched its third mercantile expedition to the Indies. Two ships, the *Dragon*, captained by William Keeling, and the *Hector*, under William Hawkins, sailed at the beginning of April 1607. A third, the *Consent*, sailed separately. Captains Keeling and Hawkins both kept journals of the voyage, and in 1625 these were published, apparently in rather dramatically edited form, by Samuel Purchas in his colossal, fascinating, and infuriating compendium of voyages, *Hakluytus Posthumus or Purchas his Pilgrimes*. In addition, the expedition's senior merchant, William Finch, who accompanied Hawkins on the *Dragon* to Surat (and who then, after many adventures, returned overland to England), kept a journal of his experiences which was likewise published in part by Purchas.

The expedition spent the last part of February 1608 at the Baie de St. Augustin, just to the south of the modern Tuléar, and Captain Keeling records that while exploring the mouth of the Onilahy River, "Here we found the beautifull beast" (in Purchas 1625, 1: 192). This part of Keeling's journal appears to have been very heavily edited by Purchas, and no more is said in print by the Captain about the local wildlife; but Purchas states in a side note that the "beautifull beast" is the animal described by Finch in his account as follows:

In the woods neere about the River, is a great store of beasts, as big as Munkies, ash-coloured, with a small head, long taile like a Fox, garled with white and blacke, the furre very fine (in Purchas 1625, 1: 417).

A clearer description for the period of *Lemur catta* would be hard to imagine.

Once the Indian route had become established, English and other ships called fairly routinely, if infrequently, at the Baie de St. Augustin, and there is evidence (see below) that a lemur had been transported alive to Surat by the early or mid-1630s. As far as I am aware, however, the next surviving record of lemurs was not composed until some decades after those of Finch and Keeling. In the journal of the Cornish traveler Peter Mundy, compiled between 1620 and 1667 but unpublished until edited by Sir Richard Temple in a series of volumes issued from 1907 to 1936, appears the following, in an entry for 1638:

Divers other creatures, etts., came to our sightt, as Bugeeas, like unto Monkyes in hands and Feete, butt sharpe snowted like a Fox, sofft, Downy haired and somewhatt bushy, long tayled, which when hee sitts he brings over all, a Dull creature . . . [and] . . . a little creature nott much bigger than a Mouse, off a fierce Nature, resembling a Kitteing. (1919:393–94).[1]

---

[1]Gmelin, editor of the 13th edition of *Systema Naturae*, listed "Simius Zambus" of Nieremberg (*Historia Naturae* 1635:176) as a synonym of *Lemur mongoz*. If this allocation were correct, then Nieremberg's would be the next known record of a lemur.

Mundy, like Flacourt a man of considerable powers of observation, and one "adicted to see [God's] wonders in strange Countries," traveled widely in Europe and Asia between 1608 and 1656, much of the time in the service of the East India Company. In June 1638, during a return voyage from China, his ship put into the Baie de St. Augustin, and stayed there for almost three months. As was his custom, Mundy diligently recorded all that he saw around him, even to the extent of compiling a Malagasy vocabulary. His "Bugeeas" (the term is probably derived from the Portuguese *bugia*, ape) must have been *Lemur catta*, and it is possible that the small animal "resembling a Kitteing" may have been *Microcebus*. The Bugeea was already familiar to Mundy, who records that he had seen one "broughtt from hence" at the East India Company's factory at Surat.

Seventeen years later, in August 1655, Mundy's ship called at Anjouan, one of the Comoro Islands, on the outward leg of his third journey to India. Although he had previously visited Anjouan, and described the island in his journal, he had up to that time made no mention of lemurs there; but on this occasion he recorded the following:

Here are munkeies and another animal called by us a bugee (which in Italian signifies a munkey). It hath a sharpe muzzell, a very long taile, a very sofft and thicke furre, the hinder part of his body much higher than the forepart. It had the foremost toe of the hinder feet unproportionable bigg like thumbs. It was exceeding nimble thatt itt would skip from rope to rope, topmast staies and uppermost lines of the ship, with such agilitie thatt it seemed rather to flie than leape. And soe famigliar to every one thatt hee would leap on their shoulders, take them fast about the next and licke their mouthes and faces (1936:45).

Mundy's reference to "munkeies" is obscure, but the bugee, which he sketched (fig. 1.1B) as well as described, is unquestionably *Lemur mongoz*, still quite abundant today in certain areas of Anjouan. The individual whose behavior Mundy described—and which he ob-

---

However, Nieremberg's work deals almost exclusively with South America (the only mention of Madagascar is in connection with the "roc" of Marco Polo), and is a curious combination of sober observation with medieval mythology. "Simius Zambus" is illustrated, but is not mentioned in the text, and the extremely stylized illustration is not convincingly that of a lemur. For instance, the extremities are shown as paws rather than as grasping organs, and Nieremberg's artist, Christoffol Jegher, acutely observed such details in his readily recognizable renditions of South American monkeys. It seems safe, then, to relegate "Simius Zambus" to the mythological aspect of the work.

viously found to be vastly more entertaining than the lemur he had seen at the Baie de St. Augustin—was taken aboard Mundy's vessel, the *Alleppo Merchant*, but the unfortunate animal "died on the way homeward."

The earliest French notice of lemurs of which I am aware is that of François Cauche, a sailor who resided in Madagascar from July 1638 to August 1642, and again from November 1643 to March 1644. Most of this time was spent in the south of the island, notably at Fanjahira (where he was when Flacourt's predecessor, the Sieur de Pronis, arrived in Madagascar); but Cauche traveled quite widely, and in addition to his engaging account of his experiences, told to C. B. Morisot and published in 1651 (and in English translation in 1711), he recorded his impressions of the customs of the Malagasy, and of the animals and plants of Madagascar. Although he dwelt most lovingly on the flora (his account of the banana tree alone is longer than that of all the lemurs), Cauche provided the following under the rubric of "monkeys":

> The Province of the *Madecasses* is pester'd with a vast number of Monkeys of several sorts. There are some brown of the colour of Bevers, their Hair Downy, the Tail long and broad, which they turn up on their backs to defend them from the Rain and Heat of the Sun, sleeping thus cover'd on the Boughs of the Trees like the Squirrels. In other Points they have a snout like the *Martin*, and round ears. This is the most innocent and harmless sort of them all. Among the Antavarres there are some of the same sort of Hair as these, with a sort of white Ruff about their Necks. There are some as white as Snow, of the same bigness with those above, long snouted, and grunting like Swine. They are no where but among the *Malegasses*, on the red Mountains, by the natives call'd *Amboimenes*. The natives believe the Monkeys can speak, but will not for fear of being made to work as men do. (Cauche 1651:127; trans. by John Stevens 1711:53).

Although these descriptions are far less recognizable and detailed than those of Flacourt, it is clear that Cauche was acquainted with several different kinds of lemur, including the ruffed lemur and the sifaka. Cauche's account was quoted on occasion by later workers (Buffon and Pennant, for instance), but despite having been translated early on, for obvious reasons it has been overshadowed by Flacourt's.

The unhappy fate which met Mundy's lemur must also have been that of any others which may have been shipped to Europe over the next half-century or so, for not until 1703 is anything more heard of these primates. In that year James Petiver, a London apothecary and naturalist, published an engraving in his *Gazophylacii Naturae et Artis* (Tab. xvii, fig. 5) of an animal "brought from Joanna [Anjouan]."[2] The figure (here fig. 1.2) undoubtedly portrays a member of genus *Lemur*; since it is uncolored the animal is not precisely identifiable, although it is very plausibly the mongoose lemur which, given its provenance, it must have been. Petiver named the animal "*Simia-Sciurus* lanuginosus fuscus," and equated it with John Ray's "*Cercopithecus Indicus* Bugée dictus" (*Synopsis Methodica Animalium Quadrupedum*, 1693:158) (see caption, fig. 2.2). But although the latter is perhaps the most poorly described of all the monkeys listed by Ray, it is nonetheless evident that it is not a lemur. The application to it of the epithet "Bugée" may have originated in confusion arising from the fact that Indiamen quite frequently called at Madagascar.

As far as I am able to ascertain, yet another forty years passed before any mention was made again of lemurs. In 1751 the zoological illustrator George Edwards, in an appendix to the fourth and last volume of his *Natural History of Birds*, published an illustration (no. 197) of "The Maucauco." Very clearly a ringtailed lemur, this individual had been brought to England from Madagascar by Captain Isaac Worth, one of the several sea captains from whom Edwards obtained many of the "nondescripts" he illustrated, and lived for some time in Edwards' house. Noting that it had "nothing of the cunning or malice of the Monkey-kind," Edwards took his lemur to be "of a genus distinct from the monkey," and indeed it was on Edwards' illustration that Linnaeus based the species *Lemur catta* in the tenth edition of his *Systema Naturae* (1758).

One should perhaps note that *catta* was only one of three species of *Lemur* named by Linnaeus in the tenth edition, the other two being *tardigradus* (the slow loris) and *volans* (the Philippine colugo). Linnaeus had introduced the genus name *Lemur* at least as early as 1754, exclusively for *tardigradus*; but common usage rapidly reserved *Lemur* for the Malagasy form, although it was not until 1911 that Thomas formally designated *catta* as the type species of *Lemur*. It is perhaps rather surprising that in 1758 Linnaeus made no reference to the genus *Prosimia* published by Brisson (*Regnum Animale*, 1st ed., p. 220) in 1756. In his first edition Brisson provided an illustration (fig. 5) of the head of a lemur, drawing particular attention

---

[2]The date of Petiver's *Gazophylacii* is usually given as 1764 (or erroneously as 1767). In fact, the separate sections ("Decades") of the *Gazophylacii*, each consisting of ten plates, originally appeared at intervals between 1702 and 1711; 1764 was the year in which these works were issued together (with some minor changes) under a single title page, some forty-seven years after Petiver's death in 1718.

Figure 1.2. Petiver's Tab. XVII, Fig. 5 of the Second Decade of his *Gazophylacii*. The complete legend to this illustration reads as follows (original 1703 version): *"Simia-Sciurus* lanuginosus fuscus, ex JOANNAE insula. *an* Cercopithecus *Indicus* Bugee *dictus* Raii *Synops. Animal.* 158. *The Wool of this is brown, soft and curled like that of a Lamb, it has several properties of a* Squirrel, *sitting often upright when it eats, which generally is done by his* Incisores, *and reflecting his bushy Tail. This strange Creature is now alive* (viz. *May* 1703) *at Mr.* Dottins *a Drugsters in* Newgatestreet, *who had him about* 12 *months since brought from* Joanna. *His Delineation was taken by that Celebrated Anatomist Mr.* William Cowper."

to the teeth both in the figure itself and in his text. Linnaeus' diagnosis of genus *Lemur*, based entirely on the teeth, is in fact very similar to Brisson's of *Prosimia*, although this is plainly coincidental since the diagnosis of *Lemur* in 1758 does not differ much from Linnaeus' description of "Lemur tardigradus" in his 1754 Catalogue of the Museum of King Adolph Frederick. In any event, by 1756 Brisson was already able to describe four species of *Prosimia*, the "Genre du Maki," all from Madagascar, all represented by specimens in the possession of M. de Réaumur, and among which there were at least three species which we would recognize today: *catta*, *mongoz*, and (probably) *fulvus*.

Similarly, even before Linnaeus' publication of *Lemur catta*, Edwards had already illustrated two more lemur species in his marvelous *Gleanings of Natural History*. The plates of which this work was composed, together with their accompanying texts, were issued individually between 1754 and 1758 and the whole was published as a unit in 1758. An early plate, already available at the beginning of 1754, was that of the "Mongooz" (his plate 216; here fig. 1.3), and in 1756 followed that of the "Black Maucauco" (plate 217). In describing the Mongooz, Edwards ventured that this animal was congeneric with his original Maucauco, but of a "less beautiful" species. Edwards had found the female Mongooz in 1752, alive, at "the house of the obliging Mrs. Kennon, midwife to her Royal Highness the Princess of Wales, who . . . informed me that it fed on fruits, herbs, and almost anything, even living fishes; and that it had a great mind to catch the birds in her cages" (p. 13). Edwards records that he subsequently saw a variety of different lemurs around London, but the only one he illustrated was the Black Maucauco, drawn in 1955 from a male individual owned by a Mr. Critington, Clerk to the Society of Surgeons. This animal "seemed to be a feeder on fruits and vegetables: while I had it, it eat cakes, bread and butter, and summer fruits, it being in July" (p. 14).

Hardly surprising, then, that in his twelfth edition of 1766 Linnaeus was able to propose two new species of *Lemur*. One of these, *mongoz*, he based uniquely on Edwards' plate of the Mongooz: the other, *macaco*, was founded partially on Edwards' Black Maucauco.[3] Whereas the identity of the Black Maucauco is unmistakable, Edwards' plate of the Mongooz (fig. 1.3) is less perfectly diagnostic, although in conjunction with his description the individual figured can be identified without too many qualms as a female mongoose lemur. Curiously enough, Linnaeus' description in this edition of *Lemur catta* is more elaborate than the one which had appeared originally in the tenth edition, and contains detail which he could not have acquired from Edwards' published work (nor does any discussion of lemurs appear in the correspondence between Edwards and Linnaeus preserved in the archives of the Linnaean Society). For instance, Linnaeus says of the eyes of this lemur: "Right, in daytime with a linear, vertical pupil; Left, in night-time my specimen

[3]Linnaeus actually based his species *L. macaco* on four references, only the last of which (erroneously given as the "Maucauco") was to the "Black Maucauco" of Edwards. Further, none of the other references ("lemur cauda floccosa, corpore fusco" of Gronovius, "Prosimia fusca" of Brisson, and "Simia-Sciurus lanuginosus fuscus" of Petiver) can be identified as a black lemur. After Linnaeus a succession of systematists moved a variety of lemurs into and out of *Lemur macaco* in an apparently random pattern; for instance, Gmelin, in the thirteenth edition of *Systema Naturae* (1788) removed Petiver's animal to *Lemur mongoz*, where it certainly belongs, but added further inappropriate taxa to *macaco*, including Flacourt's "Vericassy" (sic). However, no subsequent reviser seems to have taken the step of formally restricting the concept of the species *Lemur macaco* to that represented by the animal in Edwards' Plate 217, even though this is implied by modern usage. To end the ambiguity, I do so here.

Figure 1.3. "The Mongooz," Plate 216 of George Edwards' *Gleanings of Natural History*, 1754. The species *Lemur mongoz* is based on this animal.

had a large round pupil; was this natural or accidental?" (12th ed., p. 45). Both the content and the wording of this statement strongly suggest that by 1766 Linnaeus was able to describe the ringtailed lemur from personal experience.

It may seem remarkable, after a lapse of a hundred years during which only one lemur had been illustrated, that at least half a dozen individuals were seen in London by Edwards within the space of three years, and that four species could be described simultaneously in Paris by Brisson. But the mid-eighteenth century had seen the collection of exotic animals become a fashion amounting to a craze among the intelligentsia, not only in England but in continental Europe as well. This had considerable spin-off value for naturalists; thus, for example, in 1765 Buffon and Daubenton, in the thirteenth volume of Buffon's great *Histoire Naturelle*, were able to write at length of a variety of "makis," providing illustrations and quite elaborate descriptions, including some detail of internal anatomy (text, plus plates 23, 24, 25, 28, 29), of the "mococo" (plate 22: *Lemur catta*), the "mongous" (plate 26, *Lemur fulvus*), and the "vari" (plate 27: *Varecia variegata*, here fig. 1.4). In the same work, Buffon briefly mentioned the "petit mongous," pretty clearly a mouse lemur, but did not illustrate it. According to Étienne Geoffroy Saint-Hilaire (*Bull. Sci. Soc. Philomath.* Paris, 1795), this was the same individual Buffon illustrated and described eleven years later (*Histoire Naturelle*, Supplément, 3:1776) as the "Rat de Madagascar." If this is indeed so, a radical reassessment of its affinities by Buffon is implied, since in 1765 he had considered the "petit mongous" almost indistinguishable from the "mongous" except in point of size. In any event, the question of the affinities of the mouse lemur was soon settled; in 1776 Brown (*Illustrations of Zoology*, plate 44) illustrated and described another mouse lemur as a "maucauco,"[4] while in the following year Miller (*Various Subjects of Natural History*, 1777, plate 13) figured yet another, naming it *Lemur murinus*.

This was also a period when travelers began to furnish more substantial accounts of exotic animals observed in their native habitats. Outstanding among such travelers of this time was the Frenchman Sonnerat, who among numerous other titles enjoyed that of "Naturaliste Pensionnaire du Roi," and who visited the islands of the Indian Ocean several times between 1774 and 1781. Sonnerat devoted a chapter of the Fourth Book of his *Voyage aux Indes Orientales et à la Chine* (1782) to the peoples of Madagascar, and many of the "objects nouveaux relatifs à l'Histoire Naturelle" described and figured in the Fifth Book came from that island. The first of these was "le Aye-Aye" (plate 86), a pair of which survived two months in Sonnerat's possession. Sonnerat's description of this animal is remarka-

---

[4]Brown (1766) is usually quoted as calling this animal "the little Maucauco." However, Brown's explanation (p. 108) of his plate of *Microcebus* gives the following reference: "Genus Lemur, Lin. Syst. 44. Maucauco, Penn. Syn. Quad. 104." Linnaeus 1766 (*Systema Naturae*, 12th ed., p. 44) lists "Lemur tardigradus" (= the slender loris), *L. mongoz*, and *L. macaco*, but nothing that could equate with *Microcebus*; similarly, Thomas Pennant's *Synopsis of Quadrupeds* (1771), plate 104, illustrated "*L. tardigradus*," the "Tail-less Maucauco." It was not until 1781 that Pennant, in his *History of Quadrupeds* (p.217), applied the name "Little Maucauco" to the mouse lemur, referring to Brown 1776: 108, plate 44. Pennant also equated the animal with Buffon's "rat de Madagascar."

Figure 1.4. "Le Vari" of Buffon, Plate XXVII of Volume 13, 1765, of the *Histoire Naturelle:* the earliest illustration of a ruffed lemur.

bly accurate; and although he noted, as did many later authors, that it resembled a squirrel in various respects, he also observed that "il tient aussi du Maquis & du Singe" (p. 137). Somewhat oddly, in view of what we know of its distribution today, Sonnerat remarked that the aye-aye was only newly discovered "parce que nous fréquentons peu la côte de l'ouest, partie de cette île qu'il habite; les habitans de la cote de l'Est m'assurerent que [his captive animal, obtained on the west coast] c'étoit le premier qu'ils avoient vu" (p. 139).

All other lemurs Sonnerat referred to as "Maquis," and although he noted that many species were found in Madagascar, he described only two more: "l'Indri" (his plate 88; here fig. 1.5), and "le Maquis à bourres" (plate 89). The former subsequently derived its scientific name from Sonnerat; the latter is now known as *Avahi*. Despite certain minor inaccuracies of detail, his illustrations and descriptions provide ample evidence that Sonnerat was closely familiar at first hand with these animals; but his claim that the southern Malagasy captured young *Indri*, "les elèvent & les forment pour la chasse, comme nous dressons les chiens" (p. 142) seems somewhat fanciful, at the very least. Among the other lemurs to which Sonnerat obliquely referred, but did not describe, was *Hapalemur*. A specimen which he brought back to France was illustrated and described by Buffon (*Histoire Naturelle*, Supplément, 1789, 7: plate 34) as "le petit maki gris," and was subsequently used as the basis for *Lemur griseus* by Link and E. Geoffroy.

As empires expanded, and the thirst for knowledge of new and exotic creatures increased, there was from the late eighteenth century onwards a rapid proliferation of published references to lemurs. Most of these consisted of brief descriptions accompanied by new names, and are largely summarized by the synonymic lists provided in chapter 3. One might note, however, that this was a time when new zoological specimens were flooding into Europe from all over the world, at a rate matched only by the desire to bestow new names; and much of the literature of the period bears all the hallmarks of haste occasioned by both factors. It is certainly unfair to say that misquotation and the piracy of illustrations were raised in some quarters almost to the level of art forms; but one is nonetheless tempted to do so when trying to unravel the labyrinthine web of names produced between the late eighteenth and late nineteenth centuries.

Much of the new-found knowledge of the diversity of the animal world was due to the efforts of a new category of explorers—professional animal collectors, who provided a steady stream of specimens for the menageries, cabinets, and museums of Europe from early in the nineteenth century. Although many museums had agents in the field, sending them specimens as they came to hand, most professional collectors disposed of their specimens through dealers. One result of this is that many fine collections, such as that

Figure 1.5. "L'Indri" of Sonnerat, Plate 88 of *Voyage aux Indes orientales et à la Chine,* vol. 2, 1782. The first illustration of the babakoto.

made by the Englishman Crossley, became dispersed among institutions and private collections throughout Europe; another is that any documentation which may have accompanied specimens often failed to survive passage through the dealers' hands.

One outstanding exception to this is the collection made for the RMNH in 1864 and the years following (Schlegel 1866; Pollen and

Van Dam 1877). On behalf of that museum, François Pollen and J. C. Van Dam journeyed to the Indian Ocean in 1864, where they carried out extensive collecting in Mayotte and in the west and northwest of Madagascar. Pollen left in 1868, intending to return, but married instead; Van Dam remained to collect until 1870, when he returned to Leiden and was appointed chief technician at the Geology Museum. In keeping with the then prevalent fascination with "first" specimens, and the general lack of recognition of the value of large series of specimens from single localities, Schlegel, Director of the RMNH, did not retain all of the material sent him by Pollen and Van Dam. Substantial samples remain, however, and almost all specimens are accompanied by locality records, something less than common among older collections. Further, Pollen and Van Dam's entrancing account of their travels (1877) contains some of the best descriptions up to that time of the habitats and behavior of the lemurs.

These important collections from the west and the northwest of Madagascar were spendidly supplemented when Pollen engaged a young German, J. Audebert, to collect on the east coast of the island. Audebert's extensive collections, made between 1876 and 1879, are also accompanied by locality names, although some of his localities are hard to identify today. Unhappily, relations between Pollen and Audebert deteriorated rapidly ("you should be ashamed . . . if you think you can make a fool of me you greatly deceive yourself": Audebert in letter to Pollen September 1876), and Audebert eventually began to dispose of his material through the dealer Schneider (who sold some to Schlegel!). Nevertheless, the Pollen-Van Dam-Audebert collection in the RMNH is without doubt among the best-documented, most comprehensive, and best-preserved skin-and-skull series of lemurs in any single institution.

Unquestionably the greatest naturalist and explorer of Madagascar during the nineteenth century was Alfred Grandidier. In 1865, at the age of 29, having explored widely in South America and Asia, Grandidier found himself recovering from fever on the island of Réunion. Cured, he decided to make a brief visit to the nearby Madagascar before returning to France. Henceforth, Madagascar was to be his life's work, ultimately monumentalized in his *Histoire Physique, Naturelle et Politique de Madagascar*. Thirty-two volumes of this huge corpus appeared between 1875 and 1930, the last of them following his death in 1921, but the series is nevertheless incomplete. Of particular importance here is the fact that the section on lemurs, compiled in collaboration with his friend Alphonse

Milne-Edwards, was never finished. Several volumes of magnificent illustrations were issued, covering (in varying depth) *Lemur, Varecia, Hapalemur, Lepilemur,* and the indriids, but the accompanying text was published only for the indriid volume (1875). That no further volumes of text appeared is particularly unfortunate because, for reasons which we can now no more than guess, Grandidier did not always adhere to generally accepted nomenclature in identifying his plates. This led inevitably to an exacerbation of the confusion reigning in lemur systematics when others used these plates to establish the identity of lemur specimens.

Regrettably, this confusion was not ameliorated by any of the attempts at synthesis, such as that by Elliot (1913), made during the early part of this century. It was not until Schwarz (1931) produced his pioneering review of the Malagasy primates that the essentials of the gradually emerging modern taxonomy of the genera and species of these animals were fixed in the literature. This is not to say that Schwarz contrived to provide a definitive statement; but his work has provided the starting point for all subsequent taxonomic studies of the lemurs.

If by the early 1930s the lower-level taxonomy of the living lemurs was in relatively good order, and all the extinct genera known today had been discovered and described thanks to the labors of Lamberton and his predecessors, the behavior of the lemurs in their natural habitat was still largely a closed book. Scientists were still dependent for knowledge of the basic natural history of these animals upon the largely anecdotal reports of explorers, missionaries (among them Shaw, Baron, Sibree, and Cowan), collectors, and others. A valuable contribution was made in 1935 by A. L. Rand, a member of the Mission Zoologique Franco-Anglo-Américaine (Archbold Expedition) of 1929–31, but his published observations were all too brief. Hill (1953) records that a manuscript, *Notes on Mammal Collecting in Madagascar,* was compiled by the zoologist and collector C. S. Webb, who was stranded in Madagascar during the years of the Second World War. Webb, although "not in the habit of doing field work for museums" (Webb 1953:214) made a modest but superbly documented collection of lemurs (now in the BMNH) during his stay, and from the amount of information squeezed onto the labels accompanying his specimens, and from references made by Hill to his notes, it is clear that Webb's records contained a wealth of valuable information of many kinds. Unhappily, my efforts to locate the manuscript have so far been unavailing, and Webb's delightful memoirs (1953), which include reminiscences of his stay in Madagascar, are concerned chiefly with his search for *Macrotarsomys.*

Thus it was that the era of modern field studies of the lemurs began only in the late 1950s, when Petter (1962) undertook a survey in which he made brief records of the behavior and ecology of several species. This was shortly followed by the earliest intensive field study, performed by Alison Jolly (1966) on *Lemur catta* and *Propithecus verreauxi*. Since then, several more or less intensive field studies have been carried out on a variety of lemur species, although even now hardly anything is known about the behavior and ecology of others. Indeed, even such basic field information as precise distribution limits is unavailable for virtually all lemur species. The advances that have been made in our knowledge of the natural history of the lemurs during the past two decades are considerable; but it must be admitted that the contents of this book speak at least as clearly of what we do not know about these fascinating animals as of what we do.

APPENDIX

Since it has proved beyond my ability to convey in English the charm of Flacourt's prose, it may be worthwhile to reproduce below the original passage quoted at the beginning of this chapter.

From E. de Flacourt, *Histoire de la grande Isle Madagascar composée par le Sieur de Flacourt,* 1658, 1661 edition, pp. 153–154:

"Il y a de diuerses sortes de singes, il y en a de grands, qui sont blancs, & ont des taches noires sur les costez, & sur la teste, ils ont le museau long comme vn renard, ils les nomment à Manghabei *varicossy*: ceux-cy sont furieux comme des tigres, ils font tel bruit dans les bois, que s'il y en a deux il semble qu'il y en a vn cent. I'en ay eu deux que je fis porter dans notre barque; mais ils se ietterent dans la mer, ils sont tres-difficiles à appriuoiser; si on ne les a de ieunesse.

Il y a vn autre espece de singe gris plus petits qui a le museau fort camus, qu'ils nomment à Manghabei d'vn autre nom que vary, qui n'est pas difficile à appriuoiser, i'en ay eu vn qui tomba dans la mer & se noya en passant à l'Isle de Saincte Marie.

Il y a vne espece plus commune qu'ils nomment *vary*, qui est aisée à appriuoiser qui fait assez de singeries, il les faut auoir de ieunesse, autrement ils se laissent mourir de faim, ils sont gris, ont le museau long, & la queuë grande & velüe, ainsi que la queuë d'vn renard, ce qu'ont aussi toutes les autres especes.

Il y a encor vne autre espece de guenuche blanche, qui a vn chaperon tanné, & qui se tient le plus souuent sur les pieds de derriere, elle a la queuë blanche & deux taches tannées sur les flancs, elle est plus grande que le *vari*:

mais plus petite que le *varicossy*, cette espèce s'appelle *sifac* elle vit de feves & il y en a beaucoup vers Andriuore Damboulombe & Ranoufoutchi.

Dans les Ampatres & Maafalles, il y a encor des singes blancs en quantité qu'ils nomment *vary*, qui ont la queuë rayée de noir & blanc, ils marchent en troupe de 30. 40. & 50. ils ressemblent aux *varicossi* de Manghabei.

Il y a vne autre espece de guenuche grise qui seroit fort belle, mais impossible à appriuoiser, à qui les yeux reluisent comme feu, elle a le poil court, elle est toujours comme enragée, elle se laisse mourir de faim.

Il y a vne espèce d'escurieu gris qu'ils nomment *Tsitsihi*, qui se cache d'ordinaire dans des trous d'arbres creux qui ne sont pas beaux ny bons à appriuoiser."

CHAPTER TWO

Madagascar: The Environmental Background

Even though it no longer seems as evident as it once did that the evolutionary diversification of the lemurs took place entirely on Madagascar, it is nonetheless obvious that the lemurs can hardly be understood in isolation from the island on which they occur. In this chapter, I attempt to set the stage for discussion of these primates by providing brief outlines of the geological history, climate, and vegetation of Madagascar. Those who wish more detailed discussion of these topics will of course find it in the various references given, and consultation of the recent volume on the biogeography and ecology of Madagascar edited by Richard-Vindard and Battistini (1972) is particularly recommended. But first, some basic figures. A thousand miles (1,600 km) long and 360 miles (580 km) wide at its broadest point, with an area of 230,000 square miles (590,000 km²), Madagascar is the world's fourth largest island. Lying in the southern Indian Ocean between the latitudes of 11° 57′ S and 25° 32′ S (fig. 2.1), it falls almost entirely within the tropical zone. The long axis of the island is oriented NNE-SSW, i.e., roughly parallel to the trend of the eastern coast of Africa, from which it is separated by the Mozambique Channel. The width of the channel varies from 220 miles (350 km) to 750 miles (1,200 km).

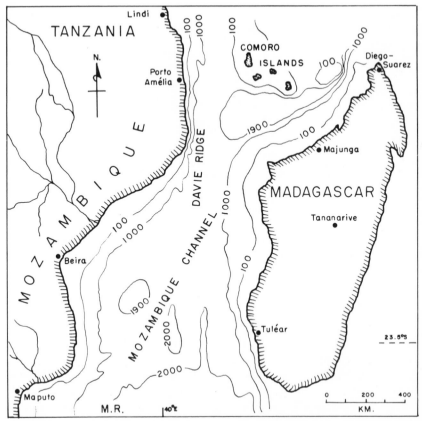

Figure 2.1. Location of Madagascar, with simplified bathymetry of the Mozambique Channel.

GEOLOGY AND TOPOGRAPHY

As a remnant of the ancient Gondwanan landmass, Madagascar is continental in origin. Its backbone consists of a heavily folded and intensely metamorphosed Precambrian basement complex which outcrops over about two-thirds of the present land area, but which is for the most part overlaid by a variable thickness of lateritic clays whose origin probably dates from the latter part of the Tertiary. The general topography of Madagascar is shown in figure 2.2; broadly, the central portion of the island consists of an elevated plateau, the mean altitude of which is in the order of 5,000 feet (1,500 m). Heavily dissected by erosion on the one hand, and built

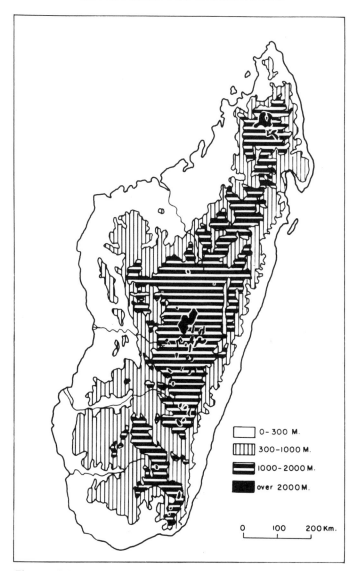

Figure 2.2. Topographic map of Madagascar.

up by volcanic activity on the other, this central massif has a rugged aspect. The greatest altitude at which the metamorphics are exposed is over 8,500 feet (2,600 m), while the highest peaks are volcanic in origin: Maromokotra, in the Massif du Tsaratanana, rises to almost 9,500 feet (2,900 m). The crystalline highlands fall off sharply to the

east, where a strongly (and in places fantastically) eroded escarp-
ment gives way to a narrow coastal strip; to the west, in contrast,
they yield more gradually to two major sedimentary basins: those of
Morondava and Majunga, which extend respectively to the south
and north of Cap St. André. A third area of extensive sedimentation
lies in the northern portion of the island, beyond the Ampasindava
Peninsula.

Although the depositional histories of the various sedimentary
regions differ considerably, these areas constitute a continuous sedimentary belt, ranging from 20 to 150 miles (30 to 240 km) in width,
along the entire west coast of the island. The oldest sediments in
Madagascar occur in the Morondava basin. These are of early Permian age, are continental in origin, and are clearly equivalent to the
earliest Karroo of southern Africa. The thick Malagasy Karroo sequence continues through the middle Jurassic, and provides evidence of repeated marine transgression starting in the middle Permian. At one end of the spectrum of possibilities, these transgressions
at the very least imply that Madagascar may have been intermittently isolated from Africa by invading seas from the middle Permian onwards (Brenon 1972); at the other, they may reflect a definitive separation of Madagascar from the East African coast by some
time in the early Jurassic (Kent 1972).

The upper Jurassic in Madagascar is entirely marine, as is the
lower Cretaceous except toward the north, where continental sands
begin to appear at the end of this time; the middle Cretaceous is
predominantly marine. The upper Cretaceous was marked by a
great deal of volcanic activity in both western and central areas of
the island, and is represented in the sedimentary areas by both marine and continental deposits. In the Majunga basin a substantial
thickness of continental sandstones from this time has yielded a varied reptilian fauna which disappears with the Maestrichtian
transgression. On the east coast, to the north of Mananjary, some
outcropping of Maestrichtian marine sediments close to the shore-
line provides the earliest evidence of transgression over the basement on that side of Madagascar (Besairie 1973; Brenon 1972).

The Tertiary of Madagascar is very poorly known. Eocene marine limestones occur in all three sedimentary regions, but most notably toward the south, and most of the known Oligocene and Miocene deposits are also of marine facies. Certain continental sediments
do, however, appear to be of later Tertiary age: these occur sporad-
ically in the Majunga and Morondava basins, but have so far proven
unfossiliferous.

The geological history of the Indian Ocean, bounded by four continental masses derived from the breakup of Gondwanaland and littered with continental debris resulting from the same event, is more complex than that of any other ocean basin. And the paleo-position of Madagascar is among the most hotly debated topics in the reconstruction of this history. Attempts to reassemble Gondwan-aland have involved placing Madagascar in one of four positions:

1. Against the East African coast from Somalia to Tanzania (e.g., Smith and Hallam 1970; Heirtzler and Burroughs 1971; McElhinny and Embleton 1976). Drift subsequent to rifting would have been to the south and east.

2. Adjacent to the coast of Mozambique (e.g., Flores 1970; Wright and McCurry 1970; Walker 1972). Drift would have been more or less directly eastward.

3. Against the Coast of Natal (Green 1972). Drift would have been to the east and north.

4. About where it is now, at least since the Paleozoic. Two variations exist on this theme: that the Mozambique Channel formed not by rifting but solely by subsidence, and that Africa and Madagascar are thus in exactly the same relative positions they always occupied (e.g., Dixey 1960; Flower and Strong 1969); and that the channel has widened somewhat since the original breakdown of the Gondwanan continental margin (e.g., Kent 1972), perhaps by a process of crustal thinning (Darracott 1974).

The choice between these alternatives, which represent the range of reasonable possibilities, is far from clear-cut. This is partly because the eastern African and western Malagasy seaboards lack large-scale, east-west trending geological structures which might be matched up (or shown not to), and because the history of the ocean floor surrounding Madagascar is still rather poorly understood. The northerly position has been favored in many best-fit reconstructions (e.g., Smith and Hallam 1970; Smith and Briden 1977) and paleo-magnetic studies (e.g., Embleton and McElhinny 1975; McElhinny and Embleton 1976), and receives support from the interpretation of the north-south trending Davie Ridge in the northern part of the Mozambique Channel as a transform fault indicating the direction of the island's movement (Heirtzler and Burroughs 1971). It suffers, however, from a number of objections, not the least among them being that seafloor sediments at least as old as the early late Creta-ceous exist to the west and north of Madagascar (Simpson, Schlich et al. 1974), and that thick sediments extend eastward from the Kenya coast which from their seismic velocity have been interpreted as an extension of the East African Karroo (Francis, Davis, and Hill 1966). Moreover, the tillites at the base of the Karroo sequence in the Morondava Basin agree in character with those of southern Africa while differing from those of East Africa.

The southern paleoposition of Madagascar, against Natal, was proposed by Green (1972) to explain certain magnetic anomaly variations at the south of the Mozambique Channel, and would necessitate spreading at the Mozambique Ridge. However, Scrutton (1973) has shown that this ridge lacks the morphology of a spreading axis and shows no symmetrical magnetic anomaly pattern over its crest, and other work has shown that the anomalies in the Mozambique Basin do not trend north-south (Darracott 1974). Further, a northerly and easterly movement of Madagascar away from the Natal coast is inconsistent with the termination of the Davie fracture zone close to the southern tip of the island.

Onshore geology does not provide a very firm base for deriving Madagascar from the Mozambique coast to the east of its present position (but see Flores 1970). Stronger evidence against such derivation comes, however, from the nature of the floor of the Mozambique Channel, where there is no convincing evidence whatever of spreading. Morphologically, the Davie fracture zone may plausibly be interpreted as a strike slip fault, but not as a spreading axis; and it completely lacks the magnetic patterns associated with seafloor spreading.

McElhinny and Embleton (1976) believe that their paleomagnetic studies of Malagasy Karroo sediments provide conclusive evidence that Madagascar occupied a position against Kenya. And indeed they have shown that, using the Smith-Hallam (1970) pole of rotation of Madagascar vis-à-vis Africa, a close agreement between the Malagasy and African-South American pole positions is achieved with the island in that position. Scrutton (1978) has, however, pointed to a variety of difficulties raised by this interpretation of the paleomagnetic evidence, and has proposed an alternative. This is based on the observation that the Davie Ridge forms part of a long curvilinear fracture zone extending from the Somali Basin and through the length of the Mozambique Channel. On the reasonable assumption that this structure does represent a strike slip fault indicating the direction of the island's motion, Scrutton used the pole of the fracture zone to rotate Madagascar back against Kenya, but found that this resulted in a poor polar agreement between Madagascar and Africa. On the other hand, using the Davie pole of rotation, a good agreement is achieved by assuming that Madagascar moved into its present position from a location only slightly to the north. Such limited southward travel has the advantage of diminishing the problems posed to a northerly derivation by the geology of the sea floor to the north and west of Madagascar. If we add to this Darracott's (1974) suggestion that Madagascar could have moved only a relatively short distance eastward by crustal thinning, it seems most likely that the island's movement relative to Africa has been rather limited. Certainly, the preponderance of the evidence now available seems to indicate that Madagascar has occupied a relatively fixed position at least since the Cretaceous; but equally certainly, the last word has yet to be said on the subject.

Almost as contentious as Madagascar's paleoposition has been the question of the timing of the rifting and drifting events, if any. At one end of the spectrum of opinion it is believed that Madagascar was isolated from Africa throughout the Mesozoic; at the other, that active drift was confined to the Cenozoic. The most plausible scenario seems to be that proposed by Kent (1972, 1974) on the basis of geological considerations, and which has recently been corroborated, at least as concerns the timing of the Gondwanan breakup, by the paleomagnetic studies of Ségoufin (1978) in the Mozambique Basin. According to this scheme, tensional stresses, leading to rifting, began to build along the Gondwanan continental margin during the Permian. By the early Jurassic, many of the major features of the eastern African coast were already developed, and by the middle Jurassic, with some subsidence along the coast, a permanent seaway had been established between Africa and Madagascar. This body of water fluctuated in width, with evidence of transgressions and regressions on both its margins, but by early in the Tertiary it had achieved oceanic depth, and was probably at or near its present width. It is reasonably certain that in the early stages of the Gondwanan breakup Madagascar remained attached to India (apparently along the latter's southwest coast: Katz and Premoli 1979). Luyendyk (1974) indicated on the basis of several Deep Sea Drilling Project cores that the separation of Madagascar from India occurred during the Paleocene, the Deccan Traps of northwest India being synchronous with this tectonic event; any movement of Madagascar would have ceased at or by that time.

In summary, it is evident that there is still a great deal of uncertainty about the history of the formation of the western Indian Ocean, although on balance of the geological evidence it appears that Madagascar has moved little if at all, at least since the beginning of the Cenozoic and probably much earlier. Eustatic sea level fluctuations of up to 500 meters or even more have been recorded in the sedimentary sequence of the continental shelves during the Tertiary (e.g., Vail 1977), and to some extent these must have affected the width of the Mozambique Channel and thus its effectiveness as a zoogeographical filter. It is evident from the modern bathymetry of the channel (fig. 2.1) that a minimum width of some 220 km exists today below the 100 meter contour; but subsidence has presumably been a long-term, continuing process in the channel, a process not entirely compensated for by the fact that in the earlier Tertiary the continental shelf would have been built out less than it is today. I discuss the possible biogeographical implications of this geological history in chapter 8.

CLIMATE

With its large size and varied topography, it is hardly surprising that climatologically Madagascar resembles a small continent rather than an oceanic island (Donque 1972). In general the Malagasy year is divisible into two seasons: the austral summer (November–April) and the austral winter (May–October), separated only by short transitional periods (Griffiths and Ranaivoson 1972); but seasonal differences and characteristics vary widely in different parts of the island. As might be expected, the mean annual temperature at sea level shows a decline from northern Madagascar southwards, e.g., from ca. 80.6° F (27° C) at Diego-Suarez to ca. 73.4° F (23° C) at Fort-Dauphin. Inland a similar trend also exists, but its effects are masked by those of altitude: temperature declines at a rate of about 2.7° F (1.5° C) per 1,000 feet (300 m). In general, the west coast is a few degrees warmer than the east at the same latitude, and the annual range in temperature increases from north to south (e.g., Diego-Suarez: 5.7° F, 3.15° C; Fort-Dauphin: 12.3° F, 6.85° C).

Diurnal temperature variation is greater in the interior and southwest than in the eastern, western and northwestern coastal regions, but in none of these regions changes very markedly with the season. Maximum temperatures are reached along most of the east coast in January or February, on the plateau in November, and in the southwest during January/March. Double maxima are experienced by the northeast in April and December, and by the northwest at about the same periods.

Rainfall, both in its quantity and in its distribution over the year, varies enormously from region to region. Annual total rainfall ranges from a maximum recorded at Tamatave of just under 200 inches (5,000 mm), to a minimum registered at Morombé of 4.7 inches (120 mm). Mean values range from 146 inches (3,700 mm) in the Baie d'Antongil to 13.4 inches (340 mm) in the southwest. Rain days vary from 30 to 250 per year. In the west and on the plateau, the austral winter may be regarded as the dry season: 90–95% of their annual precipitation falls between October and April. In contrast, the eastern slopes and the eastern coastal strip receive only about 30–50% of their annual rainfall between these months, the only approximation to a dry season being some reduction in rainfall during September and October.

Legris and Blasco ("Carton des Bioclimats" in Humbert and Cours Darne 1965) have divided Madagascar into a number of bioclimatic zones on the basis of temperature, rainfall and seasonality. These are shown in figure 2.3. The caption accompanying the figure outlines the essential climatic characteristics of each zone; briefly the plateau, normally defined as that area of the island lying above 2,300 feet (700 m), is generally temperate, with a fairly clear distinction between wet and dry seasons. The eastern coastal strip is hot and humid, with no dry season; the eastern slopes are also wet, but of varying temperature and seasonality depending on the altitude. The extreme south (and the Cap d'Ambre in the north) is hot and semiarid; rainfall is irregular, and drought not infrequent. A more distinct seasonality of rainfall occurs in the warm western lowlands, which become drier from north to south. The Sambirano area to the north of this region is hotter and more humid, its climate tending more closely to resemble that of the east coast.

VEGETATION

The vegetation which covers Madagascar today is in large part only an impoverished remnant of that which existed before the advent of man, probably less than two thousand years ago. One of the earliest monographs on the Malagasy flora (Humbert 1927) bears the subtitle: "La destruction d'une flore insulaire par le feu"; and the process lamented by Humbert over half a century ago has since proceeded, until today only a tiny fraction of the land area of the island is covered by undisturbed forest. Different forest types have suffered in different proportion: while certain floral communities have virtually disappeared, others are still quite widely represented. But all of the native Malagasy vegetal formations are threatened in

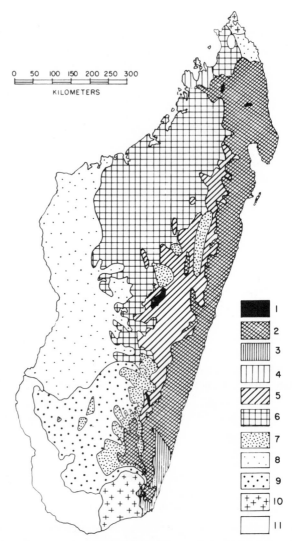

Figure 2.3. Bioclimatic zones of Madagascar, after Legris and Blasco (in Humbert and Cours Darne 1965). 1: High altitude zone: rainfall exceeds 2000 mm annually; no dry months; *t* (temperature of coldest month) is between 0° and 10° C. 2: rainfall above 2000 mm, no dry months, *t* almost invariably above 15° C. 3: rainfall 1500–2000 mm, no dry months, *t* always above 15° C. 4: rainfall above 2000 mm, dry season 3–4 months, *t* above 20° C. 5: rainfall 1500–2000 mm, dry season 1–4 months, *t* 10–15° C. 6: rainfall 1500–2000 mm, dry season 5–6 months, *t* above 20° C. 7: rainfall 1000–1500 mm, dry season 5–6 months, *t* 10–20° C. 8: rainfall 1000–1500 mm, dry season 7–8 months, *t* above 20° C. 9: rainfall 600–1000 mm, dry season 7–8 months, *t* above 15° C. 10: rainfall 400–600 mm, dry season 7–8 months, *t* 15–20° C. 11: rainfall under 400 mm, dry season 9–11 months, *t* 15–20° C.

the medium to long term: a double tragedy, for not only do the forests of Madagascar provide the habitat for a unique and fascinating fauna, but the flora itself is largely unique, with an endemism rate of over 80% at the specific level, and, depending on the region, of up to 50% at the generic (Perrier de la Bathie 1936; Humbert 1959). If trees alone are considered, endemism is even higher: 94% at the species level for all regions.

The climatic diversity of Madagascar is reflected in the richness and variety of the floral formations the island supports (or supported). In his review of 1921, Perrier de la Bathie divided Madagascar into two major floral zones, a moister Eastern and a drier Western, each containing various subzones. The schema was adopted and elaborated by Humbert (1927, 1955), who established the phytogeographic divisions shown in figure 2.4, where each major Region is divided into two or more Domains. Almost every Domain was in turn subdivided by Humbert into several Sectors, not shown in the figure. In general, vegetational differences between Domains are determined by climate, while variations within them are largely due to edaphic factors except in the very dry south, where soil type seems to have had relatively little effect on the vegetation.

The dry Western Region of Madagascar contains two Domains: the Southern and the Western. The semiarid Southern Domain is characterized by thickets or forests of strongly endemic, bushy, xerophytic vegetation, with Euphorbiaceae and Didiereaceae predominant (Perrier de la Bathie 1921; Humbert 1927). Deciduous forests similar to that of the Western Domain also occur locally in this region, usually but not invariably along watercourses.

Three types of forest are distinguished by Perrier de la Bathie (1936) in the Western Domain, their occurrence being governed largely by soil type. The deciduous forests which flourish on the moister siliceous soils are generally characteristic of river valleys and are dominated by *Tamarindus indica* (kily). In the climax condition, they exhibit a continuous canopy 25–50 feet (8–15 m) in height, with emergent trees reaching to perhaps 65 feet (20 m); the forest floor is generally quite open. The drier calcareous soils support a more xerophytic type of vegetation in which there is no predominant tree species and where no well-defined canopy exists: the stature of these forests is much lower than that of those which grow on the damper soils, and the undergrowth is dense. Where the soil is driest, xerophytic thickets are found which are generally less rich in species than those in the south. Forests of these different types may merge over short distances where local changes occur in soil

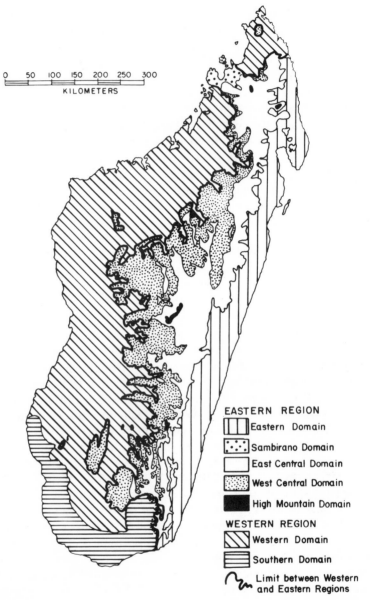

Figure 2.4. Vegetational zones of Madagascar, as established by Humbert (1955).

Figure 2.5. Views of Malagasy forest formations of three kinds. Top: medium-altitude dense rain forest near Andasibé (courtesy of Ted Schiffman); middle: transitional deciduous southwestern forest near Sakaraha; bottom: semi-arid *Didierea-Alluaudia* formation near Ambovombé.

and humidity; but, in parallel with the diminishing rainfall, there is a general trend toward the preponderance of the drier formations as one moves south.

The Eastern Region is far more humid, and supports a wider variety of forest types. The Eastern Domain is typified by dense rain forest (Koechlin 1972) which flourishes to an altitude of about 2,600 feet (800 m). This high evergreen forest, in which the trees reach heights of 100 feet (30 m) or more, is composed of vast numbers of different species. There are several distinct forest strata, a relatively open floor, and abundant epiphytes. The forest of the Sambirano domain is of similar structure, but of rather different floral composition. Between about 2,600 and 4,300 feet (800–1,300 m), the eastern portion of the Central Domain, more seasonal in climate, still supports in places a medium-altitude dense rain forest (Koechlin 1972). As tall as the Eastern rain forest, this type of formation is, however, less complexly stratified, presenting a single continuous layer at a height of about 85–95 feet (27–30 m); below this is a substantial herbaceous or shrubby undergrowth. At altitudes between 4,300 feet and about 6,600 feet (2,000 m), forest of this type gives way to a "lichen forest" composed of trees around 35–40 feet (10–12 m) in height, with small, hard leaves and twisting trunks; these support abundant epiphytes, which also cover the mossy ground (Koechlin 1972). The High Mountain Domain is restricted to those few massifs which rise above 6,600 feet; its vegetation is adapted to the more severe climatic conditions which prevail in such areas and is usually bushy or herbaceous in nature.

The western part of the Central Domain, lying below 2,600 feet, is warmer and drier than the eastern portion described earlier, but low minimum temperatures remain a limiting factor, and a sclerophylous forest with a low profile and a shrubby undergrowth is typical. In some valleys, however, vegetation of rather damper aspect occurs, characterized by the presence of some eastern species.

Outside Madagascar, wild-living lemur populations occur uniquely on three of the islands of the Comoro archipelago (fig. 2.1), which lies off the northwest coast of Madagascar. Despite their geographical situation, these islands, like the Sambirano Domain and for similar reasons, are regarded by Humbert (1955) as belonging floristically to the Eastern Region of Madagascar. There are, however, very considerable climatic and vegetational differences within each of the islands.

Sadly, these brief descriptions portray the vegetation of the various regions of Madagascar as it was, rather than as it is. Some of these climax forest types, particularly those of the central plateau, are represented today only by degraded residua or at best by isolated remnants. The central plateau is now largely treeless, apart from small patches of regrowth in steep valley bottoms and plantations of exotic trees such as *Eucalyptus*. Grasslands now characterize vast areas of the western lowlands, and in the east the humid forest has been widely replaced by *savoka*, a dense secondary formation composed mostly of heliophilous species. The evergreen forest of the Sambirano, where it has not been totally devastated for cultivation, has given way in many places to deciduous forests more typical of the drier areas to the north and south.

The question of whether Madagascar was originally entirely forested, or whether at least some of today's grasslands are of long standing, has been energetically debated (see Koechlin 1972, for a balanced discussion). In recent years, however, it has become increasingly evident that the now widespread herbaceous formations are of recent origin. For example, Battistini (1976) has analyzed the present distribution of certain butterfly species in Madagascar, particularly the genus *Charaxes*, and has shown convincingly that they could have come about only as the result of recent disruption of an effectively continuous forest cover. Certainly, the climatic changes of the Pleistocene appear at certain times to have reduced the forest cover of Madagascar (as witnessed, for instance, by evidence of ancient *lavaka*, the dramatic erosion scars which pockmark so much of the modern plateau, Petit and Bourgeat 1965), but apparently never to the extent found today; and at the time just prior to the advent of man on Madagascar, it is certain that the island was continuously covered by the climax formations already described for each phytogeographical province.

The most extensive area of primary vegetation which survives today in Madagascar is the strip of humid forest which extends from the Massif du Tsaratanana southeast to the Masoala Peninsula, then southward along the eastern slopes, becoming progressively more broken up as the south of the island is approached. A similar type of forest is also found further to the north, on the upper slopes of the Mt. d'Ambre, but the north of the island is characterized in general by a drier forest of which patches of varying extent remain. Stands of deciduous forest also occur in many areas of the Western Domain, often along river banks and in depressions. Similarly, both gallery forests and tracts of the relatively fire-resistant xerophytic vegetation may still be found scattered throughout the Southern Domain. Figure 2.6 gives an approximate idea of the areas of forest

Figure 2.6. Existing vegetal formations in Madagascar (modified after Humbert and Cours Darne 1965). 1: savanna and steppe formations, grassland; 2: dense rain forest; 3: savoka (heliophilous humid secondary forest formation); 4: montane forests; 5: deciduous and sclerophyllous forests; 6: xerophilous bush and thicket formations.

remaining in Madagascar. No reliable recent figures are available for the exact extent of the aboriginal forest which still survives; but as long ago as 1927, Humbert estimated that perhaps only 20–30,000 sq km of Madagascar were still thus forested, or a mere 3–5% of the land surface of the island. Other estimates have been slightly more optimistic, but with endemic forest disappearing at the rate of up to 200 sq km per year (Chauvet 1972), it is clear that the long-term prospects are poor indeed.

CHAPTER THREE

===

The Living Species of Malagasy Primates

In this chapter I provide a systematic account of all the genera, species, and subspecies of lemurs. I cannot claim, of course, that this review is in any way definitive; too many fundamental questions about the biology and distribution of the lemurs still remain unanswered. But where uncertainties exist, they are noted, and to that extent the account reflects the current state of our knowledge.

Perhaps the most glaring lacuna in our understanding of the primates of Madagascar lies in the fact that we know far less than we should like about the identity of the real biological units represented by the multitude of lemur populations which exist in the great island. It is at the level of the subspecies that this deficiency is most deeply felt; we are at the stage of only a rudimentary understanding of the problems involved in the identification of, variation within, and possible natural hybridization between many apparently distinct but conspecific lemur populations. The problem of identification is often particularly acute in the case of nocturnal forms, where between-population variation in pelage coloration, for instance, may be extremely subtle. It also seems realistic, if pessimistic, to conclude that some of the specimens without locality data contained in some of the older museum collections pose taxonomic problems that we shall never be able to solve, given the accelerating fragmentation and destruction of the Malagasy fauna. The species and subspecies treated below are those which appear to me now to be real units in the sense that they are discrete, although future investigation may show that some groups which now appear discrete in fact may not be so, or vice versa. Taxonomic notes are appended in cases where some elaboration as to status or nomenclature seems to be appropriate.

The synonymies that follow I believe to be complete, at least as far as new generic and specific names are concerned. I have not attempted to list new combinations. Except in those very few cases so indicated, all the original references have been consulted and checked. This was found to be necessary because of the numerous inaccuracies discovered in earlier synonymies. Thus, for instance, some names traditionally considered to be Linnaean synonyms do not appear below. Where type specimens could be located they have been examined, and in many cases synonymies could be verified on the basis of type descriptions and/or illustrations. Finally, however, I am obliged to admit that a few instances remain where identification is based upon little more than tradition. Some early names were based on several specimens or references which would not today be regarded as belonging to the same species, or even genus; discussion of these is provided where necessary.

Malagasy names for the various lemurs are given in Malagasy orthography (in which, for instance, *oo* is rendered as *o*); English orthography is used for Comorian terms. English names are given, even though in many cases they are merely notional.

Descriptions of pelage coloration and pattern (which are rather variable in most cases) are kept as brief and simple as possible, especially since no satisfactory practical terminology for coat color or pattern exists: one man's "buffy ochraceous" may be another's "medium brown." Sexual dichromatism is characteristic of only a limited number of species and subspecies, and does not exist except in those cases where it is specifically described. The description of external characteristics is kept to a minimum in order to avoid repetition of material in the next chapter. No key is provided (although one may be found in Petter et al. 1977), since the illustrations, together with the pelage descriptions, should be adequate for the purposes of identification.

A single dimension, mean maximum cranial length (most anterior point on premaxilla to opisthocranion), is given in millimeters for all subspecies and monotypic species discussed. Although body proportions obviously vary considerably between the various genera, this dimension is as reliable an index as any of the relative sizes of individuals belonging to the taxa involved, and it has the advantage of being an unequivocal measure usually available for substantial samples of individuals; thus, in arriving at the means ($\bar{x}$), I have been able to use only adult specimens the identity of which could be checked against skins. Where sample sizes (n) are adequate, I have also provided the standard deviation of the mean (s), while the overall range (o.r.) within the sample is provided in all cases. The means are calculated upon combined samples of individuals of both sexes, since there appears to be no significant difference in body size, certainly on the basis of this measure, between the males and females of any lemur species. Student's t tests were performed to test the significance of any difference between male and female maximum cranial lengths in ten subspecies or monotypic species (including representatives of all extant subfamilies except that of *Daubentonia*) in which adequately large and well-documented samples were available, and in no case did the value of t attain significance at the .05 level of probability.

Although the standard of documentation accompanying the various museum collections of lemur material is in general fairly dreadful, reliable field measurements are recorded for a substantial number of preserved specimens, notably those collected by the Mission Zoologique Franco-Anglo-Américaine (Archbold Expedition) of 1929–31 (and now housed in the AMNH, BMNH, AND MNHN). These measurements consist of head and body length (H + B, taken from the tip of the muzzle to the base of the tail), tail length (TL, taken from the base of tail to its bony tip), hind foot length (HF, taken from the back of the heel to the end of the longest claw, or occasionally to the fleshy tip of the longest digit), and ear height (EAR, usually taken from the base of the tragal notch to the furthest point on the external ear). Where available, means and ranges of these measurements are quoted in millimeters, with their sources. Because of possible differences in technique, sets of measurements for the same taxon made by different collectors are quoted separately.

W. L. Abbott was apparently the only major collector who ever troubled to weigh his specimens, so except in the few cases otherwise noted, all weights have been provided by Michael D. Stuart, of the Duke University Primate Center, where individuals belonging to the captive colony are regularly weighed. Thus several body-weight figures were available for each individual lemur, and in arriving at the figures given, I have averaged an annual series of observations for each one. Where sample sizes are adequate, I quote a mean and range for the sample based on these individual means. When consulting these figures it should be borne in mind both that they are based on captive lemurs, and that the weight of individual animals may vary substantially over short periods of time.

Although the general geographical ranges of the various lemur taxa are reasonably well understood, at least in the sense that one knows where to go to find them, our detailed knowledge of the distribution of these animals is hardly better than rudimentary. Much published information is downright misleading, and it is likely that isolated populations of a variety of taxa remain to be discovered. Perhaps most appalling is that after a century and more of exploration, during which time the habitat available to the lemurs has been steadily shrinking, precise boundaries, both geographical and ecological, remain unknown for almost all lemur populations. Intensive surveys of distributions are hence among the most urgently required of all lemur field studies. The problem of delineating ranges is exacerbated by the fact that many collectors did not trouble to record the precise localities where specimens were obtained; "Madagascar" often sufficed as an indication of locality even when the same collector, working on the neighboring continent, would never have regarded "Africa" as adequate for the purpose. In other cases, locality names provided by early collectors are no longer indentifiable; this is so, for example, for many of the localities at which Audebert obtained specimens. In yet others, several places of the same name exist today, and available evidence is inadequate to identify to which, if any, the locality record refers.

It is against this background of substantially less than total knowledge that the following ranges should be considered. In compiling the distribution maps I have tried to indicate current ranges, based upon my own surveys and on information supplied by Georges Randrianasolo, Guy Ramanantsoa and others, and on recently published sources such as Appert (1968), Andriamampianina and Peyrieras (1972), Petter et al. (1977), Sussman (1977), and Tattersall (1976b, 1977b). These are supplemented in the maps by those definitely identifiable localities at which museum specimens (limited to skins or stuffed specimens in that majority of cases where definite attribution to species-group taxon cannot be made on the basis of the skull alone) have been collected over the past century and a half. It should be kept in mind when these maps are consulted that the shaded areas represent preliminary denotations of broadly drawn range limits; they do not imply that the distributions of the species/subspecies involved are necessarily continuous, even within the forested parts of the areas shown, or that population densities are uniform. In any event, the maps should not be consulted in isolation from the text, or from figure 2.6, which shows approximately the areas of forest remaining in Madagascar today. Figure 3.1 shows the major towns and certain other features of Madagascar as an aid to placing localities; the small black circles on the distribution maps correspond to these towns.

The karyotype of each subspecies is given together with the work in which it was first correctly published. 2N indicates the diploid number of chromosomes; M, metacentric; S, submetacentric; and A, acrocentric.

To provide an indication of the order in which the various taxa are discussed, I provide here a classification of the Malagasy lemurs at and below the family level. The problems of classification are addressed in chapter 6; suffice it here to note that the classification embraces, in addition to the living taxa discussed in this chapter, the recently extinct forms considered in chapter 5. Such extinct taxa are denoted by daggers (†). As noted, the order in which the extant taxa are listed is also the order in which they are profiled.

ORDER PRIMATES Linnaeus, 1758
SUBORDER STREPSIRHINI Pocock, 1918

Family Lemuridae Gray, 1821.
  Subfamily Lemurinae Gray, 1821.
    Lemur Linnaeus, 1758. "True" lemurs.
      L. catta Linnaeus, 1758. Ringtailed lemur.
      L. mongoz Linnaeus, 1766. Mongoose lemur.
      L. macaco Linnaeus, 1766. Black lemur.
      L. fulvus E. Geoffroy, 1796.
        L. f. fulvus E. Geoffroy, 1796. Brown lemur.
        L. f. albifrons E. Geoffroy, 1796. White-fronted lemur.
        L. f. rufus Audebert, 1800. Rufous or red-fronted lemur.
        L. f. collaris E. Geoffroy, 1812. Collared lemur.
        L. f. mayottensis Schlegel, 1886. Mayotte lemur.
        L. f. sanfordi Archbold, 1932. Sanford's lemur.
        L. f. albocollaris Rumpler, 1975. White-collared lemur.
      L. coronatus Gray, 1842. Crowned lemur.
      L. rubriventer I. Geoffroy, 1850. Red-bellied lemur.
    Varecia Gray, 1863.
      V. variegata (Kerr, 1792). Ruffed lemurs.
        V. v. variegata (Kerr, 1792). Black-and-white ruffed lemur.
        V. v. rubra (E. Geoffroy, 1812). Red-ruffed lemur.
      †*V. insignis* (Filhol, 1895).
      †*V. jullyi* (Standing, 1908).

Figure 3.1. Map of Madagascar and the Comoros, showing main towns and geographical features. The heavy line delineates the boundary between the Eastern and Western phytogeographic Regions of Humbert (1955). A.K.: Ankarafantsika; B.A.: Baie d'Antongil; M.A.: Montagne d'Ambre; M.I.: Massif d'Isalo; M.T.: Massif du Tsaratanana; P.M.: Masoala Peninsula.

Family Lepilemuridae Rumpler and Rakotosamimanana, 1972.
  Subfamily Lepilemurinae Rumpler and Rakotosamimanana, 1972.
    Lepilemur I. Geoffroy, 1851. Weasel and sportive lemurs.
    L. mustelinus I. Geoffroy, 1851.
      L. m. mustelinus I. Geoffroy, 1851. Weasel lemur.
      L. m. ruficaudatus A. Grandidier, 1867. Red-tailed sportive lemur.
      L. m. dorsalis Gray, 1870. Gray-backed sportive lemur.
      L. m. leucopus Forsyth Major, 1894. White-footed sportive lemur.
      L. m. edwardsi Forsyth Major, 1894. Milne-Edwards' sportive lemur.
      L. m. septentrionalis Rumpler and Albignac, 1975. Northern sportive lemur.
  Subfamily Hapalemurinae Remane, 1960.
    Hapalemur I. Geoffroy, 1851. Gentle lemurs.
    H. griseus (Link, 1795).
      H. g. griseus (Link, 1795). Gray gentle lemur.
      H. g. alaotrensis Rumpler, 1975. Alaotran gentle lemur.
      H. g. occidentalis Rumpler, 1975. Western gentle lemur.
    H. simus Gray, 1870. Broad-nosed gentle lemur.
    †*H. gallieni* Standing, 1905.
  Subfamily †Megaladapinae Forsyth Major, 1894.
    †*Megaladapis* Forsyth Major, 1894.
    †*M. madagascariensis* Forsyth Major, 1894.
    †*M. edwardsi* (G. Grandidier, 1899).
    †*M. grandidieri* Standing, 1903.
Family Indriidae Burnett, 1828.
  Subfamily Indriinae Burnett, 1828.
    Indri E. Geoffroy, 1796.
    I. indri (Gmelin, 1788). Indris.
    Avahi Jourdan, 1834.
    A. laniger (Gmelin, 1788). Woolly lemurs.
      A. l. laniger (Gmelin, 1788). Eastern woolly lemur.
      A. l. occidentalis (Lorenz, 1898). Western woolly lemur.
    Propithecus Bennett, 1832. Sifakas.
    P. diadema Bennett, 1832.
      P. d. diadema Bennett, 1832. Diademed sifaka.
      P. d. candidus A. Grandidier, 1871. Silky sifaka.
      P. d. edwardsi A. Grandidier, 1871. Milne-Edwards' sifaka.
      P. d. holomelas Gunther, 1875. Black sifaka.
      P. d. perrieri Lavauden, 1931. Perrier's sifaka.
    P. verreauxi A. Grandidier, 1867.
      P. v. verreauxi A. Grandidier, 1867. Verreaux's sifaka.
      P. v. coquereli Milne-Edwards, 1867. Coquerel's sifaka.
      P. v. deckeni Peters, 1870. Decken's sifaka.
      P. v. coronatus Milne-Edwards, 1871. Crowned sifaka.
    †*Mesopropithecus* Standing, 1905.
    †*M. pithecoides* Standing, 1905.

†*M. globiceps* (Lamberton, 1936).
Subfamily †Archaeolemurinae Standing, 1908.
  †*Archaeolemur* Filhol, 1895.
    †*A. majori* Filhol, 1895.
    †*A. edwardsi* (Filhol, 1895).
  †*Hadropithecus* Lorenz, 1899.
    †*H. stenognathus* Lorenz, 1899.
Subfamily †Palaeopropithecinae Tattersall, 1973.
  †*Palaeopropithecus* G. Grandidier, 1899.
    †*P. ingens* G. Grandidier, 1899.
  †*Archaeoindris* Standing, 1908.
    †*A. fontoynonti* Standing, 1908.
Family Daubentoniidae Gray, 1870.
  Daubentonia E. Geoffroy, 1795.
    D. madagascariensis (Gmelin, 1788). Aye-aye.
    †*D. robusta* Lamberton, 1934.
Family Cheirogaleidae Gregory, 1915.
  Subfamily Cheirogaleinae Gregory, 1915.
    Cheirogaleus E. Geoffroy, 1812. Dwarf lemurs.
      C. major E. Geoffroy, 1812. Greater dwarf lemur.
      C. medius E. Geoffroy, 1812. Fat-tailed dwarf lemur.
    Microcebus E. Geoffroy, 1828. Mouse lemurs.
      M. murinus (J. F. Miller, 1777). Gray mouse lemur.
      M. rufus (Lesson, 1840). Brown or rufous mouse lemur.
    Mirza Gray, 1870.
      M. coquereli (A. Grandidier, 1867). Coquerel's dwarf lemur.
    Allocebus Petter-Rousseaux and Petter, 1967.
      A. trichotis (Günther, 1875). Hairy-eared dwarf lemur.
    Phaner Gray, 1870.
      P. furcifer (Blainville, 1839). Fork-marked lemur.

FAMILY LEMURIDAE GRAY, 1821

Lemur Linnaeus, 1758. "True" lemurs

As proposed by Linnaeus in the tenth edition of his *Systema Na-turae* (1758), the genus *Lemur* contained three species: *tardigradus* (the slender loris), *catta* (the ringtailed lemur), and *volans* (the Philippine colugo). Linnaeus' use of the genus name, however, goes back at least to 1754, when in his catalog of the Museum of King Adolph Frederick he published an account of "Lemur tardigradus." Nonetheless, despite the fact that Linnaeus' concept of the genus *Lemur* thus appears to have been based originally on a loris, the name has been used traditionally for the Malagasy form, while the genus name *Loris* E. Geoffroy, 1796 rapidly became adopted for the slender loris. It was only in 1911, however, that Oldfield Thomas formally designated *catta* as the type species of *Lemur*, an action later enshrined in Opinion 122 of the International Commission (January 10, 1929). The following synonymy omits the often quoted term Mococo Trouessart, 1878 (*Rev. Mag. Zool. Sér.* 3, 6, p. 163), since the name is in fact given as a vernacular term for genus *Prosimia*. Similarly, Brisson's (*Regn. Animal.*, 2 ed., 1762) species names are omitted since they do not qualify as Linnaean binomina (see discussion of *L. mongoz*).

All members of genus *Lemur* possess mental, mystacial, superciliary and genal vibrissae, but the interramal tuft is absent. Carpal vibrissae are always present, if not pronounced. In all species the face is covered by short, flat-lying hairs except at the very tip of the muzzle. Irrespective of species, the pelage is generally denser and longer in individuals from higher altitudes.

1758. *Lemur* Linnaeus, Syst. Nat., 10 ed., 1, p. 29. Type species by subsequent designation of Thomas: *L. catta* Linnaeus.

1762. *Prosimia* Brisson, Regn. Anim. in Classes IX Distrib., 2 ed., pp. 13, 155. Based on individual in Cabinet of M. de Réamur, shown in Fig. 5 of 1 ed. 1756. Species of this uncertain; type species by subsequent designation of Elliot (Rev. Primates, 1, 1913) *Lemur catta* Linnaeus.

1780. *Procebus* Storr, Prodromus Method., p. 32. Type by original designation, *Lemur catta* Linnaeus.

1806. *Catta* Link, Beschreib. Nat.-Samml. Univ. Rostock, 1, p. 7. Said to be based on *Lemur catta* Linnaeus; not seen.

1819. *Maki* Muirhead, Brewster's Edinburgh Encyclop., 13, p. 405. Type species *Maki mococo* = *Lemur catta* Linnaeus.

1895. *Eulemur* Haeckel, Syst. Phyl. Wirbelth., 3, p. 600. No species designated. Not seen.

1960. *Odorlemur* Bolwig, Mém. Inst. Sci. Madagascar, Sér. A, 14, p. 205. Type species *Odorlemur (Lemur) catta* = *Lemur catta* Linnaeus.

Lemur catta Linnaeus, 1758. Ringtailed lemur.

1778. *Lemur catta* Linnaeus, Syst. Nat., 10 ed., 1, p. 30. Based on the "Maucauco" of Edwards, Nat. Hist. Birds, 4, Pl. 197 (wrongly given by Linnaeus as Pl. 199). Madagascar.

1819. *M[aki] mococo* Muirhead, Brewster's Edinburgh Encyclop., 13,

p. 405. Muirhead gives authorship of this name to Desmarest (Nouv. Dict. Hist. Nat., 18, 1817), but the latter in fact used it as a vernacular term. Based on Desmarest and Linnaeus.

Malagasy names. Maki, hira.

Pelage and external characters. Fur dense. The back is usually a warm rosy brown, the rump and limbs a light gray or gray-brown, and the crown and neck a darker gray. The underparts are white or cream, lightly haired, with dark skin showing beneath. Tail ringed black and white, with black tip; almost invariably 13–14 black and 12–13 white rings. A brownish pygal patch is sometimes present. Forehead, cheeks, ears and throat white; there is a dark gray or black orbital ring, but the interocular area is white, the muzzle pale (fig. 3.2). Ears prominent for the genus: hairy, but only slightly tufted if at all. In males there is a large brachial cutaneous gland on the medial aspect of the upper arm near the shoulder; in females it is less well developed if present. Both sexes possess a naked elliptical area ("carpal" or "antebrachial" gland) on the palmar surface of the wrist, but only in males is this overlaid by a horny "spur." Glandular development is also present in the perianal region which is, how-

Figure 3.2. *Lemur catta* male.

ever, furred. Scrotum naked, black; most females possess two pairs of mammae, but only one pair becomes functional.

Dimensions. Cranial length, $\bar{x} = 83.9$, $s = 2.3$, $o.r. = 78.0–88.4$, $n = 27$. Field measurements (Archbold, $n = 9$): H + B, $\bar{x} = 423$, $o.r. = 385–455$; TL, $\bar{x} = 595$, $o.r. = 560–624$; HF, $\bar{x} = 108$, $o.r. = 102–113$; EAR, $\bar{x} = 44$, $o.r. = 40–48$. **Body weight** ($n = 11$): $\bar{x} = 2,760$, $o.r. = 2,295–3,488$.

Range. Restricted to southern and southwestern Madagascar, but ranges into the interior highlands farther than any other lemur (fig. 3.3). Range bounded approximately by a line connecting Belo-sur-Mer with Fianarantsoa (although the species does not appear to occur in the area immediately around Manja), and the latter with Fort-Dauphin. Within this area, however, population distribution is far from continuous.

Karyotype. $2N = 56$. Autosomes: M, 4; S, 4; A, 46. Sex chromosomes: X, M; Y, A (Hayata et al. 1971).

Lemur mongoz Linnaeus, 1766. Mongoose lemur.

The earliest name potentially available for the mongoose lemur is *Prosimia* (= *Lemur*) *fusca* Brisson, 1762 (Regn. Anim., 2 ed., p. 156) based on the animal from Anjouan illustrated by Petiver in 1703 (Gazophylacii Naturae et Artis, decade 2, Tab. 17, Fig. 5). Brisson failed, however, to apply consistently the principles of binominal nomenclature; what appears to be the species name in *P. fusca* is in fact merely a descriptive adjective, replaced in other names (e.g., in Brisson's three other species of *Prosimia*) by entire phrases. Brisson's species names must therefore be regarded as unavailable under Article 11c of the Code, and the valid name for the mongoose lemur is *Lemur mongoz* Linnaeus.

1766. *Lemur mongoz* Linnaeus, Syst. Nat., 12 ed., 1, p. 44. Based on Edwards, Gleanings of Nat. Hist., 1, p. 12, Pl. 216. Madagascar.

1812. *Lemur nigrifrons* E. Geoffroy, Ann. Mus. Hist. Nat. Paris, 19, p. 160. Based on Simia-sciurus lanuginosis fuscus of Petiver, Gazophylacii Nat. et Artis, Decade 2, Tab. 17, Fig. 5. Anjouan.

1812. *Lemur albimanus* E. Geoffroy, Ann. Mus. Hist. Nat. Paris, 19, p. 160. Based on "Maki aux pieds blancs" of Brisson (Regn. Anim., 2 ed., p. 156, 1762), and "Le Mongous" of Audebert, Hist. Nat. Makis, p. 10, Fig. 1, 1799.

1812. *Lemur anjuanensis* E. Geoffroy, Ann. Mus. Hist. Nat. Paris, 19,

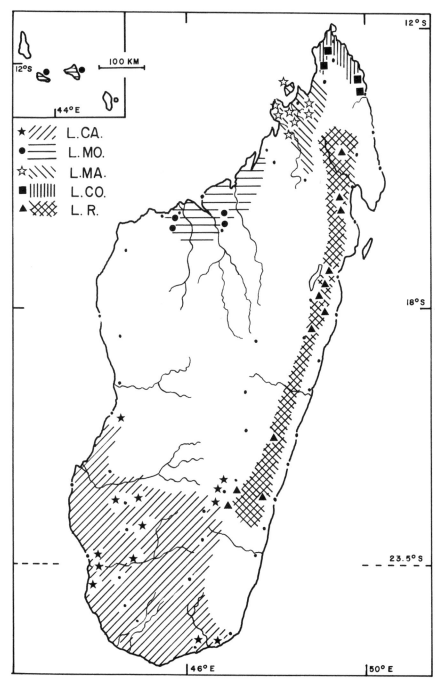

Figure 3.3. Distributions of various species of *Lemur*. Shaded areas represent approximate limits of distribution, symbols represent localities of museum specimens. L.CA.: *Lemur catta*; L.MO.: *L. mongoz*; L.MA.: *L. macaco*: L.CO.: *L. coronatus*; L.R.: *L. rubriventer*.

p. 161. Based on a specimen from Anjouan in MNHN[1]: no. 126 of Rode Catalogue.

1819. *M[aki] mongous* Muirhead, Brewster's Edinburgh Encyclop., 13, p. 405. Muirhead gives authorship of this name to Desmarest (Nouv. Dict. Hist. Nat. 18, 1817), but that author in fact used it as a vernacular term. Based on Desmarest and Linnaeus.

1840. *Prosimia micromongoz* Lesson, Spec. Mamm., p. 226. Based on Edwards, Pl. 216 (see above) and other references.

1840. *Prosimia macromongoz* Lesson, Spec. Mamm. p. 226. Based on numerous references, not all of which are *mongoz*, but the first identifiable one of which is Brisson's *Prosimia fusca*, based on Petiver's animal from Anjouan (see above).

1840. *Prosimia bugi* Lesson, Spec. Mamm., p. 227. Based on numerous references, the first of which is Petiver (see above). Anjouan.

1840. *Prosimia brissonii* Lesson, Spec. Mamm., p. 230. Based on the "Maki aux pieds blancs" of Brisson, Regn. Anim., 2 ed., p. 156, and on *Lemur mongoz* Linnaeus.

1840. *Prosimia ocularis* Lesson, Spec. Mamm., p. 231. Based on *Lemur nigrifrons* of E. Geoffroy (see above) and other references.

1870. *Lemur cuvieri* Fitzinger, Sitzb. K. Akad. Wiss. Wien, Math.-Nat., 62, 1, p. 646. Apparently based on a specimen from Anjouan.

Malagasy names. Dredrika, gidro. Comorian name: Komba.

Pelage and external characters. Fur dense, especially in individuals from high-altitude regions such as central Anjouan. Sexually dichromatic (fig. 3.4). Females generally gray on head, forelimbs, and shoulder region, grading to brown on back, flanks, rump, and hindlimbs (but hind extremities gray). Tail gray, darkening distally. Ears gray, hairy but not tufted. Face and forehead dark, but often with a white patch on the muzzle. Cheeks and beard bushy, white. Ventrum white to pale brown. Males gray, often with some brownish elements in shoulder region and on extremities. Cheeks and beard bushy, reddish brown, this coloration frequently extending to the forehead also. Ears hairy but not tufted, sometimes reddish; ventrum white to light brown. Tail gray, darkening distally. Face somewhat paler than that of the female, normally with a white patch on the muzzle. All Comorian *mongoz* conform to this dichromatic pat-

---

[1]MNHN accession numbers for Rode Catalogue entries are provided only where available.

Figure 3.4. *Lemur mongoz* male (left) and female. Female from Anjouan.

tern, but Tattersall and Sussman (1975) have observed a darker-faced, pale-bearded male variant in Madagascar. Infants uniformly exhibit the female coloration up to the age of several months. Glandular development exists in the perianal region, which is nonetheless hairy, and also in the scrotal and vulval areas. Head rubbing is common among males, but no specialized glandular structure on the crown has been described. Most females possess two pairs of mammae, of which only one pair is functional.

   Dimensions. Cranial length, $\bar{x} = 84.4$, $s = 4.0$, $o.r. = 78.4–89.6$, $n = 31$. Field measurements (Archbold, $n = 5$): H + B, $\bar{x} = 349$, $o.r. = 325–369$; TL, $\bar{x} = 482$, $o.r. = 470–510$; HF, $\bar{x} = 86$, $o.r. = 81–91$; EAR, $\bar{x} = 37$, $o.r. = 34–39$. **Body weight:** $\bar{x} = 2,023$, $o.r. = 1,950–2,190$, $n = 5$.

   Range. Largely because of numerous misidentifications (many of them stemming from the use in determinations of the plates published by Milne-Edwards and Grandidier) the range of *Lemur mongoz* has been more wildly misquoted than that of any other lemur. Indeed, it still remains uncertain in detail. The species occurs on the Comorian islands of Mohéli and Anjouan wherever there is suitable habitat (see Tattersall 1976a, 1977a), and on the mainland in a relatively circumscribed area of the northwest (fig. 3.3). The southern and western limit of the species has yet to be defined precisely. Mongoose lemurs exist in the area of Lake Kinkony, just to the south of Mitsinjo and to the west of the River Mahavavy; but the species has not been reported from the Tsingy de Namoroka Reserve, 20 km due south of Soalala. It does occur, however, both to the east and to the west of the Betsiboka River in the region of Ambato-Boéni, at the same latitude as the Namoroka Reserve. To the north, the range of the mongoose lemur extends as far as the Bay of Narinda although again, the precise limit awaits definition.

Karyotype. 2N = 60. Autosomes: M, 0; S, 4; A, 54. Sex chromosomes: X, A; Y, A (Chu and Swomley 1961).

Lemur macaco Linnaeus, 1766. Black lemur.

1766. *Lemur macaco* Linnaeus, Syst. Nat., 12 ed., p. 44. Based on four references, only one of which is to a black lemur (see p. 9). The concept of *Lemur macaco* is restricted here to the animal illustrated by Edwards, Gleanings of Nat. Hist., 1, Pl. 217, 1756. Madagascar.

1775. *Lemur macaco niger* Schreber, Von Schreber's Säugthiere, 1, p. 142, Pl. 40A. Based on Edwards, Pl. 217 (see above).

1862. *Lemur leucomystax* Bartlett, Proc. Zool. Soc. Lond., 347, Pl. 41. Based on BMNH 67.10.5.15, adult female. Madagascar.

1863. *Varecia nigra* Gray, Proc. Zool. Soc. Lond., p. 136. New name for *Lemur macaco* Linnaeus.

1867. *Lemur flavifrons* Gray, Proc. Zool. Soc. Lond., p. 596. Based on BMNH 67.10.5.19, adult female. Madagascar.

1880. *Lemur nigerrimus* Sclater, Proc. Zool. Soc. Lond., p. 451, text-fig. 2. Based on a male individual in the London Zoo, now MNHN 1882–2753, Rode Catalogue no. 127. Madagascar.

Malagasy names. Akomba, ankomba, komba.

Pelage and external characters. The most strikingly sexually dichromatic of all the lemurs (fig. 3.5). Females range in dorsal coloration from a light golden brown to a dark, almost chestnut brown. Most individuals fall toward the paler end of this spectrum, and there is a tendency for the highest proportion of paler individuals to be found at the southern end of the range of the species. In darker individuals the limbs tend to be a little paler than the dorsum, and vice versa. Tail yellowish gray through golden brown to rufous, occasionally darkening distally. Crown of head varies from russet through gray to black; the ears are usually white and luxuriantly tufted. Face reddish to blackish. Males uniformly black, sometimes with slight tints of dark brown, especially ventrally. Ears almost always lavishly tufted, as in females, although very occasion-

Figure 3.5. Above: *Lemur macaco* female; below: two *Lemur macaco* males (courtesy of Ted Schiffman).

ally individuals of either sex may be found which lack the tufts (whence, apparently, the various versions of the enduring myth of *"Lemur flavifrons"*). Poorly developed carpal glands sometimes occur; glandular development is present in the scrotal and vulval areas, although the scrotum is furred. There is also in males an extensive circumanal area of naked, wrinkled, glandular skin; this is absent in females. Females may have one or two pairs of mammae, but in the latter case only one pair is functional.

Dimensions. Cranial length: $\bar{x} = 90.5$, $s = 217$, $o.r. = 86.9$–99.4, $n = 39$. Field measurements (Archbold, $n = 8$): H + B, $\bar{x} = 412$, $o.r. = 380$–450; TL, $\bar{x} = 550$, $o.r. = 510$–640; HF, $\bar{x} = 109$, $o.r. = 102$–115; EAR, $\bar{x} = 37$, $o.r. = 34$–39. **Body weight,** $\bar{x} = 2,406$, $o.r. = 2,000$–$2,860$, $n = 11$.

Range. The western portion of northern Madagascar, southward from the region of Anivorano Nord to the area of Befandriana Nord (in the interior) and to some distance south of Maromandia (along the coast) (fig. 3.3). The precise limits remain undetermined, but this range does include the western part of the Tsaratanana Massif, the Ampasindava Peninsula, and the islands of Nosy Bé and Nosy Komba.

Karyotype. $2N = 44$. Autosomes: M, 12; S, 8; A, 22. Sex chromosomes: X, A; Y, A (Chu and Bender 1962).

Lemur fulvus E. Geoffroy, 1796. Brown lemurs.

Many recent authors, following Schwarz (1936) have placed *Lemur fulvus* in synonymy with *Lemur macaco*. Discrete populations of the two species are, however, now known to exist in sympatry west of the Galoka mountains in northern Madagascar (see Tattersall 1976a), and it is clear that separate specific status is warranted for each. But while *L. macaco* is definable with little difficulty as a biological entity, *L. fulvus,* widespread over Madagascar, consists of a series of apparently closely related populations, some of which are strikingly distinguished by the coloration of their pelage, but others of which pose exceedingly difficult problems of definition. These problems are exacerbated by the depressingly sparse field information available on distribution, geographical variation, possible secondary contact and hybridization, and so forth. The discussion which follows is of necessity based on certain simplifying assumptions: for instance, the subspecies recognized are established on the basis of certain modal types, and apparently anamalous distributions are omitted from consideration. I do not believe that this will ultimately be found to have led to any very substantial distortion, but it is essential to bear in mind, especially in the case of eastern populations of *fulvus,* that the taxonomy as well as the distribution information is provisional. One example of the kind of difficulty to which I refer may help to illustrate the situation. The Archbold Expedition collected virtually identical series of females, which look plausibly like dark-headed female *albifrons* (they are not *L. mongoz* as was thought), at sites as far apart as Maroantsetra (15° 30'S), where one would reasonably expect to find such animals, and Manombo and 20 km west of Vondrozo (both 23° S), where for many reasons one would not (this is *albocollaris* territory, as borne out by other specimens collected by the expedition). At present, there is no way to account for this odd circumstance with confidence, and rather than speculate, I feel it best to leave the matter in suspension pending a thorough survey of lemur populations in eastern Madagascar. Let us hope that someone will have the opportunity to do this before it is too late. If it is not already.

All *Lemur fulvus* subspecies possess mystacial, submental, superciliary, carpal, and usually genal vibrissae; in all the face is clothed in short hair except at the tip of the muzzle; and although females often possess a second pair of mammae, only the anterior pair is functional. In both sexes of all the subspecies the circumanal region is distinguished by an area of naked, wrinkled, glandular skin; this area is larger in males, and more highly developed histologically; but at both the gross and microscopic levels it is similar in quality in males and females. Scrotal glands are present in males, although the scrotum itself is furred.

Lemur fulvus fulvus E. Geoffroy, 1796. Brown lemur.

1796. *L[emur] fulvus* E. Geoffroy, Mag. Encyclop., 1, p. 47. Based on the "Grand mongous" of Buffon, Hist. Nat., Gén. et Part., Suppl., 7, p. 118, Pl. 33. Madagascar.

1844. *Lemur bruneus* van der Hoeven, Tijdschr. Nat. Gesch. Physiol., 11, p. 35. Based on two specimens in the RMNH: Jentinck Syst. Cat. *Lemur mongoz* v², w². Madagascar.

Malagasy names: Gidro (northwest); boromitoko (Beramanja area); varika, varikosy (eastern region).

Pelage and external characters. Upper parts gray brown to brown, sometimes with olivaceous tints (see fig. 3.6). Crown of head dark gray or black; cheeks and beard pale. Muzzle black, continuing up center of forehead, leaving lighter patches above each eye. Ears hairy, but not tufted. Underparts pale: off-white to light brown. Dark pygal patch present; extremities often tend to reddish. Tail darkens distally. Sexual dichromatism is not pronounced, but females tend to be somewhat lighter in coloration than males, and the pelage of their cheeks and beard is less luxuriant, giving the impres-

Figure 3.6. Left: *Lemur fulvus fulvus* male (courtesy of Ted Schiffman); right: *Lemur fulvus mayottensis* female.

sion of a sharper face. Within-population variation in pelage coloration is quite extensive, but in general individuals from the east tend to be darker, with denser fur.

Dimensions. Cranial length: $\bar{x} = 90.1$, $s = 2.8$, $o.r. = 85.0-95.8$, n = 22. Field measurements (Lamberton, $n = 3$): H + B, $\bar{x} = 473$, $o.r. = 430-500$; TL, $\bar{x} = 530$, $o.r. = 500-550$; HF, $\bar{x} = 61$. $o.r. = 52-70$; EAR, $\bar{x} = 30$, $o.r. = 27-35$; (Archbold, one specimen): H + B, 373; TL, 510; HF, 108; EAR, 38. **Body weight:** $\bar{x} = 2,620$, $o.r. = 2,078-4,215$, n = 9.

Range. Occurs in at least three distinct areas of Madagascar (fig. 3.7): in the northwest, to the north and east of the Betsiboka River, from south of Ambato-Boeni to Analalava; in the north, in a small area to the east of the Galoka mountains, south of Beramanja; and in the east, south of Lake Alaotra and around Andasibé. The limits of this last isolate are not well understood; in the north, towards the coast, *L. f. albifrons* occurs at Betampona, just north of Tamatave, while in the interior, at about the same latitude, *fulvus* has been collected at Didy. The southern limit of this population is even more obscure, and this lack of certainty is reflected in figure 3.7; according to Petter et al. (1977) however, this boundary is formed by the Mangoro River, which flows eastward to reach the sea at Mahanoro.

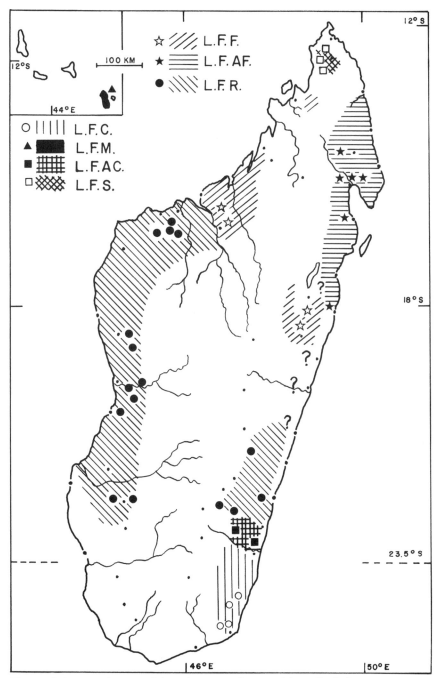

Figure 3.7. Distributions of the subspecies of *Lemur fulvus*. Shaded areas represent approximate distribution limits; symbols denote localities of museum specimens. L.F.F.: *Lemur fulvus fulvus*; L.F.AF.: *L. f. albifrons*; L.F.R.: *L. f. rufus*; L.F.C.:*L. f. collaris*; L.F.M.: *L. f. mayottensis*; L.F.AC: *L. f. albocollaris*; L.F.S.: *L. f. sanfordi.*

In the USNM collections made in 1895 by the splendid W. L. Abbott, there are several *fulvus* specimens said to have been collected in the Faraony River valley, north and east of Manakara, i.e., considerably to the south even of the Mangoro River. However, other *Lemur fulvus* specimens, including *rufus* (plausibly) and *albifrons* (implausibly) are said to have come from the same area, and some confusion must surely have occurred in the documentation of the specimens. This seems quite reasonable, for Abbott was in Madagascar at a troubled moment: he had gone there to enlist as a surgeon in the Hova army, to help defend the country against the impending (second) invasion of the French, although in common with the rest of the foreign mercenaries he resigned before the attack. Communications at the time were extremely precarious, and much correspondence was seized by the French; it seems likely that this accounts for the fact that the existing documentation of Abbott's Madagascar collections is below his normal standards. The *fulvus* and *albifrons* almost certainly came from more northerly sites; Abbott is known to have collected within the modern ranges of both.

Karyotype. 2N = 60. Autosomes: M, O: S, 4: A, 54. Sex chromosomes: X, A; Y, A (Rumpler and Albignac 1969b).

Lemur fulvus albifrons E. Geoffroy, 1796. White-fronted lemur.

1796. *Lemur albifrons* E. Geoffroy, Mag. Encyclop., 1, p. 48. Based on three individuals which are no longer traceable. Madagascar.
1840. *Prosimia frederici* Lesson, Spec. Mammif., p. 232. Based on the "Maki à front blanc" of F. Cuvier, Mammif., 1, 1819.

Malagasy names. Varika, alokasy.

Pelage and external characters. Among *albifrons* males there appear to be two distinct color phases. Most males are darkish brown or gray brown dorsally, with black faces but luxuriant white or cream forehead, crown, ears, cheeks, and throat (fig. 3.8). The tail is dark, the underparts pale. Other males, however, lack the striking white coloration of the head, the fur of this region instead being shorter, and black or dark gray. According to J. Buettner-Janusch (pers. comm.), captive white-headed males have sired drab male offspring. There is also substantial variation among the females, where the upper parts are usually gray brown but may be gray; in some females the head is dark gray, while in others it is a paler shade of the same color. Light patches may or may not be present over the eyes. Tail usually darker than dorsum, ventrum al-

Figure 3.8. Left: *Lemur fulvus albifrons* male (courtesy of Ted Schiffman); upper right: *L. f. albifrons* female with infant; lower right: *L. f. albocollaris* male.

ways lighter. In essence then, female *albifrons* also appear to fall into two groups: one more closely resembling *fulvus*, the other quite similar to female *sanfordi*. This is a complex situation which clearly merits further investigation.

Dimensions. Cranial length: $\bar{x} = 90.2$, $s = 2.4$, $o.r. = 84.4-95.4$ Field measurements (Archbold, $n = 8$, all from the Maroantsetra area): H + B, $\bar{x} = 399$, $o.r. = 380-425$; TL, $\bar{x} = 512$, $o.r. = 495-550$: HF, $\bar{x} = 106$, $o.r. = 102-110$; EAR, $\bar{x} = 34$, $o.r. = 29-37$. **Body weight:** $\bar{x} = 2,293$, $o.r. = 1888-2552$, n = 8.

Range. Humid eastern forests of Madagascar, but precise limits poorly known. If apparently anomalous records, such as that mentioned earlier, are excluded, the southern limit of the subspecies appears to be around Tamatave, at least along the coast (fig. 3.7). As already noted, *fulvus* is recorded as far south as Didy, on the western edge of the rain forest strip, while *albifrons* exists at Betampona, only slightly· to the north of this, but near the coast. It is possible that, at least toward the coast, the southern boundary of *albifrons* is marked by the Ivondro River, but this remains to be confirmed. In the interior the subspecies occurs at least as far south as the Natural Reserve of Zahamena, east of Andreba. In the north, the distribution of *albifrons* extends as far west as the Marojejy Massif, north of Andapa, but not on to the Tsaratanana Massif; along the coast it reaches beyond the northern limit of the rainforest, to the Fanambana River, near Vohémar.

Karyotype. 2N = 60. Autosomes: M, O; S, 4; A, 54. Sex chromosomes: X, A; Y, A (Chu and Swomley 1961).

Lemur fulvus rufus Audebert, 1799. Rufous or Red-fronted lemur.

1799. *Lemur rufus* Audebert, Hist. Nat. Makis, p. 12, Pl. 2. Based on an adult female in the collection of the MNHN, said by Rode to be no. 129 of his Catalogue. Madagascar.

1833. *Lemur rufifrons* Bennett, Proc. Zool. Soc. Lond., 1833, Part 1, p. 106. Based on BMNH 55.12.24.54, adult male. Madagascar.

Malagasy names. Varika (southwestern and eastern parts of range), gidro (northern part of range).

Pelage and external characters. Sexually dichromatic (fig. 3.9), and individually quite variable. Males: upper parts gray, tail darkens to tip, dark pygal patch usually present. Bushy head cap bright rusty orange, quite variable in extent. Muzzle black, this color extending up between eyes to meet orange crown. Pale gray patches above each eye; the ears, bushy cheeks, and throat are also this color. Underparts pale gray to gray brown. Scrotum is haired, dark. Females: upper parts a light to medium reddish brown; the tail is russet, darkening distally, often showing a darkish pygal patch. Crown of head gray, muzzle black, as is center of forehead and a narrower circumocular ring. A light gray or white patch occurs above each eye; the cheeks are also this color, but the thinly haired ears are reddish brown. Underparts pale golden brown or gray. Both males and females from the east tend to be darker than western ones, with, as would be expected, a denser pelage. A moderate degree of carpal

Figure 3.9. *Lemur fulvus rufus* male (above) and female. Courtesy of R. W. Sussman.

glandular development has been noted in males (Rumpler and Oddou 1979).

Dimensions. Cranial length: $\bar{x} = 88.5$, $s = 3.4$, $o.r. = 83.9$–97.7, $n = 44$. Field measurements (Archbold, $n = 18$): H + B, $\bar{x} = 414$, $o.r. = 380$–445; TL, $\bar{x} = 542$, $o.r. = 485$–600; HF, $\bar{x} = 102$, $o.r. = 95$–108; EAR, $\bar{x} = 38$, $o.r. = 33$–42; (Webb, one specimen): H + B, 405; TL,

517; HF, 74; EAR, 32. **Body weight:** x̄ = 2686, *o.r.* = 2,128–3,631, *n* = 15.

Range. Found in both western and eastern Madagascar (fig. 3.7). In the west the subspecies occurs along the southwestern bank of the Betsiboka River, from Katsepy (opposite Majunga) at least as far as Ambato-Boéni, and is found where suitable forest exists as far south as the Fiherenana River, which runs south and west to meet the sea near Tuléar. The subspecies has been recorded south of the Fiherenana only at Lambomakandro, north and east of Sakaraha, and close to the river. To the east, in common with other lemurs, *rufus* is limited by the availability of forests, which diminish progressively toward the interior. In the eastern humid forests the distribution of *rufus* is less well understood. It has been collected as far south and west as Ivohibé, at the southern end of the Andringitra Massif, only a short distance north of the most northerly occurrence of *albocollaris*. Along the coast, the most southerly collecting locality is near Manakara. The northern boundary of eastern *rufus* is highly uncertain, as indicated in figure 3.7; Petter et al. (1977) indicate that the subspecies occurs as far north as the Mangoro River.

Karyotype. 2N = 60. Autosomes: M, O, S, 4; A, 54. Sex chromosomes: X, A; Y, A (Chu and Swomley 1961).

Lemur fulvus collaris E. Geoffroy, 1812. Collared lemur.

1812. *Lemur collaris* E. Geoffroy, Ann. Mus. Hist. Nat. Paris, 19, p. 161. Probably based on MNHN 1819–73 (6A), adult male; no. 128 of Rode Catalogue. Madagascar.

Malagasy name. Varika.

Pelage and external characters. Upper parts a darkish brown or gray brown, sometimes slightly olivaceous. There is often a distinctly darker strip down the center of the back. Ventrally somewhat paler; tail dark, especially at base and tip. Face and top of head including ears and neck black in males, generally gray in females. Lighter patches are sometimes found above the eyes in males. In both sexes the cheeks are pale orange in color; in males they are notably bushy (fig. 3.10).

Dimensions. Cranial length: x̄ = 91.9, *s* = 1.2, *o.r.* = 89.8–93.7, *n* = 16. **Body weight:** x̄ = 2505, *o.r.* = 2,095–2,817, *n* = 5.

Range. Southwestern Madagascar, from the southern end of the humid forest strip, near Fort-Dauphin, north to the Mananara River, which flows in a southeasterly direction to meet the sea at Vangaindrano (fig. 3.7). The northern and western limits are not well established, however.

Figure 3.10. *Lemur fulvus collaris* male.

Karyotype. Polymorphic: individuals examined have had 2N = 50, 51 and 52 (Buettner-Janusch and Hamilton 1979). These authors suggest that the 2N = 51 karyotype is a result of natural hybridization between 2N = 50 and 2N = 52 animals. The 2N = 50 karyotype has: M, 10; S, 4; A, 34; X, A; Y, A (Hamilton et al. 1977); the 2N = 52: M, 6 or 8; S, 4; A, 40 or 38; X, A; Y, A (Rumpler and Albignac 1969a). Buettner-Janusch and Hamilton (1979) suggest that the 2N = 50 karyotype is derived from the 2N = 52 by centric fusion.

Lemur fulvus mayottensis Schlegel, 1866. Mayotte lemur.

1866. *Lemur mayottensis* Schlegel, Ned. Tijdschr. Dierk., 3, p. 76. Based on a syntype series in the RMNH: Jentinck Cat. Ost., 19, *L. collaris* s-z. Mayotte, Baie de Jongoni (= Baie de Boueni).

Comorian name. Komba.

Pelage and external characters. Essentially as in *L.f. fulvus* (fig. 3.6) from which the form is undoubtedly derived, and from which it is separated on the basis of its geographical isolation. There is no evidence whatever to support the contention of Petter et al. (1977) that the subspecies is the result of hybridization between introduced *fulvus* and *rufus*. Perhaps the most striking characteristic of this lemur is its chromatic variability, which, although it centers around a mean very close to that of *fulvus*, is greater than anything recorded from the mainland. As in *fulvus*, males tend to be darker than females and to have bushier pelage on the cheeks and jowls, giving the impression of a fuller face.

Dimensions. Cranial length: $\bar{x} = 90.8$, $s = 2.7$, $o.r = 85.9–95.7$, $n = 14$. Field measurements (J. Nicoll, two specimens): H + B, 406, 380; TL, 500, 510; HF, 97, 100; EAR, 36, 34.

Range. Island of Mayotte, Comoro group (fig. 3.7) Occurs on the island wherever there is forest, but it is rare at altitudes above 300 m.

Karyotype. 2N = 60. Autosomes: M, O; S, 4: A, 54. Sex chromomosomes: X, A; Y, A (Hamilton et al. 1980).

Lemur fulvus sanfordi Archbold, 1932. Sanford's lemur.

1932. *Lemur fulvus sanfordi* Archbold, Amer. Mus. Novitates, 518, p. 1. Based on AMNH 100585, adult male. Madagascar: Mt. d'Ambre.

Malagasy name: Varika.

Pelage and external characters. Sexes clearly distinguished (fig. 3.11). Males: upper parts gray, usually washed with brown; crown of head and bushy cheeks brownish, but ears strikingly tufted white; muzzle black, but forehead and areas beside and below orbits white or cream. Extremities sometimes reddish; tail darkest at tip; darkish pygal patch present. Scrotum hairy, black. Females: upper parts gray, sometimes gray brown; muzzle black, but rest of head a darkish gray. Ears not tufted, cheeks relatively short-haired. Tail darkens to tip, pygal patch usually indistinct. Underparts pale gray or cream. Further, it is apt in this case to quote the observation of George Shaw (in Miller, Cimelia Physica, 1796) on another animal entirely: "there is also a cast of visage, which, though not easily described in words, is very obvious in these animals" (p. 64).

Dimensions. Cranial length: $\bar{x} = 89.2$, $s = 1.7$, $o.r. = 88.0–92.5$, $n = 11$. Field measurements (Archbold, $n = 7$): H + B, $\bar{x} = 395$, $o.r. = 360–425$; TL, $\bar{x} = 512$, $o.r. = 465–525$; HF, $\bar{x} = 104$, $o.r. = 102–107$; EAR, $\bar{x} = 35$, $o.r. = 32–38$. **Body weight** (one individual): 2,234.

Range. Restricted to the immediate area of the Mt. d'Ambre in northern Madagascar (fig. 3.7), from the northern flanks of the mountain south at least as far as the Ankarana Massif, between Anivorano Nord and Ambilobé, where specimens were collected by

Figure 3.11. *Lemur fulvus sanfordi* male (left) and female.

the Archbold Expedition. Petter et al. (1977) indicate that toward the east the subspecies occurs widely down to the latitude of Sambava, but this is not substantiated by my surveys.

Karyotype. 2N = 60. Autosomes: M, O; S, 4; A, 54. Sex chromosomes: X, A; Y, A (Rumpler and Albignac 1969b).

Lemur fulvus albocollaris Rumpler, 1975. White-collared lemur.

It has been suggested by Groves (e.g., 1978) that the valid name for this subspecies is *cinereiceps* Milne-Edwards and Grandidier, 1875 (Hist. Phys. Nat. Pol. Madagascar, 5 [Atlas] Pls. 140 and 147). These plates, published without text, do not however bear much if any resemblance to the lemur under consideration here, and Schwarz (Proc. Zool. Soc. Lond., 1931, p. 142) subsequently selected two "typical specimens of *L. f. collaris* in the Paris Museum" as types, believing for reasons unclear to me that these were the individuals illustrated. This was unfortunately no solution, however, since both specimens are female and therefore not diagnostic. I thus prefer to adopt Rumpler's name, poorly proposed though it is, for this subspecies. It should be noted that the subsequent references by Rumpler and others to the authorship of the name *albocollaris* as Rumpler, 1972, are without foundation (see Tattersall 1979).

1975. *Lemur fulvus albocollaris* Rumpler, Lemur Biol., pp. 29, 37. Based on captive individuals at Tsimbazaza, Madagascar, one of them illustrated by Petter et al. (Faune de Madag., 44, p. 137, Fig. 58). Madagascar.

Malagasy name. Varika.

Pelage and external characters. Males are distinguished from those of *L. f. collaris* by possessing a white, rather than an orange, beard (fig. 3.8). I have not observed this subspecies in the wild, but captive females in Tananarive (confirmed karyotypically as *albocollaris*) resembled those of *L. f. collaris*. Most female museum specimens from sites where white-bearded males were collected are orange-bearded also; but a series collected near Vondrozo by the Archbold Expedition consists of females indistinguishable from those of *L. f. albifrons*.

Dimensions. Cranial length: 87.7 ($n = 1$). Field measurements (Archbold, $n = 15$): H + B, $\bar{x} = 407$, o.r. = 395–435; TL, $\bar{x} = 540$, o.r. = 450–585; HF, $\bar{x} = 104$, o.r. = 97–108; EAR, $\bar{x} = 35$, o.r. = 32–38.

Range. Eastern humid forests between the Mananara and Faraony Rivers (fig. 3.7). This range has not been adequately surveyed, however.

Karyotype. 2N = 48. Autosomes: M, 6; S, 10; A, 30. Sex chromosomes: X, A; Y, A (Rumpler and Albignac 1969a).

Lemur coronatus Gray, 1842. Crowned lemur.

1842. *Lemur coronatus* Gray, Ann. Mag. Nat. Hist. 10, p. 257. Based on BMNH 37.9.26.80, adult male received from Verreaux, "Madagascar." Contra Gray (1870) and Schwarz (1931), *not* collected during the voyage of the "Sulphur," and *not* from the Bay of Mahajamba.

1846. *Lemur chrysampyx* Scheuermans, Mém. Couronnés et des Savants Etrang. Acad. Roy. Sci., Lettres et Beaux-Arts Belgique, 22, Mém. 7, p. 6, Pl. 1. Based on the female individual figured; "probably Madagascar."

Malagasy names. Ankomba, varika.

Pelage and external characters. Sexually dichromatic (fig. 3.12). Females generally light gray with some brownish tipped hair on the back, especially caudally. Tail darkens distally to dark gray tip. Underparts and extremities pale, face and supraorbital region

Figure 3.12. *Lemur coronatus* male (left) and female.

light gray; cheeks and throat also pale. Head cap gray, with orange patches bilaterally above forehead; these meet in the midline and diverge caudally. Ears fairly prominent, hairy but not tufted, with some white at tip. Males tend dorsally to be of a slightly darker gray than females, but are quite variable in this. Many dorsal hairs are dichromic: basally gray, with cream to brownish tips. The limbs tend to be paler than the back, and the tail darkens distally. Underparts cream to light brown. Face white or off-white; cheeks and throat pale. As in females there is a V-shaped orange patch above the forehead, but in almost all male individuals this is filled in with black instead of gray. Ears quite prominent: hairy but not tufted, white at tip. Perianal and perineal regions hairy, but with some glandular development in both sexes. Two pairs of mammae are normally present in females, of which only one pair is functional.

Dimensions. Cranial length: $\bar{x} = 79.9$, $s = 1.7$, $o.r. = 77.4$–82.5, $n = 16$. Field measurements (Archbold, $n = 18$): H + B, $\bar{x} = 336$, $s = 13$, $o.r. = 315$–355; TL, $\bar{x} = 451$, $s = 25$, $o.r. = 415$–510; HF, $\bar{x} = 91$, $s = 3$, $o.r. = 86$–97; EAR $\bar{x} = 36$, $s = 2$, $o.r. = 34$–40. **Body weight:** ca. 2 kg.

Range. *Lemur coronatus* is the only lemur to be found in the dry forests of the arid Cap d'Ambre, the northern tip of Madagascar. To the south of this, the species occurs in the west as far as the Ankarana Massif, between Ambilobé and Anivorano Nord (fig. 3.3); in the east its distribution extends south to the Fanambana River, which reaches the sea a few miles beyond Vohémar. This range encompasses the Mt. d'Ambre; and while in 1935 Rand reported that the species was common in the dry forests of the lower slopes of the mountain but was absent from the humid forest at altitude, Petter et al. (1977) noted that it is now also found in the latter, possibly because of pressure on its preferred habitat.

Karyotype. 2N = 46. Autosomes: M, 6; S, 12; A, 26. Sex chromosomes: X, A; Y, A (Rumpler and Albignac 1969b).

Lemur rubriventer I. Geoffroy, 1850. Red-bellied lemur.

1850. *Lemur rubriventer* I. Geoffroy, Compt. Rend. Acad. Sci. Paris, 31, p. 876. Based on MNHN Rode Catalogue no. 123*a*, subadult male. Madagascar.

1850. *Lemur flaviventer* I. Geoffroy, Compt. Rend. Acad. Sci. Paris, 31, p. 876. Based on MNHN Rode Catalogue no. 124, adult female. Madagascar.

1870. *Lemur rufiventer* Gray, Cat. Monkeys, Lemurs and Fruit-eating Bats in Brit. Mus., p. 74. Lapsus for *Lemur rubriventer* I. Geoffroy.

1871. *Lemur rufipes* Gray, Proc. Zool. Soc. Lond., p. 339. Based on BMNH 44.5.14.24, adult male. Madagascar.

Malagasy names. Tongona, soamiera, barimaso.

Pelage and external characters. Fur relatively long, dense, silky (fig. 3.13). Upper parts of most individuals a dark, lustrous chestnut brown, but tinged in some with lighter, slightly olivaceous elements. Tail black. Underparts dark reddish brown in almost all males; pale, even white, in almost all females. Similarly, in some individuals the throat is dark, in others (again, mostly females) it is white, and in yet others it is intermediate in coloration. Cheeks occasionally pale. Face dark brown or black, but in most instances a naked or thinly haired area in front of the eyes reveals the pale skin beneath, creating the impression of whitish patches. Normally the ears are largely concealed in the surrounding fur; they are not tufted, but are haired on both aspects. Scrotum black, well haired; perianal region naked, wrinkled and presumably glandular, but apparently not as highly differentiated as in *Lemur fulvus*.

Dimensions. Cranial length: $\bar{x} = 86.6$, $s = 2.2$, $o.r. = 81.0–91.2$, $n = 52$. Field measurements (Archbold, $n = 8$): H + B, $\bar{x} = 386$, $o.r. = 360–420$; TL, $\bar{x} = 491$, $o.r. = 465–535$; HF, $\bar{x} = 113$, $o.r. = 105 – 117$; EAR, $\bar{x} = 40$, $o.r. = 36 – 41$; (Webb, one individual): H + B, 420; TL, 489; HF, 73; EAR, 39. **Body weight:** no data.

Range. *Lemur rubriventer* appears to occur, if only sparsely,

Figure 3.13. *Lemur rubriventer female.*

throughout the forested interior of eastern Madagascar, from the Tsaratanana Massif in the north at least as far south as Ivohibé, at the southern end of the Andringitra Massif (fig. 3.3). The species appears to be confined to forests at medium to high altitudes, and there is no reliable basis for reports that it occurs, or at least occurred, in western Madagascar.

Karyotype. 2N = 50. Autosomes: M, 2; S, 12; A, 34. Sex chromosomes: X, A; Y, n.d. (Rumpler and Albignac 1971).

Varecia Gray, 1863

1863. *Varecia* Gray, Proc. Zool. Soc. Lond., p. 136. Type species by monotypy: *Varecia variegata;* Madagascar.

Varecia variegata (Kerr, 1792). The ruffed lemurs.

Considerable variation in pelage pattern and color occurs within this species, but the status of various populations comprising it is unclear. Provisionally, I recognize two subspecies: a northern one with red and black coat coloration, and a southern one with black and white pelage. The latter, however, shows at least four distinct and consistent coat patterns, and better knowledge of the distributions of these varieties might ultimately suggest their recognition as distinct subspecies. *Varecia variegata* is the largest species of its family. For the early part of its taxonomic history this animal was regularly confused with *L. macaco*. The following synonymy omits all names also used for the black lemur.

Varecia variegata variegata (Kerr, 1792). Black and white ruffed lemur.

1792. *Lemur macaco variegatus* Kerr, Animal Kingdom, p. 86. Based on von Schreber's Säugethiere, Pl. 40B, based in turn on "Le Vari" of Buffon, Hist. Nat., Gén. et Part., 13, Pl. 27, 1765. Madagascar; the individual illustrated is I. Geoffroy's type a (see below).
1819. *Maki vari* Muirhead, Brewster's Edinburgh Encyclop. 13, p. 405. Based on le Maki Vari=*Lemur macaco* of Desmarest, Nouv. Dict. Hist. Nat. 18, p. 437, 1817, said to be based in turn on a young male illustrated by Audebert, Hist. Nat. des Makis, Pl. 5, 1797. Madagascar, Canton de Mangabey; type a.
1833. *Prosimia subcincta* A. Smith, S. Afr. Quart. Jour., 2, p. 30. Based on "Le vari à ceinture" of E. Geoffroy, in E. Geoffroy and F. Cuvier, 1796, Mag. Encyclop., 1, p. 47, 1796; Pl. 6 of Audebert (see above); and evidently also on live specimens. Madagascar; type c.
1851. *Lemur varius* I. Geoffroy, Cat. Méthod. Coll. Mammif. Mus. Hist. Nat. Paris, p. 71. Species based on Buffon (see above); three varieties recognized. Madagascar.
1863. *Varecia varia* Gray, Proc. Zool. Soc. Lond., p. 136. Based on Buffon (above) and others.
1870. *Lemur melanoleucus* Gray, Cat. Monkeys, Lemurs, and Fruit-eating Bats in Brit. Mus., p. 70. Given in synonymy with *Varecia varia*.
1953. *Lemur variegatus editorum* Hill, Primates, Comp. Anat. Tax., 1, p. 401. Based on *L. varius*, var. b, of I. Geoffroy 1851 (see above); Ambatondrazaka.

Malagasy names. Varikandana, varikandra.
Pelage and external characters. As noted above several color patterns can be distinguished. Where series exist from the same locality, they always conform to the same pattern, but it is not clear whether or not the various types distinguishable are allopatric (see below). All individuals of the species have dense, long, silky fur

which is particularly luxuriant around the ears, cheeks and throat (hence the English name). The muzzle is long and the face is only lightly haired, also bearing superciliary, mystacial, genal and mental vibrissae. Carpal vibrissae are normally present, and males have a single median gular cutaneous gland (see Rumpler and Andriamiandra 1971, for its histological structure). Three pairs of mammae present in females.

Most of the pelage variations in the black and white ruffed lemur have been recognized for a long time. Audebert illustrated two before 1800 (Hist. Nat. Makis, Plates 5 and 6), and in 1851 I. Geoffroy (Cat. Méthod.) described three, under the name *Lemur varius* Variétés a, b, and c. To minimize confusion, I shall follow Geoffroy's designations, complicating his system only by dividing variety b into two variants, b1 and b2, of which my b1 corresponds exactly to his b. Descriptions of each of these follow.

Type a. Face black except for short white hairs on muzzle below eyes. Forehead and crown black; ears, cheeks and throat tufted white. Otherwise white, except for ventrum, tail, proximal part of forelimb, lateral aspect of shoulders, extremities, and lateral aspect of thighs, all of which are black (fig. 3.14). This is the variety originally described, and it corresponds to "form 2" of Petter et al. 1977.

Type b1. Resembles type a, except that the black shoulder patches extend posteriorly, on to the flanks, and medially, to meet in the midline. This variation corresponds to *L. variegatus editorum* of Hill (1953), and to "form 4" of Petter et al. (1977).

Figure 3.14.   *Varecia variegata variegata* (type a) (left), and *V. v. rubra.*

Type b2. Pattern as in b1, except that a narrow white stripe runs forward in the dorsal midline, invading the back forequarters but not reaching the neck area. One specimen in USNM lacks black on the outer aspect of the thighs. No designation has been proposed for this variant.

Type c. Entirely black except for white cheeks, ears, and throat, a white transverse band extending across the back and sides just behind the shoulders, and another across the rump, extending down the posterior aspect of the thighs onto the lateral surface of the lower leg. White patches also occur laterally on the lower arm. This variant is that described as *Prosimia subcincta* by A. Smith (1853), and as "form 1" by Petter et al. (1977).

In all of these forms the black fur may be replaced by a very dark brown, or show brownish or grayish tints. The white fur may sometimes be flecked with gold on the shoulders, flanks and back. Some other variants have been noted, for example the "form 7" of Petter et al. (1977), based on a BMNH specimen collected by Crossley in "south Madagascar" and resembling Type b2 except that the anterior transverse white band is displaced somewhat caudally, and the fur behind it is brown, as is a longitudinal median stripe anterior to it. Such variants, however, are all represented by single specimens from unknown or unidentifiable localities, and admittedly subject to very poor sampling, do not appear to be repetitive over several localities as are the major types described above. Obviously, there is vast room for improvement in our understanding of the variation seen in *V. v. variegata*.

Dimensions. Cranial length: $\bar{x} = 104.7$, $s = 2.5$, $o.r. = 100.4–110.7$, $n = 33$. Field measurements (Archbold, $n = 3$): H + B, $\bar{x} = 553$, $o.r. = 545–560$; TL, $\bar{x} = 602$, $o.r. = 575–650$; HF, $\bar{x} = 128$, $o.r. = 125–130$; EAR, $\bar{x} = 41$, $o.r. = 39–45$. **Body weight:** $\bar{x} = 3785$, $o.r. = 3,293–4,512$, $n = 5$.

Range. The geographical distribution of this subspecies is exceptionally poorly documented, and the question of the allopatry or otherwise of the various types described is not resolved, although the balance of the inadequate available evidence suggests some sympatry. Evidently, we are not dealing with a simple cline from north to south, although there may be a tendency for the overall amount of black in the pelage to be reduced toward the south. The subspecies as a whole is found throughout most of the remaining humid forest of the eastern strip, from somewhat north and west of Maroantsetra, where the River Antainambalana seems to separate it from its red and black relative (but see below), to a point south of

Farafangana but north of the Mananara River, which flows southeast to meet the sea at Vangaindrano. The distribution of the various types within this area is somewhat confusing, at least if the lamentably few collecting records are to be believed (see fig. 3.15). In general, Type c specimens come from the region of Maroantsetra, and Type a specimens from south of this. Type b1 specimens are known to have come from south and east of Lake Alaotra, as also have b2 individuals. According to Petter et al. (1977), a population resembling Type a, but with very little black, still survives in what remains of the coastal forest south of Farafangana. Given the rate at which these extraordinarily beautiful creatures are disappearing, it is unlikely that we will ever know all that we would wish about their distribution. Subfossil bones most probably attributable to *Varecia* have been recovered at sites on the central plateau.

Karyotype. 2N = 46. Autosomes: M, 16; S, 2; A, 26. Sex chromosomes: X, S; Y, A (Hamilton and Buettner-Janusch 1977).

Varecia variegata rubra (E. Geoffroy, 1812). Red-ruffed lemur.

1812. *Lemur ruber* E. Geoffroy, Ann. Mus. Hist. Nat. Paris, 19, p. 159. Based on a specimen illustrated by Commerson in 1763. Madagascar.

1833. *Prosimia rubra* A. Smith, S. Afr. Quart. Jour., 2, p. 29. Based on E. Geoffroy (above) but Smith may have been familiar with the animal at first hand. Madagascar.

1840. *Prosimia erythromela* Lesson, Spec. Mammif., p. 236. Based on Commerson's drawings and several other references. Madagascar.

1863. *Varecia. rubra* Gray, Proc. Zool. Soc. Lond., p. 136. Based on E. Geoffroy (above); Madagascar.

Malagasy name. Varignena, varimena.

Pelage and external characters. Much more uniform in color pattern than *V. v. variegata*. Ventrum black, as are the extremities, tail, inner aspects of the limbs, the forehead, and the crown (but not the ears). A patch of white fur occurs on the neck, and very occasionally at the base of the tail; usually, however, the pygal region is merely a bit paler than the rest of the dorsum. Small patches of white may also be found in some in some individuals on the heels, or the digits, or the tip of the muzzle, although the face is normally uniformly black. Otherwise, the animal is entirely of a deep, lus-

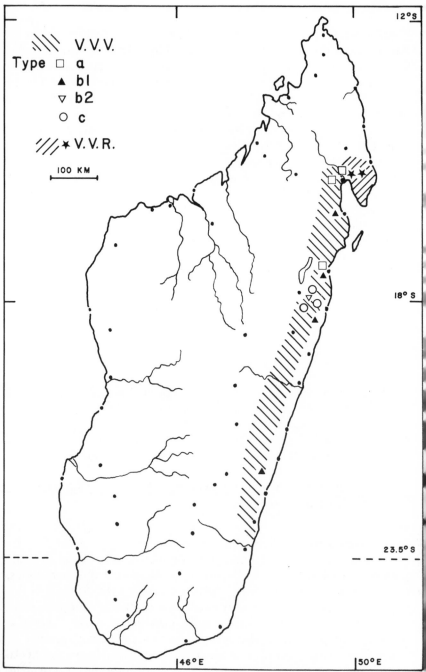

Figure 3.15. Distribution of *Varecia*. Shaded areas represent approximate range limits; symbols denote localities of museum specimens. V.V.V.: *Varecia variegata variegata*; V.V.R.: *V. v. rubra*.

trous, rusty red (fig. 3.14). Carpal vibrissae present, as is a full facial set, with the exception of the interramal tuft. Median gular gland present in males; three pairs of nipples in females.

Dimensions. Cranial length: $\bar{x} = 103.7$, $s = 2.6$, $o.r. = 97.2–106.9$, $n = 13$. Field measurements (Archbold, $n = 6$): H + B, $\bar{x} = 533$, $o.r. = 510–550$; TL, $\bar{x} = 592$, $o.r. = 560–625$; HF, $\bar{x} = 129$, $o.r. = 126–133$; EAR $\bar{x} = 42$, $o.r. = 39–43$. **Body weight** ($n = 2$): 3,961, 3,773.

Range. Confined to the Masoala Peninsula, to the east of the River Antainambalana (fig. 3.15). Very rare throughout this range, but particularly in the north; indeed, it is possible that the subspecies is extinct north of Cap Est. Several museum specimens are much lighter in color than are typical *V. v. rubra*, and resemble captive-bred hybrids between the red and black and black and white subspecies. The only locality of such a specimen that I have been able to identify with any certainty lies 40 km to the northwest of Maroantsetra (Archbold Expedition), and it seems possible that the two subspecies are in secondary contact in the region between the confluent Vohimaro and Antainambalana Rivers, north and west of Maroantsetra. Alternatively, this region may shelter a slightly albinistic local population of *V. v. rubra*.

Karyotype. 2N = 46. Autosomes: M, 16; S, 2; A, 26. Sex chromosomes: X, S; Y, A (Hamilton and Buettner-Janusch 1977).

FAMILY LEPILEMURIDAE RUMPLER AND RAKOTOSAMIMANANA, 1972

Lepilemur I. Geoffroy, 1851. Weasel or Sportive Lemurs

The status of the various populations of the widely distributed *Lepilemur* is the most vexed problem in the lower level taxonomy of the Malagasy primates. Recent classifications of the genus have ranged from a single species with five subspecies (Petter and Petter-Rousseaux 1960) to seven species, one with four subspecies (Petter et al. 1977). What is certain is that a number of distinct populations of *Lepilemur* exists; what is not clear is exactly how many or at what taxonomic level they should be separated. These nocturnal animals are often extremely difficult to survey in the field; and even where reasonable samples of adequately documented museum specimens exist, it is impossible to know how much weight to place upon the often subtle variations found. Clearly, detailed comparative craniological study is needed (but of a more comprehensive sample than that used by Drenhaus [1975]), in conjunction with thorough field surveys and reanalysis of the moderately extensive (although tax-onomically very unevenly distributed) skin collections available. Even then, however, my feeling is that the resulting classification would probably owe a good deal to intuition.

For the purposes of description, I recognize here a single species of *Lepilemur*, with six subspecies. This is the arrangement which, I think, involves the minimum number of assumptions at the present woeful state of our knowledge. I fully realize, however, that certain—possibly all—of these taxa may ultimately turn out to deserve separate specific status, or that some may require subdivision while others may in the end prove not to be distinct even at the subspecies level. In any event, in view of the fact that karyotypic polymorphism can be found within a single regional population (see *L. m. septentrionalis*), it is far from evident that karyotypic differences are a legitimate basis for separating populations at the species level as, for instance, Petter et al. (1977) have chosen to do.

In all *Lepilemur* the muzzle is thinly haired, and short buccal, genal, and superciliary vibrissae are present. A few interramal vibrissae may also occur. Females have a single pair of pectoral mammae; there is a glandular development on the scrotum of the male.

1851. *Lepilemur* I. Geoffroy, L'Institut, 19, p. 341. Type species by designation of I. Geoffroy, 1851 (Cat. Méthod. Coll. Mammif. Mus. Hist. Nat. Paris, p. 76), *L. mustelinus* I. Geoffroy.

1855. *Galeocebus* Wagner, von Schreber's Säugthiere, Suppl., 5, p. 147. Based on *Lepilemur* of I. Geoffroy; type species *L. mustelinus* I. Geoffroy.

1859. *Lepidilemur* Giebel, Säugethiere, p. 1018. Emendation of *Lepilemur* I. Geoffroy.

1875. *Mixocebus* Peters, Monatsb. K. Preuss. Akad. Wiss. Berlin, for 1874, p. 690. Type species *M. caniceps = Lepilemur mustelinus* I. Geoffroy.

1875. *Lepidolemur* Peters, Monatsb. K. Preuss. Akad. Wiss. Berlin, for 1874, p. 690. Emendation of *Lepilemur* I. Geoffroy.

Lepilemur mustelinus I. Geoffroy, 1851.

Lepilemur mustelinus mustelinus I. Geoffroy, 1851. Weasel lemur.

1851. *Lepilemur mustelinus* I. Geoffroy, Cat. Méthod. Coll. Mammif. Mus. Hist. Nat. Paris, p. 76. Based on MNHN specimen, no. 133 of Rode Catalogue; adult from Madagascar, via Goudot.

Schwarz plausibly suggests that the type locality should be taken as Tamatave.

1875. *Mixocebus caniceps* Peters, Monatsb. K. Preuss. Akad. Wiss. Berlin, for 1874, p. 690. Based on a subadult female in the MB; type locality not recorded, but Schwarz (Proc. Zool. Soc. Lond., 1931, p. 420) says that its collector, Crossley, was exploring between Tamatave and Mananjary when the specimen was collected in 1872.

1894. *Lepidolemur microdon* Forsyth Major, in Forbes, Handbook to the Primates, 1, p. 88. Based on BMNH 82.3.1.6, adult from Ankafana Forest, Eastern Betsileo.

1897. *Lepidolemur mustelinus typicus* Milne-Edwards, A. Grandidier, and Filhol, Hist. Phys., Nat., Pol. de Madagascar, 10, pt. 2, 6, Atlas III, Mammif. Pl. 255.

Malagasy names. Trangalavaka, kotrika, fitiliky, hataka, varikosy (Andapa area only).

Pelage and external characters. Relatively large *Lepilemur*, with prominent naked ears, and rather long, dense fur. Upper parts chestnut brown with lighter elements (the basal color of the dichromic hairs is dark gray), but quite variable; in some individuals the dorsum becomes slightly paler caudally, while the tail invariably darkens distally. A darker median stripe can often be distinguished on crown and/or back. Underparts gray or light brown, flecked with cream. Face dark gray or brown, with light cheeks and throat.

Dimensions. Cranial length: $\bar{x} = 58.9$, $s = 1.4$, $o.r. = 56.7–62.1$, $n = 24$.

Range. Eastern humid forests, from the Tsaratanana/Andapa region south to Fort-Dauphin (fig. 3.16). A single early collecting record suggests that *mustelinus* formerly ranged north just beyond the limits of the humid forest, to the area of Vohémar.

Karyotype. $2N = 34$. Autosomes: M, O; S, 6; A, 26. Sex chromosomes: X, S; Y, A (Rumpler 1975).

Lepilemur mustelinus ruficaudatus A. Grandidier, 1867.
Red-tailed sportive lemur.

1867. *Lepilemur ruficaudatus* A. Grandidier, Rev. Mag. Zool., 2 Sér., 19, p. 256. Based on MNHN 1867–583 (no. 134a of Rode Catalogue), adult from southwest coast: Morondava.

1872. *Lepilemur pallicauda* Gray, Proc. Zool. Soc. Lond., p. 850. Based on BMNH 72.8.19.6, adult female from Morondava.

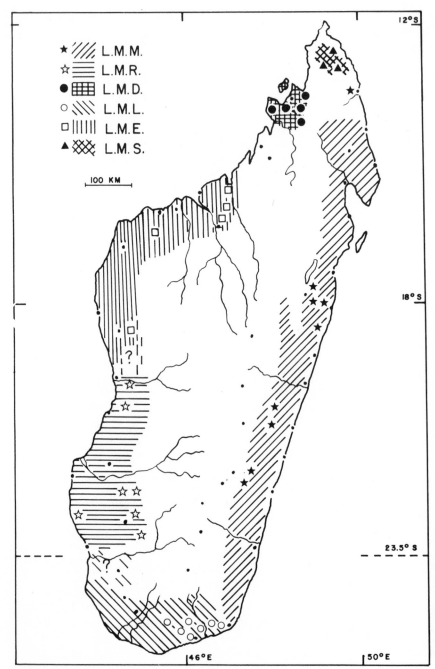

Figure 3.16. Distributions of the various subspecies of *Lepilemur mustelinus*. Shaded areas represent approximate distribution limits; symbols denote localities of museum specimens. L.M.M.: *Lepilemur mustelinus mustelinus*; L.M.R.: *L. m. ruficaudatus*; L.M.D.: *L. m. dorsalis*; L.M.L.: *L. m. leucopus*; L.M.E.: *L. m. edwardsi*; L.M.S.: *L. m. septentrionalis*.

1894. *Lepidolemur globiceps* Forsyth Major, in Forbes, Handbook to the Primates, 1, p. 89. Based on BMNH 92.11.6.1, adult male from Ambolisatra, southwest Madagascar. This synonymy is not absoutely certain. The type, the soft parts of which are preserved in spirit, and the pelage of which thus cannot at present be adequately assessed, came from a site about 25 miles north of Tuléar, and therefore within what is believed to be the range of *ruficaudatus*. The skull, however, is rather small and reminiscent of *leucopus*.

Malagasy name: Boenga.

Pelage and external characters. Dichromic hair of upper parts basally gray, light brown at tips; overall impression generally of light brown, becoming paler caudally. Tail reddish. Ears large, face pale gray to pale brown; underparts light gray flecked with cream. Throat pale. (fig. 3.17).

Dimensions. Cranial length: $\bar{x} = 56.4$, $o.r. = 52.9$–61.4, $n = 10$. Field measurements (Archbold, $n = 10$): H + B, $\bar{x} = 278$, $o.r. = 261$–304; TL, $\bar{x} = 265$, $o.r. = 240$–287; HF, $\bar{x} = 81$, $o.r. = 73$–84; EAR, $\bar{x} = 38$, $o.r. = 35$–41; (Webb, $n = 5$): H + B, $\bar{x} = 297$, $o.r. = 279$–323; TL, $\bar{x} = 246$, $o.r. = 215$–267; HF, $\bar{x} = 57$, $o.r. = 53$–61; EAR, $\bar{x} = 34$, $o.r. = 32$–

Figure 3.17. *Lepilemur mustelinus ruficaudatus* (left) and *L. m. dorsalis.*

36. **Body weight.** Average given, probably for this subspecies, by Bauchot and Stephan (1966): 915g.

Range. Western Madagascar, but limits ill defined. Range extends southward at least as far as the Onilahy River, and possibly as far as Ejeda; in the north the boundary with *edwardsi* (if indeed the two warrant separation) appears to be the Tsiribihina River (fig. 3.16).

Karyotype. 2N = 20. Autosomes: M, 2; S, 16; A, O. Sex chromosomes: X, M; Y, A (Rumpler et al. 1972).

Lepilemur mustelinus dorsalis Gray, 1870. Gray-backed Sportive lemur.

1870. *Lepilemur dorsalis* Gray, Cat. Monkeys, Lemurs, and Fruit-eating Bats in Brit. Mus., p. 135. Based on BMNH 68.9.7.5, adult male from northwest Madagascar.

1894. *Lepidolemur grandidieri* Forsyth Major, in Forbes, Handbook to the Primates, 1, p. 89. Based on BMNH 68.9.7.4, adult female from northwest Madagascar.

Malagasy name. Apongy.

Pelage and external characters. Rather small *Lepilemur*. Upper parts of most individuals are a medium to darkish brown, occasionally flecked with yellow, although in some the basal dark gray of the dichromic hair predominates caudally. Tail the color of the dorsum; that of type paler at tip. Underparts brown, only a little paler than dorsum. Face relatively blunt, somber gray to brown; ears round and quite short (fig. 3.17).

Dimensions. Cranial length ($n = 2$): 52.1, 53.0. Field measurements (Archbold, $n = 3$): H + B, $\bar{x} = 254$, $o.r. = 250–260$; TL, $\bar{x} = 266$, $o.r. = 260–278$; HF, $\bar{x} = 80$, $o.r. = 78–84$; EAR, $\bar{x} = 27$, $o.r. = 27–28$.

Range. Limited to the Sambirano region and Nosy-Bé (fig. 3.16).

Karyotype. 2N = 26. Autosomes: M, 8; S, 10; A, 6. Sex chromosomes: X, M; Y, A (Rumpler 1975).

Lepilemur mustelinus leucopus Forsyth Major, 1894. White-footed sportive lemur.

1894. *Lepidolemur leucopus* Forsyth Major, Ann. Mag. Nat. Hist., 6th ser., 13, p. 211. Based on BMNH 94.1.22.1, adult male from Fort-Dauphin.

Malagasy name. Songiky.

Pelage and external characters. Small *Lepilemur*. Upper parts, including crown and limbs, are medium to light gray, becoming yet paler caudally; hints of light brown occasionally present anteriorly. Tail is a very light brown. Underparts very pale gray or white. Ears relatively large, rounded (fig. 3.18).

Dimensions. Cranial length: $\bar{x} = 50.9$, $s = 1.2$, $o.r. = 49.1-53.5$. Field measurements (Webb, $n = 3$): H + B, $\bar{x} = 252$, $o.r. = 245-261$; TL, $\bar{x} = 236$, $o.r. = 215-261$; HF, $\bar{x} = 49$, $o.r. = 48-50$; EAR, $\bar{x} = 32$, $o.r. = 32-33$. **Body weight:** $\bar{x} = 544$, $s = 82.9$, $n = 10$ (Russell 1977). According to this author, young mature individuals weigh significantly less than older ones ($\bar{x} = 446$ vs. 600g).

Range. Dry south of Madagascar, from Fort-Dauphin westward at least to Ejeda, and possibly to the Onilahy River (fig. 3.16).

Karyotype. 2N = 26. Autosomes: M, 4; S, 14; A, 6. Sex chromosomes: X, M; Y, A (Rumpler 1975).

Lepilemur mustelinus edwardsi Forsyth Major, 1894. Milne-Edwards' sportive lemur.

1894. *Lepidolemur edwardsi* Forsyth Major, in Forbes, Handbook to the Primates, 1, p. 87. Based on BMNH 91.1.22.6, adult male collected at Betsako, near Majunga.

1898. *Lepilemur mustelinus rufescens* Lorenz, Abh. Senckenberg. Naturforsch. Gesell., 21, p. 446, Pls. 30, 31. Based on three MB specimens, adult male and two females, from Ambundubé, northwest Madagascar.

Malagasy names. Repahaka (Ankarafantsika and north), boenga (south of the Ankarafantsika).

Pelage and external characters. Quite similar in appearance to *ruficaudatus*, from which it may not in fact deserve distinction. In general, however, individuals of *edwardsi* may tend to be a little darker in coloration than those of *ruficaudatus*. Upper parts gray to gray-brown, darker anteriorly than on hindquarters. There is often a darker median stripe along the back. Tail light brown. Face gray or brownish, tending to dark. Underparts gray, flecked with cream. Ears large.

Dimensions. Cranial length: $\bar{x} = 58.4$, $s = 2.6$, $o.r. = 53.5-62.3$, $n = 21$. Field measurements (Archbold, $n = 5$): H + B, $\bar{x} = 280$, $o.r. = 272-292$; TL, $\bar{x} = 277$, $o.r. = 269-290$; HF, $\bar{x} = 80$, $o.r. = 76-85$; EAR, $\bar{x} = 37$, $o.r. = 35-38$.

Range. Western Madagascar, from the Bay of Mahajamba south

Figure 3.18. *Lepilemur mustelinus leucopus*: Berenty, southern Madagascar. Courtesy of R. D. Martin.

at least as far as Antsalova, and possibly to the Tsiribihina River (fig. 3.16).

Karyotype. $2N = 22$. Autosomes: M, 4; S, 14; A, 2. Sex chromosomes: X, A; Y, A (Rumpler, 1975).

Lepilemur mustelinus septentrionalis Rumpler and Albignac, 1975. Northern sportive lemur.

In proposing this taxon as a new species, Rumpler and Albignac (1975) created four new subspecies within it to accommodate animals of differing karyotype. However, there is no morphological or chromatic difference of any significance between individuals of any of these various karyotypes, and there is ample evidence in the form of wild-caught hybrids that all belong to a single interbreeding population. Thus if the concept of the subspecies is to retain any meaning, all of these animals must be regarded as belonging to a single subspecies.

1975. *Lepilemur septentrionalis septentrionalis* Rumpler and Albignac, Amer. J. Phys. Anthrop. 42, p. 425–29. Based on MNHN 1974–80, adult female from the Forêt de Sahafary, northern Madagascar ($2N = 38$).

1975. *Lepilemur septentrionalis sahafarensis* Rumpler and Albignac, Amer. J. Phys. Anthrop., 42, p. 425–29. Based on MNHN 1974–78, adult male from Forêt de Sahafary, northern Madagascar ($2N = 36$).

1975. *Lepilemur septentrionalis andrafiamensis* Rumpler and Albignac, Amer. J. Phys. Anthrop., 42, p. 425–29. Based on MNHN 1974–79, adult male from Chaîne de l'Andrafiamena, northern Madagascar ($2N = 38$).

1975. *Lepilemur septentrionalis ankaranensis* Rumpler and Albignac, Amer. J. Phys. Anthrop., 42, p. 425–29. Based on MNHN 194–77, youngish female from Forêts de l'Analamerana, par Ambondromifehy, northern Madagascar ($2N = 36$).

Malagasy names. Mahiabeala; songiky.

Pelage and external characters. Upper parts gray, darkest (but never extremely dark) on crown, becoming lighter caudally to pale gray rump and hindlimbs. There tends to be a darker median stripe along crown and back. On the foreparts there may be brownish, slightly olivaceous, or rosy tints. Tail pale brown, tending to darken slightly towards the tip. Underparts are gray, as are the muzzle and the cheeks. Ears moderate in size.

Dimensions. Cranial length: $\bar{x} = 53.0$, $o.r. = 49.9–56.5$, $n = 5$. Field measurements (Archbold, single specimen): H + B, 278; TL, 247; HF, 84; EAR, 30.

Range. Extreme north of Madagascar, north of Ambilobé and R.N. 5A, and to the south and east of Mt. d'Ambre (fig. 3.16).

Karyotype. Polymorphic: 2N = 34 (M, O; S, 6; A, 26; X, M; Y, A); 36 (M, O; S, 4; A, 30; X, M; Y, A), and 38 (M, O; S, 2; A, 34; X, M; Y, A). Hybrids of 2N = 35 and 2N = 37 have been found (Rumpler and Albignac 1975).

Hapalemur I. Geoffroy, 1851. Gentle Lemurs

Earlier synonymies of *Hapalemur* have included *Prolemur* Gray, 1870 as a full generic synonym. But although Gray's format is difficult to interpret, it is fairly evident that the author intended *Prolemur* as no more than a subgeneric appelation. I have therefore preferred to include *Hapalemur (Prolemur) simus* Gray, 1870 in the species synonymy of *H. simus* Gray, 1870.

1851. *Hapalemur* I. Geoffroy, L'Institut, 19, no. 929, p. 341. Based on "le Maki griset des auteurs"; type species *H. griseus* by designation of I. Geoffroy, Cat. Méthod. Coll. Mammif. Mus. Hist. Nat. Paris, p. 75, 1851.

1859. *Hapalolemur* Giebel, Von Schreber's Säugthiere, p. 1018. Lapsus for *Hapalemur* I. Geoffroy.

1913. *Myoxicebus* Elliot, Rev. Primates, 1, p. 124. Type species *Lemur griseus* E. Geoffroy; genus name based on a misinterpretation of *Mioxicebus griseus* Lesson, 1840 (= *Cheirogaleus major* E. Geoffroy, 1812).

Hapalemur griseus (Link, 1795).

For many years it was customary to recognize two subspecies of *Hapalemur griseus*, each corresponding to one of the two species given by I. Geoffroy in 1851 (Cat. Méthod. Coll. Mammif. Mus. Hist. Nat. Paris 19, p. 341): *H. griseus* and *H. olivaceus*. I. Geoffroy based the first of these on *Lemur griseus* of E. Geoffroy (Mag. Encyclop., 1, p. 48, 1796), although Link (Beyträge Naturgesch. 1, pt. 2, 1795) had proposed the same name a year earlier than had E. Geoffroy, on the basis of the same specimen (the "Petit maki gris" illustrated by Buffon, Hist. Gen. et Partic., Suppl., 7, p. 121, Pl. 34, 1789). I. Geoffroy was hesitant to propose the second species, a new one; indeed he would not have done so "sans la nécessité de donner dans le Catalogue un tableau complet de la Collection" (p. 75), since he was less than confident of the characters (of the lower jaw) separating either of his types (both immature and without locality) from *H. griseus*. This lack of confidence appears in the event to have been justified; and although the skull of the more mature of the two syntypes of *H. olivaceus* can no longer be located for comparison, it is reasonably clear that these two names apply to a single subspecies, and that all *H. griseus* from the humid eastern forests of Madagascar may be regarded as belonging to the nominate subspecies, *H. g. griseus* (Link, 1795).

Two other subspecies of *H. griseus*, long known but unrecognized as separate biological entities, do, however, exist: a large form from Lake Alaotra, and a relatively small one from the deciduous forests of the west. The first of these was published as a "new" subspecies, *H. g. alaotrensis*, by Petter (in Petter et al., Faune de Madag., 44, 215) in 1977; but in fact that name was employed at least as early as 1975 by Rumpler (Lemur Biol., pp. 28, 33). Similarly, in typical cavalier fashion, Petter et al. (1977) credited Albignac and Rumpler (J. Hum. Evol. 2, 1973) with authorship of the name *H. g. occidentalis* for the second of these subspecies, when in fact in the publication cited those authors referred to the form only as "*Hapalemur griseus* ssp. 1." The earliest publication of the name *H. g. occidentalis* of which I am aware is again by Rumpler in Lemur Biol., pp. 28, 33, 1975. I am prepared for the sake of simplicity to accept Rumpler 1975 as valid authorship of both subspecies names despite the flagrant disregard for the fundamentals of nomenclature characteristic of this trio of authors. There remains, however, some question as to whether *occidentalis* should be regarded as a junior synonym of *schlegeli* Pocock, 1917 (Ann. Mag. Nat. Hist., 8 ser., 19, 348). In proposing *H. schlegeli* on the basis of a specimen now in the BMNH, Pocock equated his new species with the material from the northwest of Madagascar collected by Pollen and Van Dam, published as *H. griseus* by Schlegel (Recherches Faune Madag. 2, Mammif. et Oiseaux, p. 6, Pl. 7, figs. 4a–3, 1868; actually, the text reads as if written by Pollen), and allocable to the western subspecies. Pocock's holotype, however, is of uncertain provenance; and although its definitive assignment may have to await a through craniological study of the populations of *Hapalemur*, it seems most likely at this point that the specimen is representative of the eastern subspecies.

Hapalemur griseus griseus (Link, 1795). Gray gentle lemur.

1795. *L[emur] griseus* Link, Beyträge Naturgesch. 1, pt. 2, p. 65. Based on the "Petit maki gris" of Buffon, Hist. Nat., Gén. et Partic., Suppl., 7, p. 121, Pl. 34. Madagascar.

1820. *Lemur cinereus* Desmarest, Mammalogie, 1, p. 101. Based on Buffon (see above) and other references, including an erroneous one to "*Lemur cinereus*" of E. Geoffroy. Madagascar.

1851. *Hapalemur olivaceus* I. Geoffroy, Cat. Coll. Méthod. Mammif. Mus. Hist. Nat., Primates, p. 75. Based on two immature specimens. According to Rode (Cat. Types Mammif., 1B, p. 59, 1939) these specimens are catalogued as MNHN 1841–117 (except for the skull of the older individual, which was registered as MNHN A 3007, but which is now lost). Madagascar.

1917. *Hapalemur schlegeli* Pocock, Ann. Mag. Nat. Hist., Ser. 8, 19, p. 348. Based on BMNH 17.3.27.2, adult from Zoo. Madagascar.

Malagasy names. Bokombolo, kotrika.

Pelage and external characters. Relatively short faced. Dorsally the dense pelage is a somber gray with olive brown elements (many hairs are banded) which are pronounced on the back and especially the crown. Tail a variably dark gray; underparts light brown to light gray. Ears hairy, rounded, and substantially hidden in surrounding fur (fig. 3.19). Cheeks, face and sometimes forehead gray, darkest between eyes. Males, in particular, possess a strongly developed glandular area high on the inside of the upper arm (a feature sometimes interpreted in females as an additional pair of nipples), and a spiny "brush" that covers another glandular structure on the inside of the lower arm above the wrist. One pair of mammae in females. Carpal vibrissae present, as is a full set of facial vibrissae, with the exception of the interramal tuft; mystacial vibrissae particularly well developed.

Dimensions. Cranial length: $\bar{x} = 64.8$, $s = 1.9$, $o.r. = 61.3 - 68.0$, $n = 24$. Field measurements (Archbold, $n = 9$): H + B, $\bar{x} = 284$, $o.r. = 270–310$; TL, $\bar{x} = 366$, $o.r. = 320–400$; HF, $\bar{x} = 85$, $o.r. = 75–93$; EAR, $\bar{x} = 32$, $o.r. = 28–39$. **Body weight:** $\bar{x} = 830$, $o.r. = 737–1,004$, $n = 4$.

Range. Occurs throughout the humid forests of eastern Madagascar, from the Tsaratanana Massif to Fort-Dauphin (fig. 3.20), although rarely in great density. Extremely sparse today in most of the northwestern end of its range.

Karyotype. $2N = 54$. Autosomes: M, 4; S, 6; A, 42. Sex chromosomes: X, A; Y, A (Chu and Swomley 1961).

Figure 3.19. *Hapaiemur* species and subspecies. Top left: *Hapalemur griseus griseus*; bottom left: *H. g. alaotrensis*; bottom right: *H. g. occidentalis*; top right: *Hapalemur simus.*

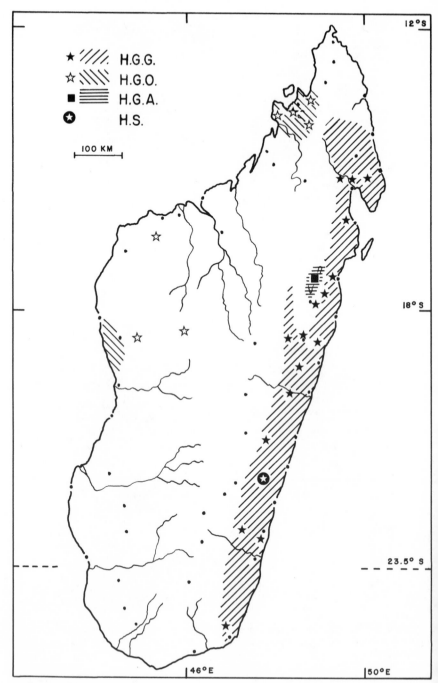

Figure 3.20. Distributions of *Hapalemur* species and subspecies. Shaded areas denote approximate range limits; symbols represent localities of museum specimens except in the case of *H. simus,* where the locality is one of capture and observation. H.G.G.: *Hapalemur griseus griseus*; H.G.O.: *H. g. occidentalis*; H.G.A.: *H. g. ala-otrensis*; H.S.: *Hapalemur simus.*

Hapalemur griseus alaotrensis Rumpler, 1975. Alaotran gentle lemur.

1975. *Hapalemur griseus alaotrensis* Rumpler, Lemur Biol., pp. 28, 33. Based on MNHN 1975–1122, male from Lake Alaotra.

Malagasy name. Bandro.

Pelage and external characters. Appreciably larger than *H. g. griseus,* slightly rounder headed and with relatively longer fur in most cases (fig. 3.19). Pelage coloration generally similar to *H. g. griseus,* but somewhat darker. Cutaneous glands and associated structures as in *H. g. griseus,* vibrissae also.

Dimensions. Cranial length: $\bar{x} = 70.4$, *o.r.* = 67.0–74.5, $n = 4$. Field measurements (probably by Lamberton, single specimen, measuring technique unknown): H + B, 380; TL, 390; HF, 50; EAR, 28; Petter et al. (1977) state that H + B is ca. 400 mm, and TL "about the same."

Range. Restricted to the reed beds of Lake Alaotra and the surrounding marshes (fig. 3.20).

Karyotype. 2N = 54. Autosomes: M, 4; S, 6; A, 42. Sex chromosomes: X, A; Y, A (Rumpler and Albignac 1973a).

Hapalemur griseus occidentalis Rumpler, 1975. Western gentle lemur.

1975. *Hapalemur griseus occidentalis* Rumpler, Lemur Biol., pp. 28, 33. Based on a syntype series in the collections at Tsimbazaza, Tananarive (Petter et al. 1977); "west Madagascar."

Malagasy names. Bekola, kofi (Antsalova region); ankomba valiha (Ambanja region).

Pelage and external characters. Again, pelage coloration close to that of the slightly larger *H. g. griseus,* but in many individuals a bit lighter. External characters as in *H. g. griseus,* except that the ears tend to be more visible and the face to be slightly sharper (fig. 3.19).

Dimensions. Cranial length: $\bar{x} = 62.1$, *o.r.* = 61.1–63.4, n = 7. Field measurements (Archbold, $n = 2$, both from the northern area): H + B, 285, 273; TL, 360, 400; HF, 81, 75; EAR, 32, 28.

Range. Two isolates are known today, both in western Madagascar (fig. 3.20). One is in the region of Antsalova/Lake Bemamba, between Maintirano and Belo-sur-Tsiribihina; the other in the Sambirano region, from Maromandia to Beramanja. In both of these areas, *Hapalemur* is confined to forests characterized by bamboo or bamboo vines ("viky"). Collecting records also exist from other localities, notably to the east of the Antsalova/Lake Bemamba isolate, and in the Namoroka area (fig. 3.20), but *Hapalemur* seems to be absent from these areas today. It is also possible that a small isolate exists in the area of Ankazoabo, but I have been unable to confirm this.

Karyotype. 2N = 58. Autosomes: M, 2; S, 4; A, 50. Sex chromosomes: X, A; Y, A (Rumpler and Albignac 1973a).

Hapalemur simus Gray, 1870. Broad-nosed gentle lemur.

1870. *Hapalemur simus* Gray, Proc. Zool. Soc. Lond., p. 828, Pl. 52, figs. 1–4. Based on BMNH 70.9.2.2, adult, "probably female." Madagascar.

1870. *Hapalemur (Prolemur) simus* Gray, Proc. Zool. Soc. Lond., p. 828, Pl. 52, figs. 1–4. Based on BMNH 70.9.2.2, adult, "probably female." Madagascar.

Malagasy names. Varibolo (eastern Betsileo); tan-tang (Antongil Bay region, according to Audebert in letter to Schlegel, March 4, 1878).

Pelage and external characters. Substantially larger than *H. griseus,* more robustly built, with a blunter muzzle (fig. 3.19). Upper parts battleship gray, washed in some areas, most notably the shoulders and top of head, with olive brown. Brownish pygal patch present; tail darkens at tip. Ears more noticeable than in *H. griseus,* and moderately tufted with white hairs. Muzzle in most cases dark; cheeks, forehead and throat brownish gray. Underparts lighter in color than dorsum, generally gray brown. A well-developed cutaneous gland is present on the distal part of the inner surface of the upper arm, and distal to the elbow joint is the homologue of the carpal gland of *H. griseus,* covered in this case by a spiny "brush" in males. The species also exhibits a pronounced gular cutaneous gland. Carpal vibrissae present, as also a full set of facial vibrissae with the exception of the interramal tuft. Single pair of mammae in females.

Dimensions. Cranial length: $\bar{x} = 80.9$, $o.r. = 78.4–81.9$, $n = 4$.

Range. Extremely rare. Known today only from the humid forest east of Fianarantsoa (fig. 3.20). A BMNH specimen (not the type) came from an unidentifiable locality in Central Betsileo, hence plausibly from the same area; the specimen in the RMNH, however, was collected by Audebert at "Passumbée," a locality not precisely identifiable but without any doubt in the region of the Bay of Antongil. The species may thus at one time have been fairly widespread in the humid forests of eastern Madagascar. Additionally, a large species of gentle lemur, possibly *H. simus* and certainly closely related, is known as a subfossil from Ampasambazimba, on the Itasy Massif in central Madagascar (see chapter 5).

Karyotype. 2N = 60. Autosomes: M, O; S, 4; A, 54. Sex chromosomes: X, M; Y, A (Rumpler and Albignac 1973a).

FAMILY INDRIIDAE BURNETT, 1828

Indri E. Geoffroy, 1796

1796. *Indri* E. Geoffroy, In E. Geoffroy and G. Cuvier, Mag. Encyclop., 1, p. 46. New genus for *L. indri* of Gmelin, based on "l'Indri" of Sonnerat (Voy. Ind. Or. Chine, 2, p. 142, pl. 88, 1782).

1805. *Indris* G. Cuvier, Leçons d'Anat. Comp., 1, Tab. Gén. Emendation of *Indri* E. Geoffroy.

1811. *Lichanotus* Illiger, Prodromus Syst. Mammal. et Av.: 72. Based on *L. indri* of Gmelin.

1815. *Indrium* Rafinesque, Analyse de Nature, p. 54. New name for *Indri* E. Geoffroy.

1827. *Lichanotes* Temminck, Monogr. de Mammalogie, 1, p. xvi. Lapsus for *Lichanotus* Illiger.

1840. *Pithelemur* Lesson, Spec. Mammif., p. 208. Based on *L. indri* of Gmelin.

Indri indri (Gmelin, 1788). Indris.

1788. *Lemur indri* Gmelin, Linnaeus Syst. Nat., 13 ed., 1, p. 42. Based on Pl. 88 of Sonnerat (Voy. Ind. Or. Chine, 2, p. 142, 1782): "l'Indri." No locality other than "Madagascar" given by either Sonnerat or Gmelin.

1796. *Indri brevicaudatus* E. Geoffroy, In E. Geoffroy and G. Cuvier, Mag. Encyclop. 1, p. 46. Based on "l'Indri" of Sonnerat, *L. indri* of Gmelin.

1799. *Indri niger* Lacépède, Tableau . . . des Mammifères, 1, p. 5. Described, but no basis given.

1825. *Indris ater* I. Geoffroy, Dict. Class. d'Hist. Nat., viii, p. 534. Lapsus for *I. niger* Lacépède.

1840. *Pithelemur indri* Lesson, Spec. Mammif. p. 208. Based on *L. indri* Gmelin.

1872. *Lichanotus mitratus* W. Peters, Monatsb. K. Preuss. Akad. Wiss. Berlin for 1871, p. 360. Based on MB 4671, collected by Crossley in northern Madagascar.

1872. *Indris variegatus* Gray, Ann. Mag. Nat. Hist., Dec. 4, 10, p. 474. Based on BMNH 72.11.8.12, collected by Crossley, possibly at the same time as MB 4671, above.

Malagasy names. Babakoto, amboanala (northern part of range); endrina (southern portion of range).

Pelage and external characters. (fig. 3.21). The largest of the Malagasy primates, *Indri indri* is perhaps most strikingly distinguished by its possession of only a vestigial tail. Pelage coloration and pattern are highly variable, but this variation is not consistent geographically, except to the extent that there is—perhaps—a tendency on the average toward a slightly lighter coloration in the south of the range. The predominant fur color is black; this is supplemented in the darkest individuals by small white or whitish patches on the crown, or the flanks, or the forelimbs and the thighs, or any combination of these; a triangular white or otherwise pale pygal patch is always present, and may be tiny or extend cranially a con-

Figure 3.21. *Indri indri*: young adult male at Analamazoatra, eastern Madagascar. Courtesy of J. I. Pollock.

siderable distance. Other individuals have in addition light gray patches of variable extent on the limbs and/or dorsum and/or flanks. Golden fur sometimes occurs in the pygal and tarsal regions. The well-haired ventral area is often dark brown, but may be much lighter, ranging to pale gray; heel always pale; tail darker at tip than at base. Face light brown or black, sometimes with pale patches on cheeks or over eyes. Ears invariably black, prominent, and tufted. Fur long, dense, silky. Facial vibrissae rather sparse although all groups except interramal are usually represented. Although "cheek marking" has been observed in *Indri* (Pollock 1975), no specialized gland of the kind found in other indriids appears to be discernible. Females have a single pair of pectoral mammae.

Dimensions. Cranial length: $\bar{x} = 102.7$, $s = 3.9$, $o.r. = 97.1–117.7$, $n = 33$. Field measurements (Archbold, $n = 10$); H + B, $\bar{x} = 607$, $o.r. = 568–695$; TL, $\bar{x} = 56$, $o.r. = 51–55$; HF, $\bar{x} = 607$, $o.r. = 568–695$; EAR, $\bar{x} = 53$, $o.r. = 51–55$. **Body weight** (n = 1): 6,250g (Bauchot and Stephan 1966).

Range. *Indri* appears to be confined today to what remains of the rain forest of eastern Madagascar, approximately between the latitudes of Sambava and Mahanoro (fig. 3.22). It does not, however, inhabit the Masoala peninsula; Schwarz (1931) believed that the type of Peters' "*Lichanotus mitratus*" came from east of the Bay of Antongil, and identified Audebert's locality of Antsompirina as being in that area, but I have been unable to confirm this. Locality data for museum specimens are poor, but Lamberton recorded in 1939 that the genus occured as far south as Maranjary, and Petter et al. (1977) note that the northern part of its range has contracted considerably during the past few decades. My own surveys in the Andapa Basin suggest that *Indri* is at best exceedingly rare in that region, and that it may well have been extirpated already from the northern extremity of its range as denoted in figure 3.22. Subfossil evidence indicates that within the past millenium or so, populations of *Indri* occupied the interior of Madagascar at least as far west as the Itasy Massif, well to the west of Tananarive.

Karyotype. 2N = 40. Autosomes; M, 12; S, 20; A, 6. Sex chromosomes: X, M; Y, A (Rumpler 1975).

Avahi Jourdan, 1834

1834. *Avahi* Jourdan, L'Institut, 2, p. 231. Based on a skull and skin supposed to be those of Sonnerat's "Maquis à bourres" (Voy. Ind. Or. Chine, 2. pl. 89). Type locality: "pays des *Betanimènes*";

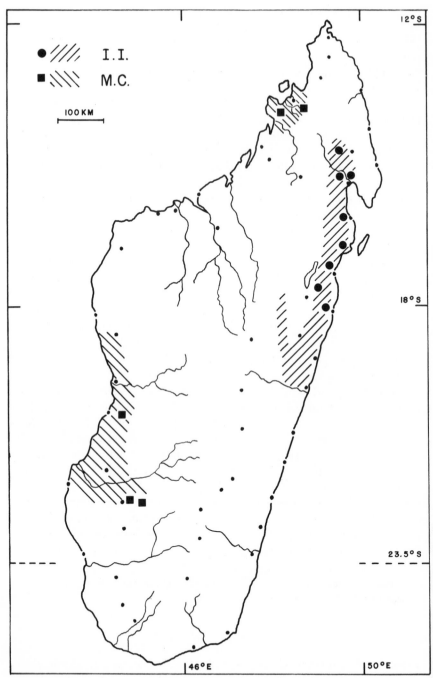

Figure 3.22. Distributions of *Indri indri* (I.I.) and *Mirza coquereli* (M.C.). Shaded areas represent approximate range limits; symbols denote localities of museum specimens.

range given as "forests of east coast of Madagascar from the mouth of the Manangara to the Baie d'Antongil."

1834. *Microrhynchus* Jourdan, Thèse, Fac. des Sciences, Grenoble. Preoccupied by Megerle, 1823 for Coleoptera. Nomen nudum. Not seen.

1835. *Avahis* I. Geoffroy, In Gervais, Résumé Leçons Mammif., p. 23. Emendation of *Avahi* Jourdan.

1839. *Habrocebus* Wagner, von Schreber's Säugethiere, Suppl. 1, p. ix. Based on *L. laniger* of Gmelin (Linnaeus' Syst. Nat., 13 ed., p. 44, 1788).

1840. *Semnocebus* Lesson, Spec. Mammif., p. 209. New name for *Avahi* Jourdan, etc.

1841. *Iropocus* Gloger, Gemein. Hand.-u. Hilfsb. Naturgesch., 1, p. 43. Based on *L. laniger* of Gmelin.

Avahi laniger (Gmelin, 1788). Woolly lemurs.

Two subspecies of this species are generally recognized, one confined to the northwest of Madagascar, the other, marginally larger, to the humid forest of the east. I am not fully convinced of the usefulness of this distinction, but nonetheless maintain it here. It is often believed that the correct name for the genus is *Lichanotus* Illiger, 1811, which certainly antedates *Avahi* Jourdan, 1834, but which is in fact a synonym of *Indri*, having been proposed by Illiger to replace *Lemur indri* and *Lemur laniger* of Gmelin. It was Jourdan who first proposed that the woolly lemur should occupy a genus apart from that of the indris. The following synonymies omit the two terms derived by Schwarz (1931) from Milne-Edwards and Grandidier (1875), as having no adequate base in nomenclature.

Avahi laniger laniger (Gmelin, 1788). Eastern woolly lemur.

1788. *Lemur laniger* Gmelin, Linnaeus, Syst. Nat. 13. ed., 1, p. 44. Based on the "Maquis à bourres" of Sonnerat (Voy. Ind. Or. Chine, 2, Pl. 89). Madagascar.

1795. *Lemur brunneus* Link, Beyträge Naturgesch., 2, p. 165. Based on Gmelin's *L. laniger,* and on Buffon (Hist. Nat., Gén. et Part., Suppl., 7, Pl. 35; "Autre espèce de Maki").

1796. *Indri longicaudatus* E. Geoffroy, Ann. Mus. Nat. Hist., Paris, 19, p. 158. Emendation of *Indri* E. Geoffroy.

1840. *Habrocebus lanatus* Wagner, von Schreber's Säugthiere, Suppl.

1, p. 258, Pl. 42A. Based on the "Maquis à bourres" of Sonnerat.

1844. *Lichanotus avahi* van der Hoeven, Tijdschr. Nat. Gesch. Physiol., 11, p. 38. Based on Sonnerat, "Maquis à bourres," and others.

1898. *A[vahis] laniger orientalis* Lorenz, Abh. Senckenberg. Naturforsch. Gesell., 21, p. 452. East coast of Madagascar, based on discussion by Milne-Edwards and A. Grandidier (Hist. Nat. Madag., 7 (1), p. 327–28).

Malagasy names. Fotsifé (northern part of range); ampongy, avahy (south of Bay of Antongil).

Pelage and external characters. Head rounded, face short, eyes large. Pelage dense (except ventrally) and quite short, but tending to form tight curls, particularly on dorsum and hind limbs (fig. 3.23).

Figure 3.23. Female *Avahi laniger laniger* carrying infant: Mandena, southeast Madagascar. Courtesy of R. D. Martin.

Many dorsal hairs dichromic, giving flecked appearance to pelage. Foreparts vary from gray brown to an olivaceous red brown; pelage becomes paler caudally and more hair is tipped with cream or yellow. Ventrum gray, with rather downy hair revealing unpigmented skin. Tail rusty red, darkening toward tip. Small pale pygal patch always present; perianal region and insides of thighs distinctly white (hence the name fotsifé). Ears small, hidden in often reddish fur but not distinctly tufted; face very short haired and usually brown; most individuals possess a pale transverse band of varying width across the forehead; some have differentiated pale supraorbital patches. Throat pale; sometimes cheeks also. In both sexes paired gular glands are present beneath the mandibular angles; those of females are whitish, those of males, brown (Bourlière et al. 1956a). No secretion could be distinguished in individuals examined by Petter (1962), who also noted cutaneous glandular development in males on the posterior part of the scrotum. Facial vibrissae reduced.

Dimensions. Cranial length: $\bar{x} = 54.6$, $s = 1.2$, $o.r. = 52.7–57.0$, $n = 19$. Field measurements (Archbold, $n = 3$): H + B, $\bar{x} = 275$, $o.r. = 265–280$; TL, $\bar{x} = 340$, $o.r. = 325–354$; HF, $\bar{x} = 29$, $o.r. = 19–34$; EAR, $\bar{x} = 26$, $o.r. = 19–34$; (Webb, $n = 3$): H + B, $\bar{x} = 285$, $o.r. = 270–292$; TL, $\bar{x} = 309$, $o.r. = 281–342$; HF, $\bar{x} = 69$, $o.r. = 64–77$; EAR, $\bar{x} = 27$, $o.r. = 25–28$. **Body weight:** 600–700 g (Petter et al. 1977); 1,279 g (n = 1, Bauchot and Stephan 1966).

Range. *Avahi laniger laniger* is found virtually throughout the eastern strip of humid forest (fig. 3.24), although infrequently at high density. Petter et al. (1977, Carte 10) indicate that the subspecies occurs in the north as far as the Tsaratanana Massif; but *Avahi* is not among the genera listed by Andriamampianina and Peyrieras (1972) for the Tsaratanana Reserve. Whatever its precise limits, it is clear that *A. l. laniger* is at the very least extremely rare at the northern end of its range. Subfossil evidence indicates that *Avahi* formerly occupied the center of Madagascar, at least as far west as Analavory.

Karyotype. 2N = 66. Autosomes: M, 4; S, O; A, 60. Sex chromosomes: X, M; Y, A (Petter et al. 1977).

Avahi laniger occidentalis (Lorenz, 1898). Western woolly lemur.

1898. *Avahis laniger occidentalis* Lorenz, Abh. Senckenberg. Naturforsch. Gesell. 21, p. 452. Based on a skin and skull collected by Voeltzkow at Ambundubé, north of Baie de Bombetoka, 1892.

Malagasy names. Fotsifé, tsarafangitra.

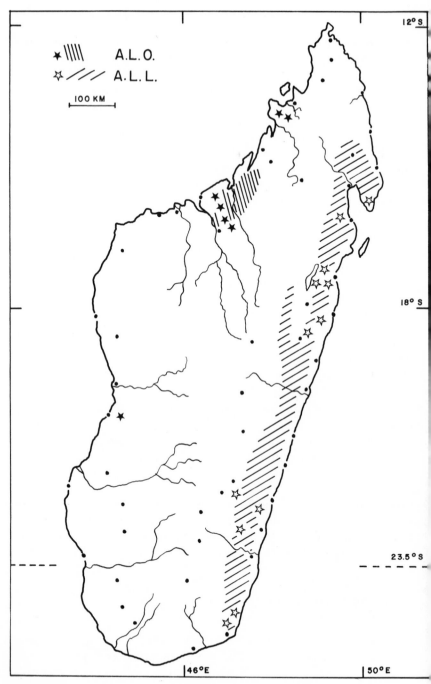

Figure 3.24. Distributions of *Avahi laniger* subspecies. Shaded areas represent approximate range limits; symbols denote localities of museum specimens. A.L.O.: *Avahi laniger occidentalis*; A.L.L.: *A. l. laniger*.

Pelage and external characters. Most specimens are distinctly lighter in color than those from eastern Madagascar. Dorsally a light to medium gray, with brown or olivaceous elements which pale caudally. Fur quite short, woolly, flecked with cream or white in the eastern form. Pygal patch small, very pale; tail usually gray but sometimes has reddish elements. In most cases the face, throat, cheeks and ventrum are light colored; the ears are small and obscured by the surrounding fur. The facial vibrissae are greatly reduced. Gular and scrotal cutaneous glands presumably present as in the eastern subspecies. Specimens from the southern (present) part of the recorded ranges are characteristically lighter in color, as described above; certain museum specimens from the north (see below) are darker, i.e., differing in the direction of the eastern form. Since the northern population on the west coast now appears to be extinct, one can only guess at its actual relationships, particularly as the entire sample of specimens from the west is tiny, and not all individuals carry precise locality data.

Dimensions. Cranial length: $\bar{x} = 50.4$, $o.r. = 49.8–51.4$, $n = 3$.

Body weight (n = 4): 859g (Bauchot and Stephan, 1966).

Range. *Avahi l. occidentalis* is confined today to a relatively restricted area to the north and east of the Betsiboka River, from the Ankarafantsika to the Bay of Narinda (fig. 3.24). Early material, however, was collected well to the north of this area, and indeed, early accounts of the range of the species (e.g., Milne-Edwards and Grandidier 1875; Kaudern 1915) placed it between Anorontsanga (west side of Ampasindava Peninsula) and Mt. d'Ambre. Whether this population was by then disjunct from that farther south (or continuous, via the Sambirano and the Tsaratanana Massif, with that further east), will probably never be known, although both possibilities seem rather unlikely despite the intriguing apparent color differences between the halves of the western population (see above). One specimen in the RMNH, a juvenile collected by Van Dam in 1868, is said to have come from Morondava. If this provenance is correct, it implies a vast earlier extension southward of the range of the western woolly lemur.

Karyotype. Unknown.

Propithecus Bennett, 1832. The Sifakas

1832. *Propithecus* Bennett, Proc. Comm. Sci. Corr. Zool. Soc. Lond., Pt. 2, p. 20. Type species *P. diadema*; Madagascar.

1833. *Macromerus* A. Smith, S. Afr. Quart. Jour., 2, p. 49. Type species *M. typicus* A. Smith.

Propithecus diadema Bennett, 1832.

I recognize, provisionally, five subspecies of this, the larger of the two species of *Propithecus*. It is not entirely clear, however, whether *P. d. edwardsi* and *P. d. holomelas* merit distinction from each other (see below). Individuals of all subspecies possess short, rather sparse, superciliary, genal, buccal and interramal vibrissae. Females have a single pair of mammae, pectoral but close to the axilla.

Propithecus diadema diadema Bennett, 1832. Diademed sifaka.

1832. *Propithecus diadema* Bennett, Proc. Comm. Sci. Corr. Zool. Soc. Lond., Pt. 2, p. 20. Based on BMNH 55.12.24.58, skin and skull presented by Telfair, Madagascar without exact locality.
1833. *Macromerus typicus* A. Smith, S. Afr. Quart. Jour., 2, p. 49. Based on a specimen owned by Verreaux.
1862. *Indris albus* Vinson, Compt. Rend. Acad. Sci. Paris, 55, p. 829. Based on an individual shot by a member of Vinson's party at Analamasoatrao, west of Andevoranto.

Malagasy name. Simpona.

Pelage and external characters. Perhaps the most strikingly beautiful of all the lemurs (fig. 3.25). Fur long, dense, silky. Face hairless, black; forehead, cheeks, throat white. Ears naked, largely hidden in bushy white hair of head. Crown almost always black, extending on to neck and shoulders. This dark fur continues a variable distance caudally; in some individuals it shades immediately behind the shoulders to a light silver gray, in others it remains dark almost to the deep golden pygal region. Hindquarters and hindlimbs usually a light gold. Tail most often white, occasionally pale golden, as are forelimbs. Extremities black. Ventral fur thicker than that of *P. verreauxi*: light silver or light golden, revealing pale skin beneath. Muzzle short; eyes have the typical "spectacled" appearance of all *P. diadema*. Males possess a large oval reddish brown gular cutaneous gland in the midline of the throat. Petter (1962) notes the presence of a perianal patch of similar color which he presumes is also glandular.

Dimensions. Cranial length: $\bar{x} = 91.6$, $s = 3.1$, $o.r. = 88.1–98.8$, $n = 15$. W. L. Abbott gave the weight of two individuals, both female, as 14 and 15 lbs. (approx. 6.3 and 6.7 kg). He noted that the latter seemed fat.

Range. The precise limits of the range of *P. d. diadema* are unknown, although the situation is clearer than for some other mem-

Figure 3.25. Unnamed variant of *Propithecus diadema* from near Daraina, northern Madagascar, provisionally ascribed to *P. d. candidus* (left; see text); right, *P. d. diadema*.

bers of the species. *P. d. diadema* occurs throughout the primary forest of the eastern humid zone between the Mangoro River and the approximate latitude of Maroantsetra (fig. 3.26), although it does not appear to exist in the immediate vicinity of Maroantsetra itself. Nowhere does it occur in high density.

Karyotype. 2N = 42. Autosomes: M, 18; S, 14; A, 8. Sex chromosomes: X, M; Y, A (Rumpler 1975).

Propithecus diadema candidus A. Grandidier, 1871. Silky sifaka.

1871. *Propithecus candidus* A. Grandidier, Compt. Rend. Acad. Sci. Paris, 72: 232. Based on observations made by Grandidier; forests northwest of Baie d'Antongil.

1872. *Propithecus sericeus* Milne-Edwards and A. Grandidier, Rev. Mag. Zool. (2) 23, p. 274. Based on MNHN 1887–55 (no. 151a of Rode Catalogue); Sambava.

Malagasy name. Simpona.

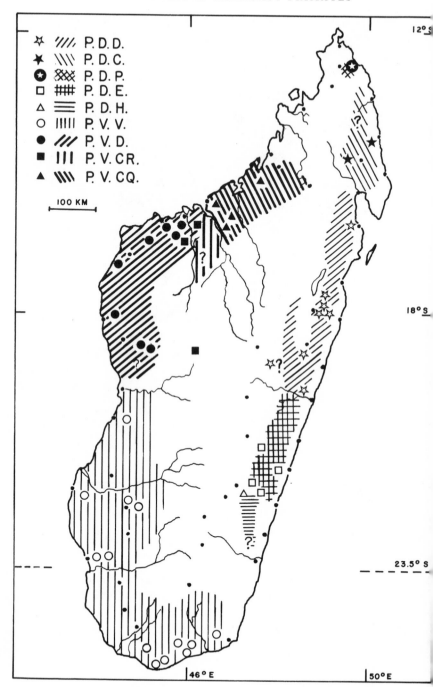

Pelage and external characters. Face black, hairless; ears mostly hidden in fur of head. Dense, silky pelage uniformly white, except that in some individuals pale to darkish silver gray tints may appear on the crown, back, and limbs, and the dorsal aspect of the anterior extremity is black or occasionally gray. An indistinct pygal coloration —pale gold or pale brown—is usually present. Gular gland present in males.

In 1974 I discovered a population of *P. diadema* living in a dry forest near Daraina, some 30 km northwest of Vohémar: an area from which the species had not previously been reported. This population, which I provisionally refer to *P. d. candidus* (because it is only very lightly pigmented and because of its location) does, however, differ in certain respects from that subspecies (fig. 3.25), and may well deserve subspecific recognition in its own right. Individuals of this population are uniformly white, except in possessing a bright orange patch between the ears (fig. 3.25). Additionally, in contrast to other *P. diadema*, they possess prominent, highly visible ears.

Dimensions. Cranial length: $\bar{x} = 90.5$, $s = 1.65$, $o.r. = 87.3-92.2$, $n = 7$. Field measurements (Archbold, $n = 3$): H + B, $\bar{x} = 522$, $o.r. = 500-545$; TL, $\bar{x} = 465$, $o.r. = 445-510$; HF, $\bar{x} = 178$, $o.r. = 170-183$; EAR, $\bar{x} = 47$, $o.r. = 46-49$.

Range. This subspecies occurs throughout the humid forest belt north of Maroantsetra to the Andapa Basin and the Marojejy Massif (fig. 3.26). Collecting records indicate that it once occurred at least as far north as Sambava. If the rather atypical population described above is properly allocable to *candidus*, then the overall range of the subspecies extends at least as far to the north as Daraina. It is unlikely, however, that this range is anything like continuous today between Marojejy (which lies to the north and the east of Andapa) and Daraina. *Propithecus d. candidus* is extremely rare today throughout its range, and it is unclear whether this range has ever included the Masoala Peninsula. It is possible that some of Audebert's collecting localities were to the east of the Bay of Antongil, although I have been unable to confirm this.

Karyotype. Unknown.

◀━━━━

Figure 3.26. Distributions of *Propithecus* species and subspecies. Shaded areas represent approximate range limits; symbols denote the localities of museum specimens. P.D.D.: *Propithecus diadema diadema*; P.D.C.: *P. d. candidus*; P.D.P.: *P. d. perrieri*; P.D.E.: *P. d. edwardsi*; P.D.H.: *P. d. holomelas*; P.V.V.: *Propithecus verreauxi verreauxi*; P.V.D.: *P. v. deckeni*; P.V.CR.: *P. v. coronatus*; P.V.CQ.: *P. v. coquereli*.

Propithecus diadema edwardsi A. Grandidier, 1871. Milne-Edwards' sifaka.

1871. *Propithecus edwardsi* A. Grandidier, Compt. Rend. Acad. Sci. Paris, 72, p. 232. Based on Grandidier's observations in Madagascar; forest west of Mananjary.

1872. *Propithecus bicolor* Gray, Ann. Mag. Nat. Hist., Dec. 4, 10, 206. Based on BMNH 72.8.19.1 (lectotype, selected by Schwarz 1931), collected by Crossley near Mananjary.

Malagasy name. Simpona.

Pelage and external characters. Face naked, black; ears hairless but largely obscured by surrounding fur. Pelage dense, almost entirely black or dark chocolate brown. Variably extensive whitish patches, grading into the black or brown, occur on the flanks and dorsum. In some individuals these lighter patches meet in the midline; in others there is a dark anteroposterior stripe in the midline of the back. Occasionally there are no distinct white patches, silvery hairs being present in their place. Tail black, ventrum brown, usually grading to whitish anteriorly. Gular gland in males a dark black brown, as also is the putatively glandular perianal skin (Petter 1962).

Dimensions. Cranial length: $\bar{x} = 88.2$, $s = 2.2$, $o.r. = 84.5$–90.9; $n = 16$.

Range. Poorly understood, but the subspecies appears to occupy an area of the eastern humid forest south of the Mangoro River to about the latitude of Manakara (fig. 3.26). It is unknown whether or not the subspecies is in secondary contact with *P. d. holomelas*, as early collecting records suggest it may be, whether the two are geographically discrete, or whether they merely represent a degree of clinal variation which extends south to Vondrozo, or possibly further. Schwarz (1931) believed that *edwardsi* was restricted to the coastal mountain range, and *holomelas* to the mountains further inland; and Petter et al. (1977, Carte 10) seem to be in agreement with this. However, such a distribution does not appear to be borne out by collecting records, which include for both localities a short distance east of Fianarantsoa. It is, moreover, difficult to discern what barrier might now, or in the recent past, have separated the two populations. Clearly a problem worthy of investigation.

Karyotype. Unknown.

Propithecus diadema holomelas Günther, 1875. Black sifaka.

1875. *Propithecus holomelas* Günther, Ann. Mag. Nat. Hist., 4 ser.,

16, p. 125. Based on BMNH 75.7.20.1 and 2, male and female collected by Crossley near Fianarantsoa.

Malagasy name. Simpona.
Pelage and external characters. Fur dense, and almost invariably uniformly black except for brown (occasionally pale) fur ventrally, and dark brown pygal patch. Face black, hairless; head hair bushy, usually obscuring ears. The status of this sifaka as distinct from *P. d. edwardsi* is uncertain; but among the rather few skins of either sifaka which are preserved in museum collections there are no really plausible intermediates. Gular gland of males reddish yellow (Petter 1962); presence of perianal skin of similar aspect unconfirmed for want of specimens.
Dimensions. Cranial length: $\bar{x} = 86.4$, $o.r. = 86.1–86.8$, $n = 4$.
Range. Uncertain. The only identifiable collecting locality is near Fianarantsoa (fig. 3.26). Petter et al. (1977) indicate that *holomelas* occurs in a narrow strip of the western part of the eastern rain forest between the latitudes of Fandriana and Vondrozo (although it may be significant that the Archbold Expedition, which collected extensively near Vondrozo, obtained no specimens); however, this is clearly a range which requires more survey before any reasonably accurate statement can be made. Moreover, the possibility of overlap with *edwardsi* requires examination (see above).
Karyotype. Unknown.

Propithecus diadema perrieri Lavauden, 1931. Perrier's sifaka.

1931. *Propithecus perrieri* Lavauden, Compt. Rend. Acad. Sci. Paris, 193, p. 77. Based on MCZ 44857, adult male from Analamera forest, southeast of Diego-Suarez.

Malagasy names. Radjako, ankomba job.
Pelage and external characters. Face naked, black; fur dense, long, silky; uniformly a deep, lustrous black except for the ventrum which is covered with short thick hair of a warm, rosy brown tint. Ears naked in some specimens, slightly furred in others, but largely concealed by the dense hair of the head. Single median gular gland in males; dark brown, as is the presumptively glandular perianal skin (Petter et al. 1977).
Dimensions. Cranial length $(n = 2)$: 86.3, 87.6. Labels accompanying the type and paratype in the MCZ (and Lavauden's published figures) give H + B as 500, 400; TL as 400, 450. These measure-

ments seem to be somewhat small, even allowing for distortion during preservation of the specimens (flat skins).

Range. Restricted to the forests to the northeast of the Andrafiamena mountain chain, just south and east of Anivorano Nord (fig. 3.26). The type site, the dry forest of Analamera, abutting on the sea, accounts for about half of this total area of distribution. Almost certainly the rarest of the sifakas, the subspecies exists at a very low population density, estimated by Petter et al. (1977) as 3–4 individuals/km².

Karyotype. $2N = 42$. Autosomes: M, 18; S, 14; A, 8. Sex chromosomes: X, M; Y, A (Rumpler 1975).

Propithecus verreauxi A. Grandidier, 1867.

I recognize four subspecies within this species, although the relationship between *P. v. deckeni* and *P. v. coronatus* is not entirely clear. Individuals of this species are distinctly smaller than those of *P. diadema*, and in general populations from higher altitudes show a thicker, longer fur than do those from lowland areas. Histological structure of the gular gland of males differs from that in *P. diadema* (Petter et al. 1977). Individuals of all subspecies possess short and rather sparse sets of superciliary (the best developed), genal, buccal, and interramal vibrissae. No carpal vibrissae. Females have a single pair of mammae, pectoral but close to the axilla.

Propithecus verreauxi verreauxi A. Grandidier, 1867.
Verreaux's sifaka.

1867. *Propithecus verreauxi* A. Grandidier, Rev. Mag. Zool. Based on MNHN 1867–580 (Rode catalog No. 147), adult male from Cap Ste-Marie, collected by A. Grandidier. Skin with skull in place.
1894. *Propithecus majori* Rothschild, Novit. Zool., 1, p. 666, Pl. 14. Type: BMNH 1939–1212A, skin of adult male collected by Last; "Antinosy Country" (see below).

Malagasy name. Sifaka.

Pelage and external characters. Face hairless, black; pelage dense dorsally but rather shorter than that of *P. diadema*; ventral fur sparse revealing black skin beneath. Fur white, except for black or terracotta cap on the head between ears and behind white forehead (fig. 3.27). This patch may be limited in size (in the holotype it barely exists at all, although this may be an artefact of fading), or it

Figure 3.27. Left: *Propithecus verreauxi coquereli*; right: Female *P. v. verreauxi* with infant. Infant is hybrid *P. v. verreauxi* x *P. v. coquereli*.

may extend back on to the neck. Ears white, slightly tufted. Light silver gray or yellow gold tints may occur on the dorsum and flanks. Pale golden pygal patch variably present. The single median gular gland of males is pale reddish brown, as is the presumably glandular perianal skin (Petter 1962).

A variant of *P. v. verreauxi* was described by Rothschild in 1894 under the name of *P. majori*. Few museum specimens of this variant exist; four in the BMNH were collected by J. T. Last in "Antinosy Country," by which he apparently meant the western part of the Onilahy River valley, in southwestern Madagascar. One specimen in the MCZ, collected by Decary, comes from Ambovombé. These specimens, few as they are, show a remarkable uniformity in pelage pattern. They are predominantly white, but possess a head cap of a dark chocolate brown, although the ears, cheeks, throat and forehead are white. The shoulders are white, but the back is brownish, as are the anterior and internal aspects of the arms and legs. The short hair of the breast and the anterior part of the ventrum is dark brown; the tail is always white-tipped but may be brown for some of its length. The type is the darkest of the preserved specimens. The recorded collecting localities of *majori* fall within the range of *P. v. verreauxi* (see below), and R. W. Sussman (pers. comm.) has found both types living in the same social groups. Francis Petter has apparently also observed this, and has in addition seen groups of each type feeding within 20 meters of each other (quoted in J. J. Petter et al. 1977). In view of this, and of the fact that *majori* occurs in widely scattered parts of the range of *verreauxi*, it seems reasonable to infer that *majori* is simply a melanistic variant within *verreauxi*: a variant which, in view of its consistency and the apparent lack of intermediates, is presumably under simple genetic control. An individual somewhat similar to *majori* was collected by the Archbold Expedition at Ambararatabé, far to the north of *verreauxi*, and in an area where the relationship between *coronatus* and *deckeni* is obscure (see below).

Dimensions. Cranial length: $\bar{x} = 80.8$, $s = 1.8$, $o.r. = 77.5–85.0$, $n = 31$. Field measurements (Archbold, $n = 4$): H + B, $\bar{x} = 437$, $o.r. = 392–465$; TL, $\bar{x} = 534$, $o.r. = 500–588$; HF, $\bar{x} = 131$, $o.r. = 127–137$; EAR, $\bar{x} = 39$, $o.r. = 38–40$.

Range. Occurs throughout the forested regions of south and southwest Madagascar from just west of Fort-Dauphin to the Tsiribihina River (fig. 3.26), flourishing in all types of forest from the arid *Didierea* formations through riverine gallery forest.

Karyotype. 2N = 48. Autosomes: M, 14; S, 16; A, 16. Sex chromosomes: X, S; Y, A (Rumpler 1975).

Propithecus verreauxi coquereli Milne-Edwards, 1867.
Coquerel's sifaka.

1867. *Propithecus coquereli* Milne-Edwards, Rev. et Mag. de Zool., sér. 2, 19, p. 314. Based on MNHN 1854–1263 (Rode catalogue No. 149): a young individual from northwest Madagascar, collected by Coquerel.
1870. *Propithecus damonis* Gray, Cat. Monkeys, Lemurs, and Fruit-eating Bats in Brit. Mus., p. 137. Madagascar.

Malagasy names. Tsibahaka, sifaka, ankomba malandy (only at extreme northeastern end of range).

Pelage and external characters. Face black, but generally with a patch of very short white hairs on the muzzle (fig. 3.27). Ears naked and quite small, but visible through surrounding fur. Pelage dense and of moderate length: predominantly white. Extensive maroon patches occur on the anterior part of the ventrum and on the anterior and internal aspects of the thighs and forelimbs. Extremities white; the back is often a pale silver gray or brownish, especially caudally. Tail white or occasionally silver gray. Gular gland and perianal skin of males dark red-brown (Petter 1962).

Dimensions. Cranial length: $\bar{x} = 82.4$, $s = 2.6$, $o.r. = 77.9–87.2$, $n = 23$. **Body weight**: $\bar{x} = 3927$, $o.r. = 3664–4306$, $n = 4$.

Range. Northwest Madagascar, north and east of the Betsiboka River (fig. 3.26). The most southerly occurrence of *P. v. coquereli* is near Ambato-Boéni; its range extends north to around Antsohihy, and its easternmost limit is near Antetemazy, a short distance to the west of Befandriana Nord.

Karyotype. $2N = 48$. Autosomes: M, 14; S, 16; A, 16. Sex chromosomes: X, S; Y, A (Rumpler 1975).

Propithecus verreauxi deckeni Peters, 1870. Decken's sifaka.

1871. *Propithecus deckenii* Peters, Monatsb. K. Preuss. Akad. Wiss. Berlin, for 1870, p. 421. Based on MB 3466, juvenile female collected at Kanatsy, northwestern Madagascar, by von der Decken.

Malagasy names. Tsibahaka, sifaka.

Pelage and external characters. Face black, ears naked but largely hidden. Muzzle relatively blunt, rounded. The fur of many individuals is completely white (fig. 3.28); in others the back, limbs, and particularly the shoulder are touched with pale yellow gold or silver gray tints. Fur short and sparse ventrally, revealing black skin. One museum specimen from Ambararatabé, near Lake Kinkony, is much darker than normal, with blackish or silver gray hairs on head, neck, back, limbs and extremities. Petter et al. (1977) note that a good deal of pelage color variation occurs between individuals on the slopes of the Bongolava Massif, northwest of Tsiroanomandidy, most being completely white but others showing brownish patches of varying darkness and extent. Median gular gland of male dark brown, as is perianal skin (Petter 1962).

Dimensions. Cranial length: $\bar{x} = 83.9$, $s = 2.6$, $o.r. = 80.4–89.6$, $n = 11$. Field measurements (Archbold, $n = 12$): H + B, $\bar{x} = 445$, $s = 18$, $o.r. = 420–475$; TL, $\bar{x} = 551$, $s = 25$, $o.r. = 510–595$; HF, $\bar{x} = 135$, $s = 13$, $o.r. = 32–44$.

Range. West coast of Madagascar, from somewhere to the south of Antsalova north to the Betsiboka River (fig. 3.26). The present range apparently does not extend southward as far as the Tsiribihina River, which marks the northern limit of *P. v. verreauxi*. There also exists inland the isolate described by Petter et al. (1977) from the Bongolava, northwest of Tsiroanomandidy. In the northwestern

Figure 3.28.   Left: *Propithecus verreauxi coronatus* female with infant. Right: *P. v. deckeni.*

part of the range of *deckeni* it is difficult to define the boundary, if there is one, of this subspecies with *P. v. coronatus* (see below); indeed, it is not clear what the status of each of these is vis-à-vis the other, particularly in view of the description by Petter et al. (1977) of rather *coronatus*–like individuals among the *deckeni* population of Bongolava. Kaudern (1915, fig. 3), who collected quite extensively in the northwest in 1911 and 1912, placed the northeastern limit of *deckeni* along the Mahavavy River, which runs roughly from south to north just to the east of Lake Kinkony. He placed *Propithecus v. coronatus* to the west of the Mahavavy between that river and the Betsiboka, which flows north into the Bay of Bombetoka and defines the western limit of *P. v. coquereli.* Petter et al. adopted these limits in their review (1977), although they noted that the Mahavavy did not appear to constitute a very effective barrier, at least in its upper reaches. However, R. W. Sussman and I, who have surveyed this region quite extensively (but not exhaustively), found pure white *deckeni* not only at Lake Kinkony, to the west of the Mahavavy, but at Katsepy, on the western shore of the Baie de Bombetoka. On the other hand, *coronatus* was found on the western side of the Betsiboka across from Ambato-Boéni, some 100 km farther to the south. Before too much is made of these rivers as barriers, one might also note that *Lemur mongoz*, among other lemurs, is found west of the Mahavavy, east of the Betsiboka, and in the region between these rivers. There is, moreover, a collecting record (Archbold Expedition) of *coronatus* slightly to the west of the Mahavavy, at Ambararatabé, and Petter et al. (1977) speak vaguely of "hybridization" around the upper Mahavavy, without giving any specifics. Clearly, this is a question which merits considerable further investigation.

Karyotype. 2N = 48. Autosomes: M, 14; S, 16; A, 16. Sex chromosomes: X, S; Y, A (Rumpler 1975).

Propithecus verreauxi coronatus Milne-Edwards, 1871.
Crowned sifaka.

1871. *P[ropithecus] coronatus* Milne-Edwards, Rev. Scient., 2 sér., 10, p. 224. Type: MNHN 1871–88 (Rode Catalogue no. 150*a*), adult from Province de Boueny, west coast of Madagascar, collected by Van Dam.
1876. *Propithecus damanus* Schlegel, Cat. Syst. Mus. Hist. Nat. Pays-Bas, 7, p. 293. Syntype series in RMNH collected by Van Dam, southern side of the Baie de Bombetoka.

Malagasy names. Tsibahaka, sifaka.

Pelage and external characters. Muzzle somewhat blunt, rounded. Face naked, black, or with some short whitish hairs on muzzle. Single central gular gland of males is dark brown, as is the perianal skin (Petter 1962). Thick fur of head, including cheeks, throat, and forehead, dark chocolate brown or black; sometimes there is some slight white tufting around the ears (fig. 3.28). Shoulders and back variably tinted, ranging from yellow gold to silver brown, paling caudally. Tail and hindlimbs white; external surface of forelimbs variably dark. Ventrum darkest on breast, which is chestnut brown; lightens caudally.

Dimensions. Cranial length: $\bar{x} = 84.3$, $s = 1.7$, $o.r. = 81.4$–85.7, $n = 7$. Field measurements (Archbold, $n = 2$): H + B, 425, 450; TL, 600, 560; HF, 155, 148; EAR, 44, 41.

Range. Unclear (see discussion of *P. v. deckeni*). Apart from the area discussed above, an isolate also exists in the region of Tsiroanomandidy (fig. 3.26), an area collected by the Archbold Expedition. As noted, the relationship between *deckeni* and *coronatus* is one which cries out for investigation; I am not convinced, as Petter et al. appear to be, that the present case is analogous to that of *verreauxi* and *majori*.

Karyotype. $2N = 48$. Autosomes: M, 14; S, 16; A, 16. Sex chromosomes: X, S; Y, A (Rumpler 1975).

FAMILY DAUBENTONIIDAE GRAY, 1870

Daubentonia E. Geoffroy, 1795

1795. *Daubentonia* E. Geoffroy, Décad. Philos. Litt. Pol., 4, no. 28, p. 195. Based on *Sciurus madagascariensis* Gmelin, Linnaeus, Syst. Nat., 13 ed., p. 152, 1789, based in turn on "l'aye-aye" of Sonnerat, Voy. Ind. Or. Chine, p. 138, Pl. 86, 1782. Specimen now in MNHN (see below).

1795. *Scolecophagus* E. Geoffroy, Décad. Philos. Litt. Pol., 4, no. 28, p. 196. Alternative name for *Daubentonia* (Geoffroy says this is the first name he thought of giving the aye-aye).

1799. *Aye-aye* Lacépède, Tab. Mammif., p. 6. No references given but presumably based on Sonnerat (see above; his specimen was the only one available up to 1844).

1803. *Cheyromis* E. Geoffroy, Cat. Mammif. Mus. Hist. Nat. Paris, p. 181. Based on Sonnerat (see above) and derivative references.

1811. *Chiromys* Illiger, Prodromus Syst. Mammal. et Av., p. 75, 1811. Type species: *Sciurus madagascariensis* Gmelin.

1816. *Psilodactylus* Oken, Lehrb. Naturgesch., 3, Zool., 2, pp. ix, 1164. Based on *Lemur psilodactylus* Shaw, but unavailable following Opinion 417 of the Commission (March 1956).

1817. *Cheiromys* G. Cuvier, Regn. Anim., 1 ed., 1, p. 207. Type species *Sciurus madagascariensis* Gmelin.

1839. *Myspithecus* Blainville, Ostéog. Mammif. 1, sect. E, p. 34. Based on Sonnerat (see above) and derivative references; alternative name for *Cheiromys* G. Cuvier.

1846. *Myslemur* Blainville, Dict. Univ. Hist. Nat., 8, p. 559. Published in synonymy with *Myspithecus* Blainville (see above).

Daubentonia madagascariensis (Gmelin, 1788). Aye-aye.

1788. *Sciurus madagascariensis* Gmelin, Linnaeus, Syst. Nat., 13 ed., 1, p. 152. Based on Sonnerat, Voy. Ind. Or. Chine, p. 138, Pl. 86; individual preserved in MNHN as no. 153 of Rode Catalogue. Western Madagascar.

1800. *Lemur psilodactylus* Shaw, Gen. Zool., 1, pt 1: Mammal., p. 109.

Based on Sonnerat (see above) and derivative references. Madagascar.

1929. *Cheiromys madagascariensis* var. *laniger* G. Grandidier, Bull. Acad. Malgache, n. s., 11, p. 106, one plate. Based on the specimen illustrated, in the collections of the Académie Malgache.

Malagasy names. Hay-hay, ahay, aiay.

Pelage and external characters. Unique among Malagasy primates in the quality of its fur. On the upper parts there is a rather dense layer of relatively short and soft white off-white hair, overlaid by a layer of extremely long, coarse guard hairs which are blackish brown for most of their length, but white in their distal part (fig. 3.29). The overall impression is of a very dark brown pelage, flecked, and in places suffused, with white. The hair of the limbs is coarse, decreasing in length distally, and blackish brown; the bushy tail is the same color, but is clothed entirely in very long, coarse monochromic hairs. The underparts are more sparsely covered by hair equivalent to the dorsal underlayer, mostly dichromic white and brown, but white alone on anterior part of breast and throat. Such minor variations in the condition of the pelage as have been

Figure 3.29. *Daubentonia madagascariensis.*

observed seem to be due to a propensity to molt. The extremely short face is quite thinly haired; the pale skin shows beneath the darkish interocular hair, and there is short white hair on the cheeks and above the eyes, although there are narrow dark circumocular rings. Ears naked, elongate, mobile and very large; the superciliary vibrissae are highly developed, and buccal, genal, and interramal vibrissae are also present. Nails of all digits except the hallux are laterally compressed and rather claw like; the third manual digit is elongated and extremely thin; the fourth is longer but more robustly made. Scrotum naked, with glandular skin; the single pair of mammae is inguinal in position.

Dimensions. Cranial length: $\bar{x} = 87.2$, $s = 2.3$, $o.r. = 82.4\text{--}90.1$, $n = 14$. Field measurements (Archbold, single specimen): H + B, 395; TL, 410; HF, 117: EAR, 99. **Body weight** $(n = 1)$: 2,800g (Bauchot and Stephan 1966).

Range. Has been reported much more widely than is generally realized. Museum specimens are rather badly documented; several are from "Tamatave," but this is probably where they were purchased, not captured or killed. One partial exception to this is a specimen "found in the forests west of Tamatave" (Kaudern 1915: 1). Other museum localities in eastern Madagascar are Beforona, about 30 km east of Andasibé (Périnet), in the center of the humid eastern forest strip; near Sahatavy, upper Maningory River, about 50 km west of Fénérive; and Seranantsera, several km inland, about 50 km SSW of Tamatave. I have myself found a dead aye-aye just north of Sambava, on the northeast coast (dead aye-ayes are easier to find than living ones; the animals are often killed on sight by villagers who believe them to be harbingers of grave misfortune), and others have been reported at several localities on the east coast. Decary (1950) records having found an aye-aye as far south as the Matitanana River, south of Manakara. In the west, the only collecting record is from Ampasimena, at the northern tip of the Ampasindava Peninsula (although the type locality is "west coast"); but the presence of aye-ayes has been reported at sites from the Mt. d'Ambre to Ankobakabaka, near Befandriana Nord. Kaudern (1915) stated that he had been told on several occasions that aye-ayes occurred in the Ankarafantsika region, although he was unable to find one himself; Decary (1950) claimed that the animal was found in the mangroves around Majunga, and implied its presence to the south of this, which lends weight to a report by Lamberton (1934), who quoted Hourcq to the effect that he had seen a fresh aye-aye skin near Andranomaro, south of Soalala. Subfossil remains of a large western *Daubentonia* have also been recovered, from the sites of Lamboharana and Tsiravé, near Morombé, and of Anavoha, in the far south (fig. 5.1).

Taken together, these reports imply that the original range of the aye-aye was extensive. Certainly it is found widely through the eastern forests, or was until not too long ago, and maybe was widespread over the western part of Madagascar as well. Its status today, however, is doubtful. Apparently a highly adaptable form, the aye-aye is (or was) found in areas of primary rain forest, deciduous forest, secondary growth, cultivation (particularly coconut groves), and conceivably even in mangrove swamps and dry scrub forest. It is unlikely, however, that the animal ever existed in high density, and its present rarity seems to be due to a combination of this factor with the hostility its unusual appearance provokes among villagers, and the rapid disappearance of its habitat.

Karyotype. 2N = 30. Autosomes: M, 16; S, 8; A, 4. Sex chromosomes: X, M; Y, A (Petter et al. 1977).

FAMILY CHEIROGALEIDAE GREGORY, 1915

Cheirogaleus E. Geoffroy, 1812. Dwarf lemurs

1812. *Cheirogaleus* E. Geoffroy, Ann. Mus. Hist. Nat. Paris, 19, p. 172. Type species (fixed by Elliot, Publ. Field Columbian Mus., Zool., 8, p. 548, 1907): *Cheirogaleus major* E. Geoffroy, 1812.

1840. *Cebugale* Lesson, Spec. Mammif., pp. 207, 213. Type species by monotypy, *Cebugale commersonii* Lesson.

1840. *Mioxicebus* Lesson, Spec. Mammif., pp. 207, 218. Type species by subsequent designation (Elliot, Rev. Primates, 1, p. xxx), *Mioxicebus griseus* Lesson.

1841. *Chirogale* Gloger, Gemein Hand.-u. Hilfsb. Naturgesch., 1, p. 44. Emendation of *Cheirogaleus* E. Geoffroy.

1842. *Myspithecus* F. Cuvier, in E. Geoffroy and F. Cuvier, Hist. Nat. Mamm., 4, Tab. Gén. Method., p. 2.

1846. *Myoxicebus* L. Agassiz, Nomenclator Zool., Index Univ., p. 243; Emendation of *Mioxicebus* Lesson.

1846. *Myoxocebus* L. Agassiz, Nomenclator Zool., Index Univ., pp. 235, 243. Emendation of *Mioxicebus* Lesson.

1872. *Opolemur* Gray, Proc. Zool. Soc. Lond., p. 853. Type species by monotypy, *Opolemur milii* Gray.

1913. *Altililemur* Elliot, Rev. Primates, 1, pp. xlvii, 111. Type species, *Altililemur medius* (E. Geoffroy).
1928. *Altilemur* Weber, Säugethiere, 2 ed., 2, p. 736. Lapsus for *Altililemur* Elliot.

Cheirogaleus major E. Geoffroy, 1812. Greater dwarf lemur.

Most recent authors (e.g., Petter et al. 1977) have recognized at least two subspecies of greater dwarf lemur, of which one (*C. m. crossleyi*), with more reddish fur, is said to exist primarily to the north of the Masoala Peninsula, while the other, browner, is found to the south (*C. m. major*). The picture is far from clear, however, since the browner and redder variants do not appear to be discrete geographically, and are united by individuals of intermediate coloration. Although future research may well show that discrete populations of *C. major* do exist which deserve recognition as separate subspecies, I prefer for the moment to regard the species as variable but monotypic.

1812. *Cheirogaleus major* E. Geoffroy, Ann. Mus. Hist. Nat. Paris, 19, p. 172, Pl. 10, fig. 1. Based on the drawing by Commerson, published for the first time by Geoffroy. Madagascar.
1822. *Lemur commersonii* Wolf, Abbild. u. Beschr. Merkw. Naturg. Gegenst., 1 ed., 2, p. 9, Pl. 4. Based on Commerson's drawing (see above).
1828. *Cheirogaleus milii* E. Geoffroy, Cours Hist. Nat. Mammif., 11ᵉ leçon, p. 24. Based on MNHN A3954 and Rode Catalogue no. 135, figured by F. Cuvier (in E. Geoffroy and F. Cuvier, Hist. Nat. Mamm., 2, 1824 [livr. 32, dated 1821]) as "le maki nain." Madagascar.
1833. *Cheirogaleus typicus* A. Smith, S. Afr. Quart. Jour., 2, p. 50. Based on BMNH 37.9.26.77, a male brought from Madagascar by Verreaux. No specific locality.
1840. *Cebugale commersonii* Lesson, Spec. Mammif., p. 213. Based on Commerson's drawing (see above).
1840. *Mioxicebus griseus* Lesson, Spec. Mammif., p. 218. Based on F. Cuvier's "maki nain" (see above).
1868. *Cheirogaleus adipicaudatus* A. Grandidier, Ann. Sci. Nat., 5 sér., Zool., 10, p. 378. Based on the author's observations; Tuléar.
1870. *Cheirogaleus major crossleyi* A. Grandidier, Rev. Mag. Zool. (2) 22, p. 49. Based on the author's observations; "forêts est d'Antsianak."

Figure 3.30. Above: *Cheirogaleus medius,* Mandena, southeast Madagascar, just after emerging from dormancy. Courtesy of R. D. Martin. Below: *Cheirogaleus major,* Sambava, northeast Madagascar.

1894. *Chirogale melanotis* Forsyth Major, Novit. Zool., 1, p. 25, Pl. 2, fig. 10. Based on BMNH 70.5.5.25, adult coll. Crossley, said by Major to come from Vohémar, but bearing the provenance "Antsianak" on its label.

1895. *Chirogale sibreei* Forsyth Major, Ann. Mag. Nat. Hist. 6 ser, 18, p. 325. Based on BMNH 97.9.1.160, adult from Ankeramadinika, "one day's journey to the east of Antananarivo."

Malagasy names. Tsitsihy, tsidy, hataka.

Pelage and external characters. Body hairs dichromatic; basally gray, paler tipped. Face covered with short, flat-lying hair; naked only at distal end of muzzle (fig. 3.30). Eyes large; ears moderate in size, naked, but partly concealed by surrounding fur. Dorsal pelage dense but not of great length; varies in color from a dull gray brown to slightly reddish. Top of head and tail uniform in color with dorsum; underparts paler, with the gray basal color apparent beneath the white or creamy tips of the downy hair. Dark rings around the eyes; the interorbital area is pale, as (usually) are the cheeks; the throat is darker. Tail swells seasonally. Mammae: usually one pectoral and one inguinal pair. Full set of facial vibrissae: superciliary, buccal, genal, and (sparse) interramal. No carpal vibrissae. Anus below first caudal vertebra.

Dimensions. Cranial length: $\bar{x} = 54.7$, $s = 3.2$, $o.r. = 46.6$–58.3, $n = 30$. Field measurements (Archbold, $n = 2$): H + B, 250, 219; TL, 310, 255; HF, 57, 55; EAR; 23, 27; (Webb, $n = 2$): H + B, 264, 221; TL, 305, 242; HF, 42, 30; EAR, 23, 25. **Body weight** is highly variable seasonally; Petter et al. (1977) quote a range of 340–600 g; Bauchot and Stephan (1966), a mean of 450 g.

Range. Throughout the forested areas of eastern Madagascar from Fort-Dauphin to Mt. d'Ambre, and extending westward to include the Tsaratanana Massif and the Sambirano region (fig. 3.31). Petter et al. (1977) also indicate that a population of *C. major* exists on the Bongolava Massif, at the far western edge of the Eastern Region. Evidently, the isolation of this population is a relatively recent phenomenon; subfossil evidence shows that *C. major* was present within the last millenium at Ampasambazimba, near Analavory, while zoological collecting records demonstrate that until only a few decades ago the range of the species extended well onto the central plateau (fig. 3.31).

Karyotype. 2N = 66. Autosomes: M, O; S, O; A, 64. Sex chromosomes: X, M; Y, A (Rumpler and Albignac 1973b).

Cheirogaleus medius E. Geoffroy, 1812. Fat-tailed dwarf lemur.

The species is often divided into two subspecies, *C. m. medius* and *C. m. samati*, but the distinction does not appear to be warranted.

1812. *Cheirogaleus medius* E. Geoffroy, Ann. Mus. Hist. Nat. Paris, 19, p. 172, Pl. 10, fig. 2. Based on the drawing by Commerson, published for the first time by Geoffroy. Madagascar.

1812. *Cheirogaleus minor* E. Geoffroy, Ann. Mus. Hist. Nat. Paris, 19, p. 172, Pl. 10, fig. 3. Based on Commerson's drawing (see above).

1868. *Chirogaleus samati* A. Grandidier, Rev. Mag. Zool., (2), 20, p. 49. Based on syntypes MNHN 1868–229/230, male and female from west coast of Madagascar, Tsidsibon River, probably south of Morondava.

1872. *Opolemur milii* Gray, Proc. Zool. Soc. Lond., p. 854, fig. 1, Pl. 70. Based on two individuals from Morondava.

1894. *Opolemur thomasi* Forsyth Major, Novit. Zool., 1, p. 20, Pl. 1, fig. 1; Pl. 2, fig. 2. Based on BMNH 91.11.30.3, female from Fort-Dauphin.

1931. *Cheirogaleus medius samati* Schwarz, Proc. Zool. Soc. Lond., p. 405. Based on *C. samati* A. Grandidier.

Malagasy names. Matavirambo (northwest), kely be-ohy (Morondava region), tsidy, tsitsihy (far south).

Pelage and external characters. Pelage dense, in most cases relatively short. Body hairs dichromic, much darker at base than at tip. Upper parts, including tail, light gray or silver gray, often with rosy or brownish tints. Underparts including throat paler, varying from light brown through cream to white, with gray base color also showing. Facial hair short; black or dark maroon rings around eyes, with pale stripe between; rest of face usually light gray or brown, with cheeks paler yet. Ears moderate in length, naked, but partly concealed (fig. 3.30). Tail swells seasonally; normally one pair of pectoral mammae, one of inguinal. Superciliary, genal and buccal facial vibrissae present; interramal rare. Anus below first caudal vertebra.

Dimensions. Cranial length: $\bar{x} = 40.6$, $s = 1.5$, $o.r. = 38.2$–44.0, $n = 41$. Field measurements (Archbold, $n = 4$): H + B, $\bar{x} = 193$, $o.r. = 167$–220; TL, $\bar{x} = 209.5$, $o.r. = 195$–230; HF, $\bar{x} = 44$, $o.r. = 38$–50; EAR, $\bar{x} = 20$, $o.r. = 18$–22; (Webb, $n = 6$): H + B, $\bar{x} = 184$, $o.r. = 172$–188; TL, $\bar{x} = 181$, $o.r. = 177$–185; HF, $\bar{x} = 26$, $o.r. = 23$–28; EAR, $\bar{x} = 18$, $o.r. = 17$–19.

Body weight. Varies greatly with season: $x = 333$, $o.r. = 274$–283, $n = 11$. Bauchot and Stephan (1966) give a mean value for two wild-caught specimens of 177 g.

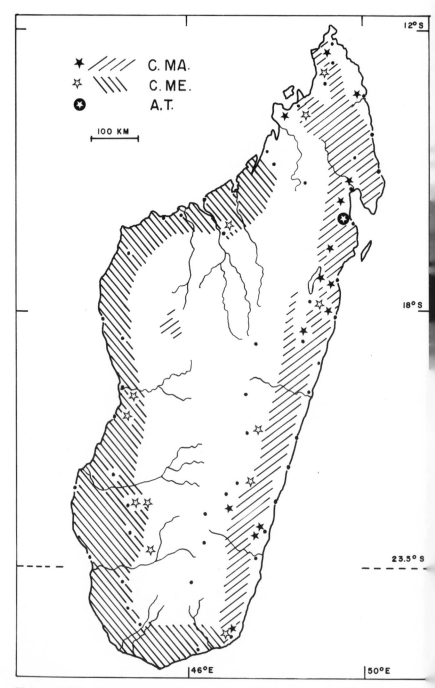

Figure 3.31. Distributions of the species of *Cheirogaleus* and of *Allocebus*. Shaded areas represent approximate range limits; symbols denote localities of museum specimens. C.MA.: *Cheirogaleus major*; C.ME.: *C. medius*; A.T.: *Allocebus trichotis*.

Range. Found in most of the forested areas of the west, south-west and south of Madagascar, from the Bay of Narinda to Fort-Dauphin (fig. 3.31). According to the locality information accompanying certain specimens collected by Van Dam, by Forsyth Major, and by the Archbold Expedition, the species also occurs, or occurred, in eastern and northern Madagascar and in the Sambirano region (fig. 3.31), in sympatry with *C. major*. The present status of such populations is problematical.

Karyotype. 2N = 66. Autosomes: M, O; S, O; A, 64. Sex chromosomes: X, M; Y, A (Rumpler and Albignac 1973b).

Microcebus E. Geoffroy, 1828. Mouse lemurs.

Two forms of *Microcebus* are normally recognized (if, as is done here, Coquerel's dwarf lemur is excluded from the genus): a gray, long-eared form from the west of Madagascar, and a brown/rufous, short-eared form from the east. Most recent students have considered these to be conspecific subspecies, but of late a tendency has been growing to regard them as separate species (e.g., Petter et al. 1977). R. D. Martin (pers. comm.) has been able to demonstrate that consistent differences exist between them in cranial morphology, in addition to the more traditional distinctions in pelage coloration and ear length; and it is now known that mouse lemurs with rufous pelage do occur sympatrically with the gray western form, if only sparsely. The western rufous is closer to the eastern rufous than to the gray mouse lemur in its cranial characters, but does have rather long ears (R. D. Martin, pers. comm.); we do not as yet know enough about it to decide whether it represents a subspecies of the eastern form or whether it deserves specific recognition in its own right. In any event, there is evidence to suggest that the classic gray and rufous forms are indeed separate species, and they are treated as such here. The question of the distinction between eastern and western rufous forms is left in abeyance pending better knowledge of them.

The nomenclature of the species of *Microcebus* is, unhappily, horribly entangled, and since my usage does not exactly follow current convention I shall attempt to outline the historical situation as briefly as possible. Almost all of the early names applied to mouse lemurs are based on one or more of three original sources:

1. The specimen illustrated by Brown (New Illustr. Zool., p. 107, Pl. 44, 1776), who gave as references "Genus Lemur, Lin. Syst. 44. Macauco, Penn. Syn. Syn. Quad. 104." Both of these references are in fact to "L. tardigradus," the "tail-less macauco," i.e., the slen-

der loris; but the animal figured by Brown is unquestionably a gray mouse lemur, and was subsequently dubbed the "little macauco" by Pennant (Hist. Quadrupeds, 1, 217, 1781).

2. The "Rat de Madagascar" illustrated and inadequately described by Buffon (Hist. Nat., Gén. et Partic., Suppl., 3, p. 149, Pl. XX, 1776). It is impossible to be sure of the species of this animal, although without any doubt whatsoever it is a *Microcebus*. The ears of this individual *appear* in the figure to be rather short, but there is nothing in the text to support (or deny) this supposition, although Buffon compares his specimen with the "rat" which was reported from the southwest coast of Madagascar by Dutch sailors, and which can hardly have been other than the gray form. In any event, Buffon's description, even with the illustration, constitutes an insufficient basis for subsequent Linnaean species designations. The same is true for Buffon's "petit mongous" (Hist. Nat., Gen. et Partic., 13, p. 177, 1765), which is even less adequately described, but which is said by E. Geoffroy (Bull. Sci. Soc. Philomath. Paris, 1, 1795) to be the same animal as the "Rat" described by Buffon 11 years later. In the same work, Geoffroy proposed the name *Lemur pusillus* for this primate, describing it as "Cinereo-fulvus," and as having relatively long ears, both of which features might well better fit a gray mouse lemur than a brown one. Rode (Cat. Types Mammif. Mus. Natl. Hist. Nat. 1B, p. 61, 1939) states that a specimen in the MNHN (no. 138 of his catalogue) is the type of *M. rufus*, and is the individual described successively by E. Geoffroy as the "maki nain" (=*L. pusillus*); as *Galago madagascariensis* (Ann. Mus. Hist. Nat. Paris, 19, p. 166, 1812), the pelage of which is described as "roux" and which is based on Brown and on *Lemur murinus*, misattributed to Pennant, as well as on Buffon's "Rat"; and as the "Microcèbe roux" (a designation accompanying Geoffroy's publication of the genus *Microcebus*: Cours Hist. Nat. Mammif., 11, p. 24, 1828). It is hard to detect any rufous elements in the pelage of this specimen, which still exists; but after its long sojourn in the Salon des Types of the MNHN, its fur is by now so deeply impregnated with dust that it is perhaps unwise to try. This may well be the specimen that was illustrated as "le maki nain, *Lemur pusillus*" by Audebert (Hist. Nat. Makis, p. 19, Pl. 8, before 1800). If so, it is almost certainly a gray mouse lemur.

3. The individual illustrated by J. F. Miller (Various Subjects of Nat. Hist., Pl. 13) in 1777. Later republished, with a description by G. Shaw (Cimelia Physica, p. 25, Pl. 13), there can be no doubt that Miller's plate depicts a gray mouse lemur.

Most subsequent early names were based on one, two, or all of these at a time when it was not realized that more than one species

(or even race) of *Microcebus* existed. We have seen that there is sufficient uncertainty about the specific identity of the "Rat de Mada-
gascar/petit mongous" to render inadequately-based any species
name, such as E. Geoffroy's *Lemur pusillus*, founded solely upon it.
All other combinations may be regarded as applying to the gray
mouse lemur—and indeed, have in effect been so restricted by
Schwarz (Proc. Zool. Soc. London, 1931, p. 402). Current convention, following Schwarz, applies the name *Microcebus rufus* (or *M. murinus rufus*) Wagner, 1840 to the brown mouse lemur; but in fact it
is not possible to justify the use of Wagner's name for the brown
mouse lemur. That author (Von Schreber's Säugthiere, Suppl. 1, p.
278, 1840) included Geoffroy's *Galago madagascariensis* in his synonymy of *M. murinus*, the only species of the genus he recognized in
his discussion of *Microcebus;* but in a footnote to his discussion of
Otolicnus (= Galago), Wagner somewhat obscurely stated (p. 291) that
"Der *Galago madagascariensis* von Geoffroy and Desmarest [Mammal-
ogie, 1, p. 103] ist unser *Microcebus rufus*." Thus, contra Schwarz, *M.
rufus* Wagner is based not on the "microcèbe roux" of Geoffroy, but
is instead an absolute synonym of *Galago madagascariensis* which
Schwarz himself regarded as a synonym of *M. m. murinus*. However,
by the exercise of what it would be uncharitable to label sophistry, it
is nonetheless possible to retain the species name *rufus* for the brown
mouse lemur. For in the same year, Lesson (Spec. Mammif., p. 217,
1840) described two species of "Makiloir" (his *Gliscebus*): *G. murinus*,
the "makiloir gris," and *G. rufus*, the "makiloir roux." Although in
his synonymy Lesson quotes in part *Cheirogaleus commersonii* of Vig-
ors and Horsfield (Jour. Zool. 4: 111, 1828), which is not a mouse
lemur, and although he assigned certain *Microcebus* to another genus, the "makirat," *Myscebus*, it is clear that his intention was to distinguish between gray and rufous species of "Makiloir." I therefore
take the valid name of the brown mouse lemur to be *Microcebus rufus*
(Lesson, 1840). The synonymy which follows omits all species names
based solely on Buffon's illustration and description.

1828. *Microcebus* E. Geoffroy, Cours. Hist. Nat. Mammif., leçon 11,
      p. 24. Based on the "Rat de Madagascar" of Buffon (Hist.
      Nat., Gén. et Part., Suppl. 3, p. 149, pl. 20); genus named
      without species, but "Le Microcèbe roux" given.
1835. *Scartes* Swainson, Nat. Hist. Classif. Quadrupeds, p. 352.
      Based on Brown, Illustr. Zool., Pl. 44, 1776; type species
      given as "*S. murinus*", presumably = *L. murinus* of J. F. Miller,
      Various Subjects of Nat. Hist., Pl. 13, 1777.

1840. *Myscebus* Lesson, Spec. Mammif., pp. 207, 214. Type species by monotypy *Myscebus palmarum*, based on numerous references.

1840. *Gliscebus* Lesson, Spec. Mammif., pp. 207, 216. Type species by subsequent designation (Schwarz, Proc. Zool. Soc. Lond., 1931, p. 401), *Gliscebus murinus = Microcebus murinus* (J. F. Miller).

1841. *Myocebus* Wagner, Arch. Naturg., 7, no. 2, p. 19. New name for *Myscebus* Lesson.

1870. *Murilemur* Gray, Cat. Monkeys, Lemurs, and Fruit-eating Bats in Brit. Mus., pp. 132, 135. Based on *Lemur murinus* of J. F. Miller (ref. given, Cimelia Physica, Pl. 13, 1796; same figure as published by Miller in 1777, see above).

1870. *Azema* Gray, Cat. Monkeys, Lemurs, and Fruit-eating Bats in Brit. Mus., pp. 132, 134. Type species by monotypy *A. smithii = Cheirogaleus smithii* Gray, Ann. Mag. Nat. Hist. 10, p. 257, 1842.

Microcebus murinus (J. F. Miller, 1777). Gray mouse lemur.

1777. *Lemur murinus* J. F. Miller, Various subjects of Nat. Hist., Pl. 13; Cimelia Physica, Pl. 13, and text (by G. Shaw), p. 25, 1796. Based on the specimen figured: Madagascar.

1785. *Prosimia minima* Boddaert, Elenchus Animalium, 1, p. 66. Type by subsequent designation (Schwarz, Proc. Zool. Soc. Lond., 1931, p. 403) Pl. 44 of Brown, New Ilustr. of Zool., 1776 = *Microcebus murinus* (J. F. Miller). Madagascar.

1792. *Lemur prehensilis* Kerr, Animal Kingdom, p. 88. Based on Pl. 44 of Brown, also by subsequent designation of Schwarz (above).

1812. *Galago madagascariensis* E. Geoffroy, Ann. Mus. Hist. Nat. Paris, 19, p. 166. Based (?) on MNHN Rode Catalogue no. 138; Madagascar.

1835. *S[cartes] murinus* Swainson, Nat. Hist. Classif. Quadrupeds, p. 352. Based on Brown, Pl. 44 (above); species name presumably from Miller, 1777 (above).

1840. *Myscebus palmarum* Lesson, Spec. Mammif., p. 215. Based on numerous sources; restricted by Schwarz (above) to Miller's specimen. Madagascar.

1840. *Gliscebus murinus* Lesson, Spec. Mammif., p. 216. Based on Brown, pl. 44 (above) and other references; Brown's illustration designated here.

1840. *Microcebus rufus* Wagner, Von Schreber's Säugthiere, Suppl. 1, p. 278 = *Galago madagascariensis* E. Geoffroy.

1842. *Galago minor* Gray, Ann. Mag. Nat. Hist. 10, p. 257. Based on BMNH 37.9.26.79, adult. Madagascar.

1852. *Microcebus myoxinus* Peters, Naturwiss. Reise nach Mossambique, Zool., 1, p. 14, pl. 3, pl. 4, figs. 6–9. Based on MB 319/14655 (lectotype, selected by Schwarz, Proc. Zool. Soc. Lond., 1931, p. 402), female from Baie de St. Augustin.

1868. *Chirogaleus gliroides* A. Grandidier, Ann. Sci. Nat. 5: Zool., 10, p. 378. Based on specimen collected at Tuléar, identified by Rode (1939) as MNHN 1868–1441.

1870. *Murilemur murinus* Gray, Cat. Monkeys, Lemurs, and Fruit-eating Bats in Brit. Mus., p. 135. New name for *Lepilemur murinus* Gray, Proc. Zool. Soc. Lond. p. 143, = *Lemur murinus* J. F. Miller.

1910. *M[icrocebus] minor griseorufus* Kollman, Bull. Mus. Natl. Hist. Nat. Paris, 16, p. 304. No type designated; "côte sud-est, sud, et sud-ouest."

Malagasy names. Tsidy (northern part of range); koitsiky, tilitilivaha, vakiandri (Morondava region); pondiky (Fort-Dauphin region).

Pelage and external characters. Fur dense, moderately long, hairs dichromic: basally a dark gray, with lighter tips. Upper parts are a light silver or rosy brown, darkest on the head; only the extreme tips of the hair bear this color so the darker basal color tends to show through. Ventrally the hairs are tipped with white or off-white; in some individuals they are this color for almost all their length, giving a uniformly pale effect; in others the basal gray shows through. A fairly distinct median dorsal stripe is frequently present. Tail brown, sometimes darkening somewhat distally. Face covered with short hairs. Forehead often light russet; some individuals show a trace of a dark circumorbital ring, especially medially. A white stripe descends from the forehead to the tip of the muzzle (fig. 3.32); the cheeks are often rather lighter than the dorsal color; the throat matches the ventrum. Ears long, naked, highly visible. Superciliary, buccal and genal vibrissae present; no interramal or carpal. Two pectoral, two abdominal mammae.

Dimensions. Cranial length: $\bar{x} = 32.0$, $s = 0.9$, $o.r. = 29.7–34.2$, $n = 193$. Field measurements (Webb, $n = 18$): H + B, $\bar{x} = 124.7$, $s = 5.5$, $o.r. = 115–134$; TL, $\bar{x} = 134.5$, $s = 6.6$, $o.r. = 118–139$; HF, $\bar{x} = 23.6$, $s = 3.2$, $o.r. = 20–25$; EAR, $\bar{x} = 23.5$, $s = 1.2$, $o.r. = 20–25$; (Archbold, $n = 11$): H + B, $\bar{x} = 130$, $o.r. = 120–138$; TL, $\bar{x} = 147$, $o.r. = 140–157$; HF,

Figure 3.32. Left: *Microcebus murinus,* Analabé, western Madagascar; courtesy of R. D. Martin. Right: *Microcebus rufus.*

$\bar{x} = 32.8$, *o.r.* $= 32–34$; EAR, $\bar{x} = 26.9$, *o.r.* $= 25–28$. Martin (1972a) gives the following ear height, measured on live specimens from the bottom of the external auditory meatus to the tip of the pinna: 22 mm (*o.r.* 21–23 mm). **Body weight.** Martin (1972a) notes that weight varies markedly with season, but suggests on the basis of extensive trapping and weighing at a variety of localities that "on the whole it is probably correct to take a figure of 60 g for the average annual body weight of adult Lesser Mouse Lemurs" (p. 52). Petter et al. (1977) estimate the annual weight variation for an adult gray mouse lemur to be in the region of 55 to 90 g. The figures supplied by the DUPC are consistently higher, overlapping with the high end of this range rather infrequently and occasionally extending up to over 180 g. Petter et al. note that one of their captive individuals attained a weight of 170 g; it seems, then, that in the case of this lemur the weights of captive animals consistently and considerably overestimate those normally achieved in the wild.

Range. Occurs throughout the forested areas of western, southern and southwestern Madagascar, from Fort-Dauphin to the Sambirano region (fig. 3.33). Martin (1972a) notes that in the Fort-Dauphin area the gray mouse lemur's area of distribution includes the littoral forest to the north and east of the town, while the brown mouse lemur occurs in the rain forest which extends southward, to the west of Fort-Dauphin, almost to the coast. There is thus in this region a sharp environmental demarcation between the two species, for although the two areas receive similar rainfall, the littoral is much better drained, and supports a vegetation of distinctly less humid aspect than does the interior. The northern boundary of the

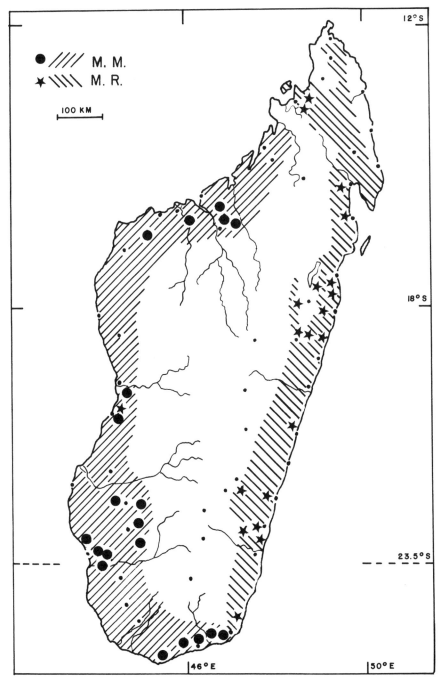

Figure 3.33. Distributions of *Microcebus murinus* (M.M.), and *Microcebus rufus* (M.R.). Shaded areas represent approximate range limits; symbols denote localities of museum specimens.

gray mouse lemur at the other extremity of its range is not precisely known, but the species appears not to occur north of the Sambirano River.

Karyotype. 2N = 66. Autosomes: M, O; S, O; A, 64. Sex chromosomes: X, M; Y, A (Rumpler and Albignac 1973b).

Microcebus rufus (Lesson, 1840). Brown or rufous mouse lemur.

1840. *Gliscebus rufus* Lesson, Spec. Mammif., p. 217. Based on several references (see above); Madagascar.

1842. *Cheirogaleus smithii* Gray, Ann. Mag. Nat. Hist., 10, p. 257. Based on BMNH 37.9.26.78, adult without locality other than "Madagascar."

Malagasy names. Tsidy, tsitsihy.

Pelage and external characters. Hairs of dense pelage dichromic: gray at base, brown or rufous at tips. Dorsally brown to reddish brown with the rufous element most marked on the head. Ventrally the gray base color shows beneath the white, off-white, or pale rufous tips. Tail uniform in color with the upper parts, but may darken distally. Hairs on face short; white or off-white median stripe down muzzle between the eyes, darkish hairs medial and rostral to eyes (fig. 3.32). Cheeks rufous, usually somewhat paler than crown; throat pale. Buccal, genal, and superciliary vibrissae present; no interramal or carpal. Ears naked; short relative to those of *M. murinus* but often rather broader; there is some variation in ear length, however, and occasionally a rufous specimen from the east coast, as well as from the west, will exhibit longish ears. Two pairs of mammae: one pectoral, one inguinal.

Dimensions. Cranial length: $\bar{x} = 32.5$, $s = 1.5$, $o.r. = 29.5$–35.4, $n = 40$. Field measurements (Archbold, $n = 5$): H + B, $\bar{x} = 124$, $o.r. = 113$–142; TL, $\bar{x} = 153$, $o.r. = 137$–168; HF, $\bar{x} = 34$, $o.r. = 33$–35; EAR, $\bar{x} = 18.8$, $o.r. = 18$–20; (Webb, $n = 3$): H + B, $\bar{x} = 123$, $o.r. = 120$–126; TL, $\bar{x} = 122$, $o.r. = 120$–125; HF, $\bar{x} = 23$, $o.r. = 21$–24; EAR, $\bar{x} = 15.3$, $o.r. = 14$–18. **Body weight.** Subject to fluctuation similar to that noted for the gray mouse lemur (above). Although the gray and brown mouse lemurs are exceedingly close in cranial and H + B lengths, Petter et al. (1977) note that the brown form tends to weigh less, about 10 g on average; they estimate normal adult variation in body weight to be from 45–80 g. They also note that red individuals from the west tend to be lighter yet, averaging about 55 g; again, this is not reflected in cranial length, two rufous individuals obtained near Morondava having values of 34.0 and 32.8 mm.

Range. Brown mouse lemurs are found throughout the humid forests of the eastern region of Madagascar, in secondary as well as in primary formations, from Fort-Dauphin (see above) to the Mt. d'Ambre, and including the Sambirano region, at least to the north of the Sambirano River (fig. 3.33). A brown mouse lemur also exists (albeit very sparsely) in western Madagascar south of the Sambirano; specimens have been collected near Morondava and have been reported from the Ankarafantsika (Petter 1962). Mouse lemurs are absent from the arid northern tip of Madagascar.

Karyotype. $2N = 66$. Autosomes: M, O; S, O; A, 64. Sex chromosomes: X, M; Y, A (Rumpler and Albignac 1973b).

Mirza Gray, 1870

Coquerel's dwarf lemur was initially described by Alfred Grandidier (1867) as a species of *"Cheirogalus,"* *C. coquereli*, largely on the basis of certain resemblances to the fork-marked lemur, known to him as *Cheirogalus furcifer*. However, following Schlegel and Pollen (1868), who believed that they were describing a new species, and who apparently chose by chance the same species name as had Grandidier, the form has generally been assigned to genus *Microcebus* (although in describing the skull of their specimen, Schlegel and Pollen remarked that, among the "Microcèbes," in which they included *Cheirogaleus*, "elle est plus voisine du Mycrocebus [sic] typicus de Smith que les autres espèces," by this referring to the form now known as *Cheirogaleus major*). The primary morphological reason given by subsequent investigators for assigning *coquereli* to *Microcebus* has been the relatively small size of its P^2, which is subequal in height with P^3. In both taxa there is some variation in this feature, however, and in its large, trenchant anterior lower premolar *coquereli* is closer to *Cheirogaleus*. Further, in its upper molars a similar observation applies: in M^{1-2} the hypocone tends to be positioned relatively anteriorly, opposite the metacone, as in *Cheirogaleus* species and in contrast to *Microcebus*, where the hypocone is displaced posteriorly. These dental differences are, however, relatively minor (and the anterior hypocone is presumably primitive), while in cranial construction and other dental characters *Cheirogaleus* and *Microcebus* are remarkably uniform.

Perhaps, then, a better approach to the assessment of the affinities of *coquereli* lies in its locomotor habit, which is much closer to the rapid scurrying of the mouse lemurs than to the slower, more deliberate locomotion of the dwarf lemurs. Indeed, R. D. Martin (pers. comm.) believes that this divergence in locomotor pattern reflects a fundamental phylogenetic dichotomy within this group of lemurs. A polytypic genus is a monophyletic grouping of closely related species, and it seems plausible that a genus *Microcebus* embracing *coquereli* would not violate the criterion of monophyly. However, the number of species—and branching events—within a genus is arbitrary, and although deference to current usage might seem desirable from the point of view of stability, I feel it is more useful to resurrect Gray's genus *Mirza* to contain *coquereli*, reserving *Microcebus* for the two vastly smaller species which in every way resemble each other much more closely than either does *coquereli*.

1870. *Mirza* Gray, Cat. Monkeys, Lemurs, and Fruit-eating Bats in Brit. Mus., p. 136. Type species *Mirza coquereli* = *Microcebus coquereli* Schlegel and Pollen, 1868.

Mirza coquereli (A. Grandidier, 1867). Coquerel's dwarf lemur.

1867. *Cheirogalus coquereli* A. Grandidier, Rev. Mag. Zool. (2), 19, p. 85. Based on seven individuals acquired by the author at Morondava; Rode (1939) gives as "holotype," i.e., presumably designates as lectotype, MNHN 1867–603, adult with skull in skin.

1868. *Microcebus coquereli* Schlegel and Pollen, Recherches Faune Madag., 2, p. 12, Pl. 6; pl. 7, fig. 2. Based on an individual "tué dans les forêts de Congony, dans l'intérieur de la Baie de Passandava [Ampasindava]."

1870. *Mirza coquereli* Gray, Cat. Monkeys, Lemurs, and Fruit-eating Bats in Brit. Mus., p. 136. New name for *Cheirogalus coquereli* of A. Grandidier (see above) and for *Microcebus coquereli* of Schlegel and Pollen (see above).

Malagasy names. Tsiba, tilitilivaha (southern area): setohy, fitily (northern area).

Pelage and external characters. Hair dense, relatively short, dichromic; basally a dark gray, paler at tips. Dorsally a warm brown or gray brown, sometimes with rosy or yellowish elements; ventrally the gray base color of the downy hair shows through beneath the yellowish or slightly russet tips. The long hair of the tail darkens distally to the black or dark chestnut tip. The face is covered with short hair of the same color as the dorsum; the cheeks are sometimes a bit lighter. No dark ring such as occurs in *Cheirogaleus* is present around the eyes in most individuals, but in some one is faintly visible, especially medially. Relatively well developed superciliary, buccal, genal and interramal vibrissae present; carpal also. The ears are long, hairless, and highly visible (fig. 3.34). Mammae: one pair pectoral, one abdominal.

Dimensions. Cranial length: x = 50.3, s = 2.1, $o.r.$ = 46.4–53.0, n = 20. Field measurements (Archbold, n = 1): H + B, 212; TL, 333; HF, 59; EAR, 34. **Body weight:** ca. 300 g (Pagès 1978).

Range. Restricted to parts of western Madagascar (fig. 3.22), apparently in a scattered series of disjunct isolates. Occurs from the region of Ankazoabo northwards to Belo-sur-Tsiribihina, or a little beyond. Also occurs on the Ampasindava Peninsula and in the adjoining Ambanja region. Petter et al. (1977) suggest that *coquereli* may occur in coastal forests in the area intervening between the southern and northern populations, but are unable to confirm this.

Karyotype. 2N = 66. Autosomes: M, O; S, O; A, 64. Sex chromosomes: X, M; Y, A (Rumpler and Albignac 1973).

Figure 3.34. *Mirza coquereli.* Drawing by Nicholas Amorosi after photograph published by Petter et al. (1977).

Allocebus Petter-Rousseaux and Petter, 1967

1967. *Allocebus* Petter-Rousseaux and Petter, Mammalia 31 (4), p. 574. Based on *Cheirogaleus trichotis* Gunther, 1875.

Allocebus trichotis (Günther, 1875). Hairy-eared dwarf lemur.

1875. *Chirogaleus trichotis* Günther, Proc. Zool. Soc. Lond., p. 78, figs. 1, 2; pl. 15. Based on BMNH 75.1.29.20, an individual collected by Crossley "on his way from Tamantave to Murundava."

Malagasy name. Unknown.

Pelage and external characters. Only two skins known: the type, and the male specimen (MNHN 135–170–37–11) described by Petter-Rousseaux and Petter (1967). Pelage dense; hair dichromic, basally a darkish gray, paler at tips. Upper parts a rosy brown, darkest along median line; tail darkens distally. Face covered with short hair: light brown, as is the forehead, which shades into the darker brown crown. Narrow dark ring around eyes; interorbital area pale. Ears relatively short, but extravagantly tufted, brown (fig. 3.35). Ventrally gray, with some cream-tipped hair. The MNHN specimen is a little grayer, paler, than the type.

Dimensions. Cranial length ($n = 2$): 37.6, 36.2 Günther (1875) gives the measurements of the type as follows: H + B, 152; TL, 149. Petter-Rousseaux and Petter (1967) give the following for the MNHN specimen: H + B, 133: TL, 170: HF, 37; and remark that possibly the tail of the type may have been damaged.

Figure 3.35. *Allocebus trichotis,* to show tufted ears of stuffed Brunoy specimen, MNHN 135–170–37–11, male.

Range. The holotype was collected by Crossley in 1874, "on his way from Tamantave to Murundava" (Günther 1875), but bears the locality "S. Madagascar" on its label. The MNHN specimen was captured in 1965 by Peyrieras, in the forest of Andranomahitsy, to the west of Mananara (fig. 3.31). No locality information accompanies the two other specimens known, collected by Humblot around 1880. This animal is unquestionably the rarest of surviving lemurs, and presumably has never existed at high density, although the collecting records, such as they are, do suggest at one time the genus occurred quite widely in the eastern humid forests. It is worth noting that an effort made in 1975 to find *Allocebus* in the Andranomahitsy forest was unsuccessful (Petter et al. 1977).

Karyotype. Unknown

Phaner Gray, 1870

1870. *Phaner* Gray, Cat. Monkeys, Lemurs, and Fruit-eating Bats in Brit. Mus., p. 135. Based on *Lepilemur furcifer* of author in same work, = *Lemur furcifer* Blainville.

Phaner furcifer (Blainville, 1839). Fork-marked lemur.

This species has a rather spotty distribution within Madagascar, but although there seem to be some minor size and pelage differences between individuals from different areas (specimens from the east coast, for instance, are larger and darker than those from the west, while Russell and McGeorge [1977] have recently reported a reddish population from the far south), it would not appear to be particularly helpful at this stage to distinguish subspecies.

1839. *L[emur] furcifer* Blainville, Ostéog. Mammif., Primates, Makis, p. 35, Pl. 7, Fig. 2. According to Rode (1939), based on MNHN 1834–136, female brought from Madagascar by Goudot. Guessed by Schwarz (1931) to come from around the Baie d'Antongil, but this cannot be confirmed.

Malagasy names. Tanta, tantaraolana, vakiandrina, vakivoho.

Pelage and external characters. Upper parts of most individuals a light brown, although the basal color of the dichromic hair is a darkish gray. Some specimens show reddish or grayish elements. Dorsal pelage dense, relatively short; that of underparts more downy; cream, white, or pale brown with gray basal color apparent. Face covered with short hair of the same color as the crown, which is often somewhat paler than the back. The dark brown rings around the eyes are sometimes expanded down the muzzle, and are always extended backwards, in stripes which meet between the large, naked ears (hence the fork mark of the English name). From this point a highly distinct median dorsal stripe of the same color extends caudally to the rump (fig. 3.36). The proximal half of the long and somewhat bushy tail is usually the color of the dorsum, but the distal moiety is very dark in most individuals, although occasionally the tip of the tail may be white. Russell and McGeorge (1977) have described individuals from the far south as being red orange, with a white throat patch and a red brown face. Males, especially, have a well-developed single median gular cutaneous gland. Nails strongly keeled, pointed. Superciliary, buccal, and genal vibrissae always present; interramal occasionally. Carpal vibrissae present.

Dimensions. Cranial length: $\bar{x} = 53.6$, $s = 1.6$, $o.r. = 51.0–56.6$, $n = 25$. Field measurements (Archbold, $n = 6$): H + B, $\bar{x} = 240$, $o.r. = 215–260$; TL, $\bar{x} = 365$, $o.r. = 350–375$; HF, $\bar{x} = 70$, $o.r. = 67–75$; EAR, $\bar{x} = 34$, $o.r. = 27–37$; (Webb, $n = 4$): H + B, $\bar{x} = 228$, $o.r. = 227–229$; TL, $\bar{x} = 341$, $o.r. = 333–348$; HF, $\bar{x} = 41$, $o.r. = 40–42$; EAR, $\bar{x} = 30.5$, $o.r. = 30–31$. **Body weight:** 350–500 g. (Petter et al. 1977).

Range. *Phaner* has a wide, if now discontinuous, distribution in the west of Madagascar (fig. 3.37). It occurs in an area extending from about the latitude of Tuléar, northward to the region of An-

Figure 3.36. *Phaner furcifer.*

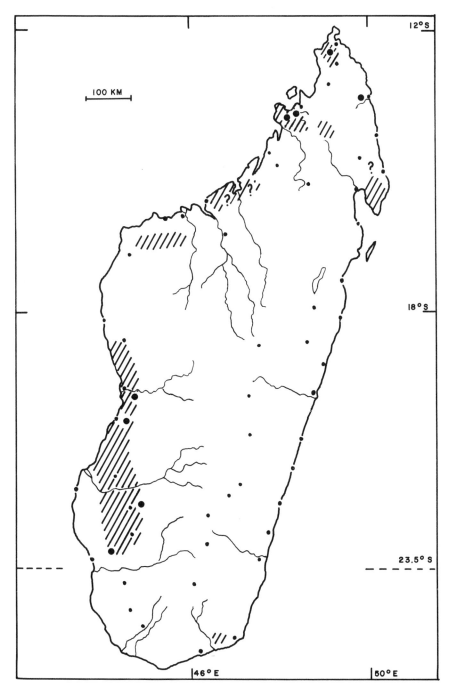

Figure 3.37. Distribution of populations of *Phaner furcifer*. Shaded areas represent approximate range limits; symbols denote localities of museum specimens.

tsalova. A population also occurs to the south of Soalala, and another in the Ampasindava Peninsula and the adjoining region. In the far north of Madagascar an isolate exists on Mt. d'Ambre, while in the east *Phaner* occurs on the Masoala Peninsula. A specimen in the MB, collected by Crossley, bears the locality "Vohémar," but it is dubious whether *Phaner* exists that far to the north and east today. Andriamampianina and Peyrieras (1972) reported that *Phaner* is found on the Tsaratanana Massif. The distribution map published by Petter et al. (1977) fails to show this, but does indicate that the genus occurs in the coastal area to the north and east of the Bay of Bombetoka. Those authors, do not, however, mention the existence of such a population in their text, and no specimens have been recorded from the area despite visits by several collectors. In the far south of Madagascar, *Phaner* is found in Réserve Naturelle 11 on the Mananara River, northeast of Amboasary (Russell and McGeorge 1977).

Karyotype. 2N = 48. Autosomes: M, 4; S, 12; A, 28. Sex chromosomes: X, M; Y, A (Rumpler and Albignac 1973b).

CHAPTER FOUR

Morphology and Adaptation

Remnant though it is of a once much larger and more diverse primate fauna, the surviving ensemble of Malagasy primates presents us with a very considerable morphological and adaptive variety. In this chapter I review various aspects of that variety, although without attempting to be exhaustive. The chapter is in no way a rounded overview of the biology of the extant lemurs; rather, its contents reflect the unevenness of our current knowledge of these fascinating creatures.

CRANIAL AND MANDIBULAR MORPHOLOGY

Cranial Osteology

In terms of overall cranial gestalt, the extant lemurs divide along family lines with remarkable neatness. It is also notable, however, that certain proportions of the skull appear to be so specifically dictated by mechanical factors (Roberts and Tattersall 1973) that, at least on a superficial level, two long-faced members of different families may resemble each other more closely in their longitudinal splanchnocranial/neurocranial proportions than do either of their shorter-faced relatives (Tattersall and Schwartz 1974). The general cranial characteristics of the various genera will be evident from figures 4.1 to 4.4.

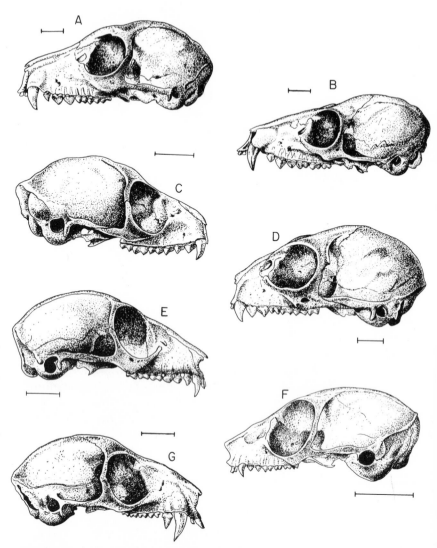

Figure 4.1. Crania of various lemurids, lepilemurids, and cheirogaleids in lateral view. A: *Lemur fulvus*; B: *Varecia variegata*; C: *Lepilemur mustelinus*: D: *Hapalemur griseus*; E: *Cheirogaleus major*; F: *Microcebus murinus*; G: *Phaner furcifer*. All scales represent 10 mm.

The cheirogaleids show a particular degree of conformity in overall cranial morphology, with only *Phaner* departing to any extent from the general pattern. The rostrum of the cheirogaleids is relatively long, narrow, and pointed, and in all genera of this family except *Phaner*, it is very distinctly "pinched" anterior to the lacrimal foramen. In *Lemur* and *Varecia* the muzzle is also quite pointed, but in the short-faced *Hapalemur* and *Lepilemur* it is rather truncated, giving it a squarer aspect. In all of these forms except the last the nasal bones are narrow and elongated. In contrast, the rostrum of the indriids is relatively broad, and in particular high, an effect due largely to the considerable deepening of the maxilla. In all indriids the nasals are relatively short and broad—but least of all in the long-faced *Indri*—and the premaxillae large. The size of the premaxilla is especially striking in *Daubentonia*, where this element is greatly expanded, making contact above with the frontal, from which it separates the maxilla. In lemurids, cheirogaleids, and lepilemurids (especially *Lepilemur*, which lacks upper incisors), the premaxilla is relatively small. All of the Malagasy primates show considerable expansion of the lacrimal anterior to the orbital margin; the lacrimal foramen lies in this portion of the bone. The nasal aperture is normally of inverted piriform shape, and in most cases is quite broad

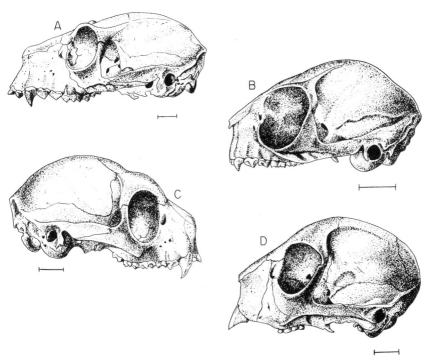

Figure 4.2. Crania of various indriids and *Daubentonia* in lateral view. A: *Indri indri*; B: *Avahi laniger*; C: *Propithecus verreauxi*; D: *Daubentonia madagascariensis*. All scales represent 10 mm.

relative to its height. Internal structure of the nasal cavity is described later in this chapter under "olfaction."

In all lemurs the palate is shallow, although it tends to be vaulted transversely somewhat more in the indriids and *Daubentonia* than in the others. Large anterior palatine fenestrae are present between the premaxilla and the palatine, reflecting the presence of broad nasopalatine ducts. In the cheirogaleids large, open posterior palatine foramina are present, while in the other lemurs these foramina are much smaller, and are generally incorporated into the posterior rim of the hard palate. The posterior margin of the bony palate falls behind the third molar in the cheirogaleids, but is more anteriorly located in the other lemurs, notably *Lepilemur*, where in the midline it extends posteriorly only as far as the anterior portion of M^2. Bony palatal rugae are absent in all genera, although in life the palatal ridges extend far posteriorly; and generally indriids and *Daubentonia* lack the bilateral longitudinal grooves for the anterior palatine nerve and the greater palatine artery which are characteristic of most other lemurs.

The orbits in all lemurs are reasonably well frontated and convergent (see "vision" in this chapter), and unsurprisingly, are relatively larger in the specialized nocturnal forms. In all species a fully formed post orbital bar is present, usually as a flat bony strut; this structure is most strongly developed among the indriids. The upper margins of the orbits in longer-faced forms such as *Indri* and *Varecia* are more projecting than are those of shorter-faced genera such as *Propithecus* and *Lemur*. This is largely because in the latter genera inflated frontal sinuses are developed at the superior junction of the neurocranium and splanchnocranium, thus creating a convex frontal profile in lateral view. In *Indri* and *Varecia*, on the other hand, sinusial development is greatly inferior; hence the profile of this region is more depressed and the structural limits of the superior orbital margins are exposed. The frontal bones are paired, with the metopic suture commonly persisting into advanced age. The interparietal (membrane-derived) portion of the occipital bone is large in all forms; its suture with the supraoccipital portion occasionally remains unclosed into adult life in most lemurs, frequently in *Daubentonia*. Superior temporal lines are usually distinctly marked, especially in larger bodied genera, but in no extant lemur do they meet in the midline.

The zygomatic arch, a rather slender strut in most lemurs, possesses a large vertical dimension in the indriids, *Daubentonia*, and *Hapalemur*. Malar foramina are either absent or tiny in the indriids,

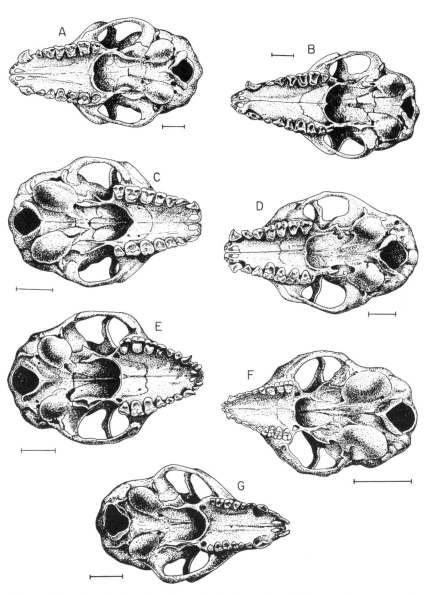

Figure 4.3. Crania of various lemurids, lepilemurids, and cheirogaleids in ventral view. A: *Lemur fulvus*; B: *Varecia variegata*: C: *Lepilemur mustelinus*; D: *Hapalemur griseus*; E: *Cheirogaleus major*; F: *Microcebus murinus*; G: *Phaner furcifer*. All scales represent 10 mm.

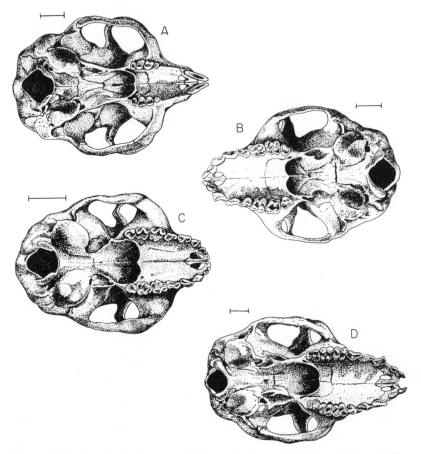

Figure 4.4. Crania of various indriids and *Daubentonia* in ventral view. A: *Daubentonia madagascariensis*; B: *Propithecus verreauxi*; C: *Avahi laniger*; D: *Indri indri.* Each scale represents 10 mm.

cheirogaleids, *Lepilemur* and *Daubentonia*; they are moderate to larger sized and sometimes multiple in the lemurids; and are very large and occasionally multiple in *Hapalemur*.

The structure of the medial orbital wall in the lemurs has attracted considerable attention; the most exhaustive recent treatment of the topic is that of Cartmill (1978), who has emphasized both the amount of intraspecific variability to be found in the composition of the orbital mosaic, and the fact that no lemur family exhibits an entirely distinctive and uniform structure of the medial orbital wall. The reader is referred to Cartmill's account for a detailed discussion of the morphological variability of the region; suffice it here to note Cartmill's finding in his sample of 200 lemur skulls that a distinct ethmoid element is invariably present in the medial orbital wall only in *Microcebus* (and in the *Allocebus* sample of two), and that a majority of specimens of *Lepilemur* also show this condition, while in *Indri*, *Avahi*, and *Hapalemur griseus* the element is only occasionally exposed in the orbital wall. Normally, the exposed ethmoid contacts the lacrimal anteriorly and the orbitosphenoid posteriorly, but a frontopalatine contact sometimes divides the ethmoid from the orbitosphenoid in *Microcebus, Cheirogaleus medius*, and *Allocebus*. Where the ethmoid exposure is of limited size, the element is occasionally separated from the lacrimal by a frontomaxillary contact in *Lepilemur* and *Avahi*. Only in *Lepilemur* is a distinct ethmomaxillary fissure found.

In all lemurs except *Daubentonia*, a floor is provided to the orbit by a lateral expansion of the maxilla, normally accompanied by at least some degree of sinusial inflation, at the root of the zygomatic arch. At the apex of the orbital cone is found a variable number of major foramina. The cranio-orbital foramen, which transmits the superior ramus of the stapedial artery, most often lies on the suture joining orbitosphenoid and alisphenoid in all lemurs except *Microcebus*, where it is usually enclosed by the alisphenoid. In almost all genera, the superior orbital fissure and the foramen rotundum are normally confluent at the rear of the orbit, forming a single large foramen ventral to the orbitosphenoid/alisphenoid suture, thus within the latter bone; the optic foramen lies dorsomedial to this, i.e., within the orbitosphenoid.

Ventrally, the longitudinal proportions of the palate and cranial base are affected by the functional relationships of the splanchnocranium and neurocranium already noted. The structures of the cranial base in indriids are, however, considerably enlarged superoinferiorly relative to those of the other Malagasy primates. In all genera there is a distinct presphenoid element which is sometimes overlapped anteriorly by the vomer; the anterior sphenoid synchondrosis which separates the presphenoid from the basisphenoid often remains unsynostosed well after the fusion of the spheno-occipital synchondrosis. Both synchondroses normally remain unfused until well into adult life. Cranial kyphosis in most lemurs is usually said to be minimal (e.g., Hofer 1969), flexure within the basicranial axis itself being no more than 10°; there is, however, as Hershkovitz (1977) has pointed out, notable variation in the angle formed by the palatal and basicranial axes. Hershkovitz gives this angle as ca. 33°

in *Lemur* (*Homo* = ca. 37°), where declination of the axis of the face on that of the cranial base is relatively small; the angle is substantially larger in many other lemurs, particularly the indriids and *Hapalemur*.

In all lemurs, as among primates in general, the lateral pterygoid plate is larger than the medial. However, the medial plate is usually well developed and bears a prominent hooked hamulus. In the indriids and, in particular, *Daubentonia*, the plates are enlarged supero-inferiorly in correlation with the greater facial depth of these forms. Uniquely in *Lemur rubriventer*, deep bilateral posterior sphenoid fossae occur anterior to the pterygoid processes. In all lemurs the lateral pterygoid plate normally extends caudally to contact the anterior wall of the auditory bulla, leading to the formation of a variably sized pterygospinous foramen (of Civinini) close to the bullar wall. Lateral to the anterior part of the bulla lies the glenoid fossa; this is generally better excavated in the indriids than in the other Malagasy primates, in which it is essentially a flat surface. In *Daubentonia* this surface is well defined, posteriorly expanded, and angled down and backward to overlap the auditory meatus slightly; furthermore, its posterior boundary is not marked by a pronounced postglenoid process as it is in other lemurs. In *Lemur* and *Varecia* this descending process remains distinct from the anterolateral wall of the bulla; in the other Malagasy primates the medial margin of the process tends to be confluent with the bulla wall. In *Lepilemur* the anterior face of the postglenoid process comprises an articular surface. Immediately posterior to the postglenoid process lies a large postglenoid foramen connecting the petrosquamous sinus with the external jugular vein; in *Microcebus* this canal debouches on the tip of the process itself.

In almost all lemurs the exoccipital region is rugose and well defined, and in many it bears a distinct paroccipital process on its medial border. This process is a rather slender, bladelike lamina in forms such as *Hapalemur*, *Avahi*, and *Cheirogaleus*, while in *Indri* it is long and robust. The mastoid region is generally rather weakly pneumatized in lemurs, although some pneumatization is almost invariably present (see discussion by Saban 1975). Among the indriids, *Avahi*, in particular, shows pneumatization in this area, while such development is most marked of all in *Allocebus*, where the mastoid is inflated by a mass of tiny bone cells. Other cheirogaleids share this tendency, but in a less elaborate and extensive way (Cartmill 1975).

In the Malagasy primates the nuchal region is usually defined above by a well marked if low nuchal crest, and is characterized by a broad median vermis. Among the lemurs, body posture is not clearly correlated with the orientation of the foramen magnum. Indeed, the axis of this foramen in the more pronograde *Varecia* is more downwardly directed than it is in the habitually upright *Indri*. Far more suggestive of posture is the nature of the occipital condyles, the articular surfaces of which are relatively expanded in the indriids and *Lepilemur*. A relatively large hypoglossal canal is present at the base of each condyle in all genera.

The area of the lemur cranial base which has traditionally attracted the most attention from systematists is the ear region. All living lemurs possess inflated auditory bullae, derived from the petrous portion of the temporal bone and wedged between the basioccipital and the alisphenoid. In almost all lemur families the size of the bulla bears a marked negative relationship to body size; but in the small *Allocebus* bullar inflation is minimal, rather as in lorisines. In *Microcebus* on the other hand, the bulla is absolutely as large as that of the much larger-bodied *Cheirogaleus major*. The characteristic most commonly cited as distinguishing Malagasy from mainland strepsirhines is the intrabullar ectotympanic ring, attached above and behind by its anterior and posterior horns to the squamosal and petrosal; the posterior horn contributes in considerable measure to the formation of the posterior part of the auditory meatus. The "free" anteroventral portion of the ring does indeed lie within the bulla (having rotated downward from a more or less horizontal foetal orientation), but is attached to the bony accoustic meatus by a cartilaginous annular membrane, which may on occasion become ossified as an apparent extension of the petrosal in older specimens of *Lemur* (Van Valen 1965), and in *Allocebus* (Cartmill 1975). Saban (1963) has observed that in *Microcebus* the annulus comes to lie so close to the meatus that it causes a ring-shaped deformation of the bullar wall. Cartmill (1975) has also noted in some *Microcebus* that the annulus membrane is replaced almost entirely by bony laminae extending medially from the tympanic and laterally from the petrosal. Cartmill suggests that in such specimens of *Microcebus* the condition of the ectotympanic is not vastly dissimilar from that seen in *Loris*, where the middle portion of the ring, although in contact with the inner bullar wall, is enclosed by the petrosal; he would thus limit the diagnostic characters of the Malagasy strepsirhines in this region to only two: extension of the tympanic air space laterally below the lower edge of the ring, and the absence of a complete longitudinal septum dividing the hypotympanic sinus from the lateral part of the intrabullar air space. The usefulness of such di-

agnostic characteristics depends, however, on one's view of relationships within Strepsirhini (see chapter 6), and it remains true that the
anatomical relationships of the ectotympanic in all lemurs (even *Microcebus*) are quite distinctive. The reader is referred to Saban (1963,
1975) for detailed descriptions of the osteology of this region.

The auditory bulla, roughly hemispherical or ovoid in shape, is
surrounded by a number of major and minor foramina concerned
mostly with the transmission of blood vessels and nerves. Laterally
lies the large postglenoid foramen already noted; posterolaterally,
just behind the auditory meatus, lies the stylomastoid foramen,
while posterior or inferior to this is found the carotid foramen,
which conducts the internal carotid artery to the interior of the
bulla. In lemurids, indriids, and *Daubentonia*, this small orifice is
somewhat laterally situated, except in *Avahi* where, as in *Lepilemur*,
it is more medially positioned. The carotid foramen lies posteromedially in cheirogaleids. The so-called posterior lacerate foramen lies
at the posterior margin of the bulla, within a capacious depression
formed at the junction of the petrosal and the occipital; in *Daubentonia* it is tightly sandwiched between the bulla and the occipital condyles. In *Lemur*, *Hapalemur*, and *Phaner* a lamina of the occipital extends across this depression, dividing it; the inferior petrosal sinus
thus debouches in these forms in the large fossa accomodating the
hypoglossal foramen (Saban 1963). Anterior to the bulla, just lateral
to the point at which it contacts the lateral pterygoid plate, lies the
oval foramen; medial to this lies the exit of the auditory tube. Yet
further medially, and slightly anterior, there is found in the cheirogaleids an "anterior carotid foramen," probably homologous with
the foramen lacerum of human anatomy (Cartmill 1975).

Carotid Circulation

The cheirogaleids share with the mainland strepsirhines a pattern of carotid circulation which sharply distinguishes them from
the other members of the Malagasy primate fauna. In both groups
the vertebral artery predominates in the blood supply to the brain.
But in the other lemurs this artery is supplemented, if only in a very
minor way, by blood derived from what is nonetheless a reasonably
well developed internal carotid system, while in the cheirogaleids
the internal carotid is feebly developed at best, and the brain is also
supplied via an "anterior carotid," which penetrates the braincase
anterior to the bulla.

Carotid circulation in *Lemur* has been described in some detail by Saban (1963) and Bugge (1974). Below and behind the bulla the common carotid trunk divides into internal and external carotid arteries (fig. 4.5). The internal carotid first gives off a fine branch which runs across the bulla to enter the orifice of the auditory tube after emitting inferior tympanic and posterior meningeal branches; it then sends off an occipital branch before entering the bulla via the carotid foramen and a bony carotid canal. Once on the promontorium, near the fenestra cochleae, the internal carotid is usually said to divide into stapedial and promontory (entocarotid) portions. Much has been made of the relative sizes of the bony promontory and stapedial canals in fossil forms, but Conroy and Wible (1978) have demonstrated that despite the presence of a substantial bony promontory canal (housing the internal carotid nerve) in a specimen of *Varecia*, the promontory artery was vestigial at best; and Bugge (1974) notes that in a specimen of *Lemur catta* the internal carotid was entirely obliterated distal to the origin of the stapedial artery.

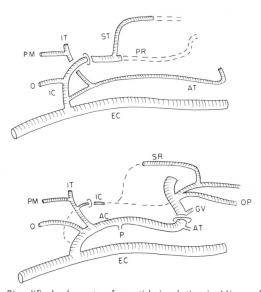

Figure 4.5. Simplified schemata of carotid circulation in *Microcebus* (above; after Cartmill [1975] and *Lemur* (below; modified after Saban [1963]). AC: anterior carotid; AT: branch to auditory tube ("rameau tubaire"); EC: external carotid; GV: branch to ganglion of fifth cranial nerve; IC: internal carotid; IT: inferior tympanic; O: occipital; OP: ophthalmic; P: pharyngeal branch(es); PM: posterior meningeal; SR: superior ramus of stapedial; ST: stapedial.

Saban (1963) has, however, reported the presence of subequally sized promontory and stapedial branches in *Lemur* and *Daubentonia*. After the split, the stapedial artery runs past the vestibular window in a bony canal and emerges to pass through the crura of the stapes before penetrating the endocranium and emitting branches to the dura and the orbit.

The internal carotid pattern described for *Lemur* is said by Saban (1963) to apply essentially equally to the indriids and *Daubentonia*. In *Lepilemur* the stapedial artery is greatly reduced.

The most detailed study of carotid circulation in a cheirogaleid is that of Cartmill (1975) on *Microcebus*. In this animal the common carotid trunk bifurcates behind the bulla into large anterior and external carotid arteries (fig. 4.5). The former runs forward and upward, giving off a tiny posterior meningeal branch which enters the jugular foramen. This branch in turn emits a minute internal carotid artery which goes to the tiny carotid foramen (in Cartmill's specimen this artery was present only on the right side), and also an inferior tympanic branch. On its way forward, the anterior carotid gives off various minor pharyngeal branches, and at the front of the bulla gives rise to slender medial pterygoid, terminal pharyngeal, and meningeal branches. The main trunk continues through the anterior carotid foramen to join the cerebral arterial circle after emitting various small intracranial branches including superior and inferior ophthalmic arteries. Inside the neurocranium, no trace of the internal carotid or any derivatives could be discerned in Cartmill's specimen. Cartmill argues convincingly, following the suggestions of Adams (1957) for *Nycticebus* and Bugge (1972, 1974) for lorises in general, that the anterior carotid in *Microcebus* is homologous with the ascending pharyngeal artery of human anatomy; he also surmises that the branch of the internal carotid to the auditory tube in *Lemur* (the "rameau tubaire" of Saban [1963]) is homologous with the anterior carotid of cheirogaleids.

Saban (1963) has described the carotid circulation of *Cheirogaleus* in somewhat less detail; according to his report, the carotid pattern of *Cheirogaleus* is essentially similar to that which Cartmill later described in *Microcebus*, except that the internal carotid is relatively larger, and following its separation from the external carotid artery stems independently from the common carotid instead of from the anterior carotid. Further, Saban indicates that in this form the internal carotid maintains an anastomosis with the anterior carotid artery, and also emits a stapedial branch; the stapedial is said to be lacking in *Phaner*.

Mandibular Morphology

The lower jaws of the indriids are sharply distinguished from those of all other lemurs in the great depth of the corpus, the rounding out and posteroventral expansion of the angle, the long symphysis which shows a deep genial fossa and large inferior tubercles, and in a variety of less obvious characteristics. Among these is the very extensive area of attachment of the anterior belly of the digastric muscle, which in *Lemur*, *Varecia* and *Lepilemur* leaves no scar at all, and in *Hapalemur* leaves only a limited impression. In the indriids the anterior digastric shows a well marked area of insertion, which increases in relative extent with body size; it is expressed in *Indri* as a bony protuberance. The mandible of *Daubentonia* is highly modified, but does resemble that of the indriids in possessing a deep, narrow corpus with a swelling of its ventral margin in the area of digastric insertion.

The horizontal rami of the lower jaws of *Lemur* and *Varecia* closely resemble each other in being long and shallow, with a hooked angle (fig. 4.6); *Hapalemur* possesses a more convex gonial angle with considerable ventral protrusion, and shows a narrow, posteriorly hooked coronoid. *Lepilemur* is intermediate in possessing a *Lemur*-like hooked angle which protrudes ventrally like that of *Hapalemur*, and also resembles the latter in showing some deepening of the mandible at the symphysis. The symphysis in lemurids, lepilemurids, and cheirogaleids is rather ovoid in section, lacking the genial pit and pronounced lateral tubercles of the indriines; it is inclined at an oblique angle comparable to that seen in the indriids, but fails to reach as far posteriorly because of the shallowness of the mandibular body. The symphyses of lemurs are usually described as unfused, but synostosis may occur in individuals of many genera.

Among the indriids there is a tendency towards increasing subcircularity of the mandibular condyle with size. Thus in *Avahi* the condyle remains relatively broad transversely but is strongly curved in the coronal plane, while in *Propithecus* the curvature is retained but the relative width decreased. In *Indri* the condyle is virtually globular. The articular surface appears to be carried on to the posterior aspect of the condylar neck in all indriines; this region conforms very closely to the shape and curvature of the anterior face of the postglenoid process. In contrast, among the lemurids the mandibular condyle is broad transversely, the articular surface is more restricted anteroposteriorly, and although the axis of the surface points more posteriorly than does that of the indriines, the condyle

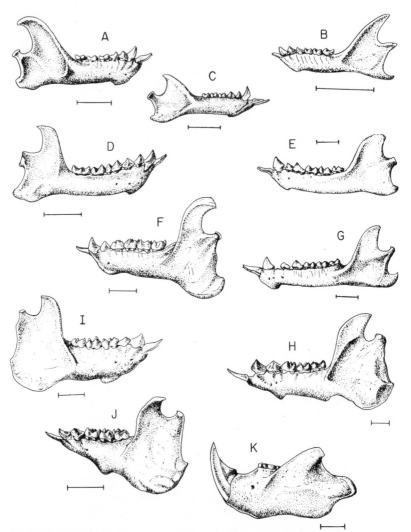

Figure 4.6. Lateral views of mandibles of various lemur species. A: *Cheirogaleus major*; B: *Microcebus murinus*; C: *Phaner furcifer*; D: *Lepilemur mustelinus*; E: *Lemur fulvus*; F: *Hapalemur griseus*; G: *Varecia variegata*; H: *Indri indri*; I: *Propithecus verreauxi*; J: *Avahi laniger*; K: *Daubentonia madagascariensis*. Each scale represents 10 mm.

is better differentiated from the neck. In *Hapalemur* condylar morphology is comparable to that found in *Lemur*, although there is more posterior lipping on the medial aspect of the articular surface; but in *Lepilemur* a definite posterior articular surface descends vertically from the medial portion of the condyle. This surface is matched by the anterior face of the postglenoid process, and is found elsewhere among the lemurs only in the extinct *Megaladapis*. The mandibular condyle of *Daubentonia* is greatly modified in being anteroposteriorly elongate and laterally compressed; it most closely resembles a drawn-out version of the condyle of *Indri*.

THE DENTITION

Morphology of the Permanent Dentition

Upper and lower permanent dentitions of representative individuals of most extant lemur genera are depicted in figures 4.7 and 4.8. The following brief notes are intended primarily as a supplement to these illustrations.

In the diurnal indriines the upper incisor teeth are relatively large, showing a pronounced lateral flare; in *Avahi* they are small and peglike. I^2 (I follow Schwartz [1974] in identifying the pair of upper incisors as I^{2-3}) is the larger tooth of the pair in *Propithecus*, while the reverse is true of *Indri* and *Avahi*. Among the lemurids the upper incisors are greatly reduced; the same is true of *Hapalemur*, while in *Lepilemur* they are absent altogether. In *Cheirogaleus*, *Microcebus*, and *Mirza*, these teeth are relatively somewhat larger than in the lemurids, and I^2 is very distinctly larger than I^3; in *Phaner* I^2 is robust, greatly elongated, and projects anteriorly.

Schwartz (1974, 1975; Tattersall and Schwartz 1974) has identified the anterior teeth of *Daubentonia* as canines, and I maintain this usage here. These teeth are large, laterally compressed, and open rooted. Among the other Malagasy strepsirhines, the upper canine is least high crowned and most mesiodistally elongate in the indriines.

In *Lemur* and *Varecia* P^3 is the most salient tooth of the upper premolar row, and P^4 is not molariform despite the development lingually of a substantial protocone; in *Hapalemur* and *Lepilemur*, on the other hand, P^2 is the highest crowned of the upper premolars, and in *Hapalemur* P^4 is large and distinctly molariform. Among the cheirogaleids, P^2 and P^3 are both bladelike and subequal in crown height in *Microcebus* and *Mirza*, whereas in *Cheirogaleus* the crown of the latter is lower but more elongate mesiodistally than that of P^2. *Phaner* is distinguished by its "caniniform" P^2, as also is *Allocebus*. In none of the cheirogaleids is P^4 molariform, although a small protocone is present in most genera. The indriines are clearly distinguished by their possession of only two upper premolars; in *Indri* these are subequal in length, while in *Avahi* and *Propithecus* P^3 is of

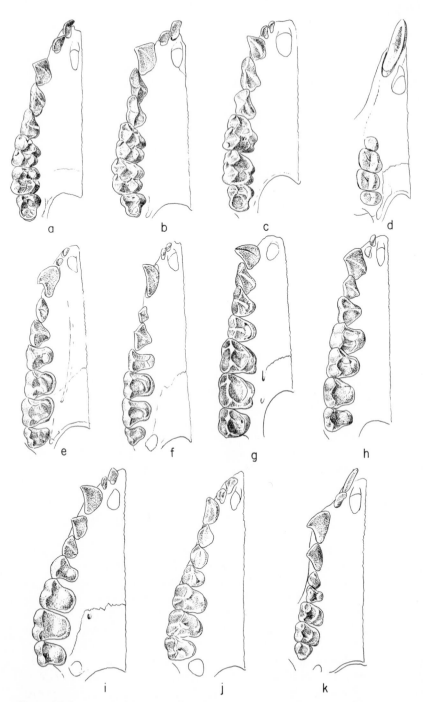

Figure 4.7. Upper dentitions of a: *Indri indri*; b: *Propithecus verreauxi*; c: *Avahi laniger*; d: *Daubentonia madagascariensis*; e: *Lemur fulvus*; f: *Varecia variegata*; g: *Lepilemur mustelinus*; h: *Hapalemur griseus*; i: *Cheirogaleus major*; j: *Microcebus murinus*; k: *Phaner furcifer*. Not to scale.

somewhat greater length than P⁴. In all indriines these teeth are compressed laterally, lacking any significant lingual development, although P⁴ is less compressed than is P³. Only a single, small, peg-like upper premolar is present in *Daubentonia*.

M¹ and M² of all the cheirogaleids except perhaps *Mirza* are reminiscent of those of the lorisids and galagids in possessing a more or less notched posterior margin, partially reflecting the posterior displacement of the hypocone except in the low-cusped *Cheirogaleus*, where this cusp is absent. The hypocone is always quite small, and arises from a lingual cingulum, in *Microcebus* and *Mirza*; the cingulum is also present, if faint, in *Cheirogaleus*, in which the molar cusps are low and rounded. In *Phaner* the hypocone is better differentiated, and there may be traces of cingulum anterior to it. The cheek teeth of *Phaner* are exceptionally small relative to body size. M³ is relatively reduced and is subtriangular in all cheirogaleids; the lingual cusp pair, as in the anterior upper molars, is higher than the protocone. In *Phaner* the paracone of M³ is somewhat larger than the metacone, rather than subequal with it, as in M¹⁻².

There is some variation in the conformation of the upper molars within genus *Lemur*. In all species, M¹⁻² have subequally sized buccal cusps faced lingually by a lower but large and crescentic protocone; in *L. catta* and *L. rubriventer* this is bounded internally by a distinct lingual cingulum. In M² of *L. catta*, and M¹⁻² of *L. rubriventer*, the cingulum gives rise to a large protostyle mesially, and a small hypocone distally. The protostyle and hypocone are normally present in M¹⁻² of other species of *Lemur* also, but do not originate from a cingulum. M³ of all *Lemur* species is greatly reduced in size; its paracone is much larger than the metacone and there is rarely any lingual development, except very occasionally in the form of a small protostyle. In *Varecia* the paracone and metacone of M¹⁻² are subequally sized, and the protocone, while still crescentic, is relatively small, shifted anteriorly, and more transversely oriented, the trigon basin thus being open distally. A broad lingual cingulum exists, and is particularly large mesiolingually. M³ is reduced and simple, its protocone lying far anteriorly. The upper molars of *Lepilemur* and *Hapalemur* show far less lingual development than do those of the lemurids. In the former, the lingual cingulum of M¹⁻² is much reduced, although a tiny hypocone may be hinted at, while in the latter there is only the merest trace of a cingulum. In both genera, the buccal cusps are somewhat compressed buccolingually, with some suggestion of a buccal cingulum, and are more distinctly interconnected than are those of *Lemur* and *Varecia*. In *Hapalemur* the buccal cingulum may give rise to a small mesostyle on M^1, and in both genera a distinct preprotocrista sweeps forward to connect with the sharp paracrista anterior to the paracone. Similarly, in both genera the protocone is rather forwardly shifted, leaving the trigon basin open distally. M^3 is relatively less reduced than in the lemurids, particularly in the case of *Hapalemur*.

The anterior upper molars of the indriines are distinctly quadricuspid. In M^1 the paracone dominates, displacing the mesiobuccal border of the tooth anteriorly; this effect is emphasized by the reduction of the protocone. In *Indri* the metacone and hypocone are in transverse alignment and the buccal cusp is larger, as in the anterior pair; in *Propithecus* and *Avahi* the hypocone is displaced somewhat posteriorly. Although lacking a buccal cingulum, *Indri* possesses a distinct parastyle and mesostyle on M^1, as do *Propithecus* and *Avahi*, in which a buccal cingulum is present; the latter genera also exhibit small metastyles, as well as paraconules and metaconules. The paracone of M^2 lies anteriorly as does that of M^1, in both *Propithecus* and *Avahi*; in *Indri* M^2 is squarish, the buccal cusps lying almost opposite the lingual cusps. Again, the trigon is better defined in *Propithecus* and *Avahi*, although the hypocone is displaced less far posteriorly. Tiny metaconules and paraconules may be present on M^2 of all three genera, as may also a small transverse crease representing the anterior fovea. Well-defined mesostyles are present on M^2 in *Avahi* and *Propithecus*, and a trace of this structure may be seen in *Indri*. The parastyle is better developed in M^2 of *Propithecus* and *Avahi* than in that of *Indri*; in M^3 of each genus the paracone and protocone are prominent and subequal in size. The posterior margin of this latter tooth is marked by a rim whose composition of a distobuccal metacone and a lingual hypocone is far more clearly evident in *Indri* than in *Propithecus* or *Avahi*. The talon basin is relatively much larger in *Indri* than in the other indriines.

The upper molar teeth of *Daubentonia* are greatly reduced in size, and are almost devoid of cuspal detail, although the differentiation of a mesostyle may be detected between the rudimentary paracone and metacone of M^{1-2}. In their generally squarish outline, however, these teeth approach the indriine rather than the lemurid condition. They decrease in length posteriorly, although M^2 is broader than M^1.

With the single exception of *Daubentonia*, all extant Malagasy primates possess procumbent anterior dental combs, or scrapers, the primary function of which has long been a subject of debate although clearly the structure is used in both grooming and feeding.

In all lemurs the lateral teeth of the comb possess a pronounced lateral flare, whereas those positioned more centrally are narrower, and are virtually parallel edged. A longitudinal ridge runs along the center of the dorsal surface of each tooth of the group, bounded on each side by a shallow groove; in the outer teeth it is the lateral groove which becomes expanded to produce the flare. In all cheirogaleids, lemurids, and lepilemurids, as in the mainland strepsirhines, the comb consists of six teeth, i.e., two incisors and a canine bilaterally; but in the indriines only four teeth are present. Traditionally, the latter have all been interpreted as incisors, but here I follow the suggestion of Schwartz (1974, Tattersall and Schwartz 1974) that the outer pair be regarded as canines. The single, large, laterally compressed, open-rooted anterior tooth of *Daubentonia* is similarly interpreted as a canine. Among the cheirogaleids the dental comb of *Phaner* is particularly notable for its elongation and for the great dorsoventral depth of the lateral teeth; these characteristics are presumably related to the specialized use of this structure as a gum-scraper. The comb of *Microcebus*, on the other hand, is relatively the shortest among all the extant lemurs.

All genera except *Daubentonia*, in which lower premolars are entirely wanting, have caniniform anterior premolars. In the indriines the low anterior premolar tends to be less salient but more elongate mesiodistally than in the others; this is especially true of *Avahi*. The cheirogaleids, like the galagids, are typified by particularly slender and somewhat procumbent caniniforms. In all lemurs, including the indriines in which only two lower premolars are present, the anterior lower premolar is identified here as P_2, following the usage of Schwartz (1974).

P_3 and P_4 in the cheirogaleids resemble these teeth in galagids; P_3 is subcaniniform except in *Phaner* and to a lesser extent in *Microcebus*, while P_4 shows some talonid development. P_3 and P_4 of the lemurids show increasing complexity posteriorly, although the paracone invariably predominates; all species of *Lemur* except *L. catta* show the development lingually of a small metaconid. In *Varecia* P_4 is clearly more salient than P_3; in *Lemur* the two teeth are generally subequal in height, although P_4 is longer. In *Lepilemur* P_3 may be slightly higher than P_4; and in this genus the latter tooth possesses a less distinct, more laterally compressed posterior fovea than does its homologue among the lemurids. P_3 in *Hapalemur* is higher than P_4, but the latter is very strongly molarized, bearing five distinct cusps and well-defined anterior and posterior foveae. Both extant lepilemurids lack the diastema between P_2 and P_3 found in *Varecia*

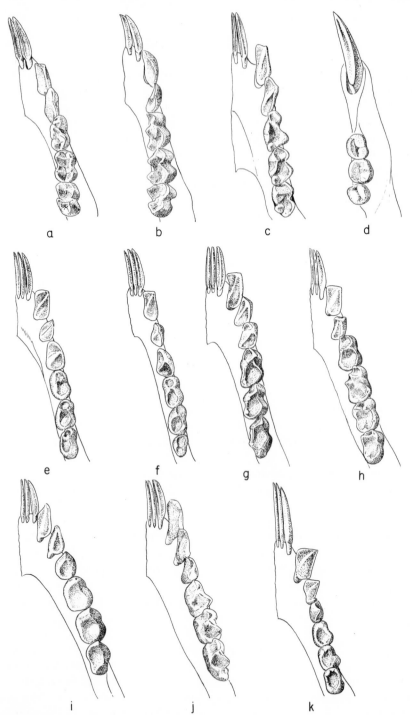

Figure 4.8. Lower dentitions of a: *Indri indri*; b: *Propithecus verreauxi*; c: *Avahi laniger*; d: *Daubentonia madagascariensis*; e: *Lemur fulvus*; f: *Varecia variegata*; g: *Lepilemur mustelinus*; h: *Hapalemur griseus*; i: *Cheirogaleus major*; j: *Microcebus murinus*; k: *Phaner furcifer*. Not to scale.

and *Lemur*. The laterally compressed posterior lower premolar of the indriines (P_3) is simple, consisting of a single cusp with crests extending mesially and distally. In *Avahi* this tooth is exceptionally elongated, and in both this genus and in *Propithecus* there is some differentiation of a metaconid.

The two anterior lower molars of the lemurids are subequal in size, although the talonid basin is wider and more open in M_1 than in M_2. The protoconid is the highest and most anterior cusp, whereas the metaconid and hypoconid are only slightly raised from the encircling rim of the talonid. A distinct fovea is present, with the suggestion of a paraconid in some *Lemur* species. The hypoconid projects somewhat laterally, contributing to a lateral waisting of these teeth which is rather more marked in *Lemur* than in *Varecia*. *Lemur* shows no discernible entoconid, and the talonid basin is open lingually; in *Varecia*, however, a small entoconid is present, and the talonid basin, relatively larger than in *Lemur*, is open posteriorly. M_3 in the lemurids, although relatively smaller than the anterior molars, nonetheless possesses four distinct cusps. In *Lemur* the talonid basin remains closed, while in *Varecia* the entoconid is displaced buccally to contact the hypoconid, so that the talonid basin is open lingually.

The anterior lower molars of the lepilemurids are quite similar to each other although they are more compressed laterally in *Lepilemur*, and the cusps are distinctly inflated in *Hapalemur*. Four distinct cusps are present, in contrast to the lemurid condition; the entoconid is well developed, especially in *Hapalemur*. But while in *Lepilemur* the metaconid and entoconid are connected by sharp crests and the entoconid and hypoconid are not, in *Hapalemur* the reverse applies. The talonid basin is thus open posterolingually in *Lepilemur*, while the opening is lingual in *Hapalemur* (although it is constricted in M_1 by the presence of a metastylid). In *Lepilemur* alone, the anterior fovea remains unclosed. M_3 of *Lepilemur* is characterized by the presence of an elongated talonid heel, due largely to the presence of a very large, posteriorly displaced hypoconulid. M_3 of *Hapalemur* is reduced, but is reminiscent morphologically of M^{1-2}.

M^{1-2} of the cheirogaleids possess four distinct cusps, and in all genera of the family except *Phaner* there exists a hypoconulid on M_3. The anterior molars are relatively broader than are those of the lemurids, and the buccal cusps are displaced further mesially. An anterior fovea is lacking, although a slight paraconid shelf may be present, as is the case among lorisids. In *Cheirogaleus* the molar cusps are notably low and rounded.

Among the indriids the lower molars, like the uppers, are strongly distinct from those of the other lemurs. Four distinct cusps are present on all lower molars, and M_1 additionally shows a paraconid while M_3 exhibits a small hypoconulid. In M^{2-3} of *Avahi* and *Propithecus* each buccal cusp lies slightly anterior to its lingual partners; in *Indri* the mesial and distal cusp pairs are in transverse alignment. The lingual cusps of the lower molars of *Indri* are subequal in size, and are somewhat rounded; in the other indriines the hypoconid predominates and like the protoconid is more angular. In striking contrast to the lower molars of the other lemurs, in which the reverse applies, the lingual cusps of the indriines are higher than the buccal ones. The simple, reduced lower molars of *Daubentonia* resemble those of the indriines in being raised lingually.

Morphology of the Deciduous Dentition

In all extant strepsirhines except for *Daubentonia* and *Lepilemur*, the deciduous upper incisors are small, but are morphologically reminiscent of their permanent replacements. In *Lepilemur* deciduous incisors appear, but are unreplaced. The deciduous upper canine similarly resembles its permanent successor, but is much smaller in size.

In both cheirogaleids and galagids dp^2 is caniniform, but is less so than in the lorisids; dp^3 is a roughly triangular tooth in which, with the exception of *Cheirogaleus*, the single major cusp may be supplemented by a small parastyle and traces of a cingulum. dp^4 is small, but morphologically resembles M^1; thus in *Microcebus* it bears four cusps, and in *Cheirogaleus* and *Phaner*, three.

In *Lemur* and the lepilemurids dp^2 is small, simple, and single-cusped; so also is dp^3 in *Lemur*, but in the lepilemurids this tooth is elaborated by distinct lingual development. The dp^4 is small but similar morphologically to M^1; in *Lemur* this tooth possesses three cusps of which the protocone is the largest, while in the lepilemurids the metacone predominates. Among the indriines the anterior deciduous upper premolar is unicuspid and laterally compressed; the posterior is broader, possessing a distinct lingual bulge, is banded by a cingulum and is centrally waisted; lingual styles are common on this tooth, a parastyle always being present. *Daubentonia* exhibits three deciduous premolars, of which only the most posterior is replaced. The first is small and peglike, and separated by a wide diastema from the two behind it; of the latter, the anterior is also small and peglike, while the posterior is somewhat triangular in outline but has undifferentiated crown relief.

The lower anterior deciduous teeth in all the lemurs resemble their successors in morphology, except in the case of *Daubentonia* where the lower anterior teeth are continually growing and are persistent throughout life. The lower deciduous premolars of the cheirogaleids and the galagids are very similar: dp_2 is caniniform, with a small posterior heel; dp_3 is long and low crowned, banded by a lingual cingulum which gives rise to small anterior and posterior stylids, and possesses a smaller secondary cusp in association with the large primary cusp; dp_4 is highly molariform, exhibiting four distinct cusps.

In both lemurids and lepilemurids dp_2 is caniniform; in *Lemur* it is the least salient deciduous premolar, but in *Lepilemur* and *Hapalemur* it is distinctly higher than the others. In both families dp_3 is long, low, laterally compressed, and essentially unicuspid with a posterior basin; in *Lemur* (except *L. catta*) however, a small secondary cusp may be associated with the primary cusp, and in the lepilemurids the tooth exhibits an anterior stylid. In *Lemur* and *Hapalemur* dp_4 is distinctly molariform, possessing a paraconid in addition to the four primary cusps; in *Lepilemur* this is not the case, but the disposition of the anterior moiety of the tooth resembles that in M_1 while the posterior part of the tooth forms a large basin opening distally and buccally.

Uniquely among the lemurs, the indriines possess four lower deciduous premolars (Schwartz 1974). The dp_1 is a small, peglike tooth and in very young individuals lies close to the much larger dp_2, which is caniniform. Echoing dp_1, dp_3 is reduced and peglike, lying close to the rather triangular dp_4, which is expanded distally. *Daubentonia* shows two unreplaced deciduous lower premolars; these are found well back in the jaw and somewhat resemble dp_{3-4} of the indriines, although with a less differentiated cusp morphology.

Dental Development and Eruption

Schwartz (1974; Tattersall and Schwartz 1974) has determined the sequences of development and initial and completed eruption in a variety of lemur species, including representatives of almost all living genera. Table 4.1 summarizes Schwartz's findings.

THE BRAIN AND CRANIAL SENSORY ORGANS

External Morphology of the Brain

Elliot Smith's monograph of 1903 still stands as the most comprehensive comparative work on the morphology of the brain of le-

Table 4.1. Sequences of dental development and eruption in various lemur species.

	Development	Initial Eruption	Completed Eruption

Lemur mongoz
```
Development:
M1  C  P2 I2 I3   M2   P4   M3   P3
          M2     P4   M3   P3
M1 I2I3C P2

Initial Eruption:
M1  I2I3 C   M2   P2   M3P4   P3
M1 I2I3C    M2   P2   M2     P4   P3

Completed Eruption:
M1 I2 I3  M2 M3 P2   C P4   P3
M1 I2I3C  M2 M3      P2   P4   P3
```

Lemur fulvus
```
Development:
M1   C   M2 I2I3   P2   P4 M3   P3
M1 I2I3C M2        P2   P4   M3P3

Initial Eruption:
M1     I2I3   M2 C P2 P4 M3   P3
M1 I2I3C      M2   P2   P4  M3 P3

Completed Eruption:
M1 I2 I3  M2 P2  C M3 P4 P3
M1 I2I3C  M2 P2    M3 P4 P3
```

Lemur macaco
```
Development:
M1  I2 C M2 I2I3  P2   P4 M3   P3
M1 I2I3C M2       P2   P4   M3 P3

Initial Eruption:
M1     I2I3 M2 C P2 ?M3 P4 P3
M1 I2I3C    M2   P2 ?M3 P4 P3

Completed Eruption:
M1 I2 I3  M2 ?C   P2   ?
M1 I2I3C  M2      ?P2   ?
```

Lemur rubriventer
```
Development:
M1   C   ?
M1 I2I3C P2   ?
```

Lemur catta
```
Development:
M1 I2 I3 C M2 P4 M3 P3 P2
M1 I2I3C M2   P4 M3 P3 P2

Initial Eruption:
M1 I2 I3   M2 C M3 P4 P3 P2
M1 I2I3C M2     M3 P4 P3 P2

Completed Eruption:
M1 I2 I3  M2 M3 P4 ?P3 P2 C
M1 I2I3C  M2 M3 P4 ?P3 P2
```

Varecia variegata
```
Development:
M1 I3I2 C P2 M2 P4 M3 P3
I2I3C M1    P2 M2 P4 M3 P3

Initial Eruption:
M1 I3I2 M2 ?C P2 P4 M3 P3
M1 I2I3C M2 ?P2 P4 M3 P3

Completed Eruption:
M1  I2I3      ?
M1 I2I3C      ?
```

Hapalemur griseus
```
Development:
M1 M2 I2I3 C P4 P3 M3 P2
M1 I2I3C M2   P4 P3 M3 P2

Initial Eruption:
M1     I2I3   M2 P4 M3 P3 C P2
M1 I2I3C      M2   P4 M3 P3   P2

Completed Eruption:
(same as initial)
```

Lepilemur mustelinus
```
Development:
M1 M2   M3 C P2 P4 P3
M1 M2 I2I3C M3   P2 P4 P3

Initial Eruption:
M1 M2       M3 P4 C P3 P2
M1 M2 I2I3C M3 P4   P3 P2

Completed Eruption:
M1 M2       M3 P4 P3 P2 C
M1 M2 I2I3C M3 P4 P3 P2
```

Avahi laniger
```
I2 M1 I3  P4 M2 M3 P3 C
   M1 I3C P3 M2 M3 P2
```

Propithecus verreauxi
```
M1    I2 M2 I3   C P4 M3 P3
M1 I3C M2       P3   M3    P2
```

Propithecus diadema
```
M1 I2  M2 M3 I3 P4 C P3
M1 I3C M2 M3    P3 P2
```

Indri indri
```
I2I3 M1 M2 C P4 M3 P3
I3C  M1 M2   P4 M3 P3
```

Daubentonia madagascariensis (development and eruption)
```
                   M1 M2 P4 M3
                   M1 M2
```

Cheirogaleus major
```
(M1 I2I3   )P2 C  M3 P4 P3
(M1 I2I3C M2)P2   M3 P4 P3
```

Mirza coquereli
```
M1    I2 M2 I3 C P2 M3 P4 P3
M1 I2I3C  M2     P2    M3 P4 P3
```

Microcebus murinus
```
(M1    I2I3 ) M2   M3 C P2 M3 P4 P3
(M1 M2 I2I3C)      P2 M3 P4 P3
```

Phaner furcifer
```
M1    I2 M2 I3 C P2 P4 M3 P3
M1 I2I3C  M2      P2 P4 M3 P3
```

```
I2I3 M1        P4 P3 M2 C M3
     M1 I3C P3 P2 M2    M3
```

```
M1 I2  M2 I3 P4 M3 P3 C
M1 I3C M2 I3 P3 M3 P2
```

```
M1 I2I3 M2 P4 P3 M3 C
M1 I3C  M2 P3 P2 M3
```

(See text)

```
(M1 I2I3  M2) P2   C M3 P4 P3
(M1 I2I3C M2)    P2   M3 P4 P3
```

(same as development)

```
(M1    I2I3) M2 C P2 M3 P4 P3
 M1 M2 I2I3C      P2   M3 P4 P3
```

(same as development)

```
M1       I2 M2        ?
      I2I3C  M2        ?
```

murs, and his citations provide a thorough guide to the earlier literature on the subject. More recently, the endocast studies of
Radinsky (e.g., 1968, 1974) have excellently supplemented this pre-
vious work, particularly in using the results of cortical mapping to
aid in functional interpretation. Figure 4.9 shows lateral views of endocranial casts of several lemur species. Certain general characteristics of the external morphology of the brain appear within related
groups to vary closely with size; for instance, smaller brains exhibit
fewer sulci (and are thus less useful for comparative analysis), while

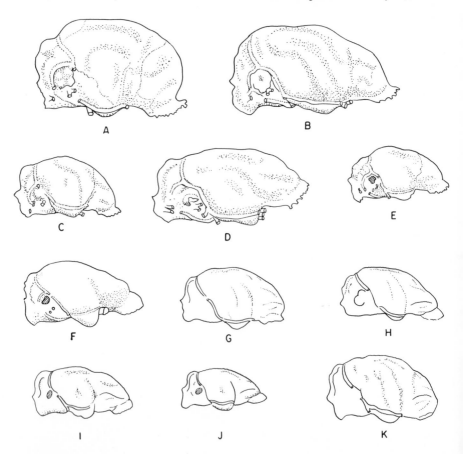

Figure 4.9. Lateral views of cranial endocasts of A: *Daubentonia madagascariensis;* B: *Indri indri;* C: *Avahi laniger;* D: *Varecia variegata;* E: *Lepilemur mustelinus;* F: *Microcebus murinus;* G: *Lemur rubriventer;* H: *Hapalemur griseus;* I: *Phaner
furcifer;* J: *Cheirogaleus major;* K: *Propithecus verreauxi.* Redrawn after illustrations
published by Radinsky (1968, 1974); not to scale. Radinsky (1968, 1974); not to scale.

the frontal lobes of smaller animals tend to be more convex than those of larger related forms.

Endocasts of *Lemur* and *Varecia* reveal the presence of coronal, postcruciate, lateral, postlateral, sylvian, postsylvian, and orbital sulci; a fronto-orbital sulcus is sometimes discernible. The posterior rhinal fissure is obscured by an overlying dural sinus. The brain of the smaller *Lemur* is relatively shorter and somewhat higher than that of *Varecia,* with more rounded frontal lobes. Radinsky (1974) notes that his lemurine sample varies considerably in certain features, however, and is reluctant to identify any diagnostic characteristics. *Hapalemur* and *Lepilemur* both possess brains broadly reminiscent of those of the lemurines, except that the endocast of *Hapalemur* is rather narrower, lacks a postlateral and sometimes a fronto-orbital sulcus, and has more ventrally positioned olfactory bulbs than is typical of most lemurines. *Lepilemur* fails to show postsylvian, postlateral, and fronto-orbital sulci, and its coronal sulcus is not well developed. This lesser sulcal development, as noted, is what would be expected empirically in a smaller animal, and the lack of folding of the cortex which it reflects is presumably due as suggested by Clark (1947) to the maintenance of a relatively constant proportion of internal white matter (volume) to gray cortex (surface area).

In contrast to that of the similarly sized *Lepilemur,* the brain of *Phaner* exhibits a postsylvian sulcus (albeit a relatively short one), but lacks a postcruciate. Otherwise the two genera are mutually fairly reminiscent in proportions and in sulcal expression. The endocast of *Cheirogaleus* resembles that of *Phaner* but does not show an orbital sulcus. The tiny *Microcebus,* whose brain has been described at some length by Clark (1931), shows very little sulcal detail, possessing only a sylvian and a faint postsylvian sulcus. The lateral compression of the olfactory bulbs in the mouse lemur is attributed by Radinsky (1974) to the combination of its small size with its relatively large orbits.

The endocasts of the indriids are not greatly dissimilar to those of the lemurines, although they are generally slightly higher and wider, especially anteriorly. In *Indri* the frontal lobe is distinctly broader than that of, for instance, *Lemur*; the postcruciate is better defined and possesses a transverse branch which intersects the coronal sulcus. Radinsky (1968, 1974) relates these latter features to expansion of the motor area for the forelimb, and suggests that this indicates finer muscular control of the hand in *Indri* than in *Lemur.* *Propithecus* shares this complex of features with *Indri,* but its relatively shorter brain is more rounded in the area of the frontal lobes, and its olfactory bulbs are further rotated ventrally. In *Propithecus* the sylvian fissure extends further caudally than in *Indri*; in both but especially in the latter, the postlateral sulcus is unusually well developed, suggesting some expansion of the indriid visual cortex. *Avahi* lacks the postcruciate and fronto-orbital sulci of its larger relatives, but displays the long sylvian fissure also characteristic of *Propithecus*.

As might be expected *Daubentonia* is unusual in the external proportions of its brain. The endocast is high and narrow, with a well rounded frontal lobe rotated a long way ventrally for an animal of its size. Radinsky notes that no orbital impressions are present, that the coronal and postcruciate sulci are discontinuous, that the postsylvian is variably developed, and that the anterior suprasylvian and pseudosylvian sulci are separated, exposing a part of the cortex normally buried in strepsirhines. This last characteristic may be due to an expansion of the auditory cortex, which is found in this region, and might then be interpreted as an adaptation reflecting the aye-aye's specialized auditory detection of its larval prey. No postlateral sulcus is discernible.

The external morphology of the brain of *Daubentonia* provides an unusually favorable situation for assessing the relative contributions of brain and bone to final skull form. In development, the membrane bones of the skull vault ride outward passively atop the expanding brain mass within; but the skull as a whole has to accommodate not only the brain but the organs of olfaction, audition, and vision, and also must provide an appropriate structure for generating and absorbing the stresses of mastication. In cases where the discordance between the requirements of the brain and of the structures external to it is considerable, air sinuses form; but usually it is extremely difficult to gauge to what extent the brain has accommodated itself to the needs of the external structures, or vice versa. In *Daubentonia*, however, it is clear that the proportions of the brain owe a great deal to the modification of the skull in response to the demands of an unusual masticatory system. The extreme downward rotation of the facial skeleton to bring the long axis of the rostrum in line with the upper incisal axis, the high, narrow conformation of the braincase, and a variety of other features seem without doubt to be related to the need for efficient stress transmission from the gnawing incisors to the braincase (Cartmill 1974; Radinksy 1974). The rather globular brain of *Daubentonia*, with its ventrally rotated frontal pole and olfactory bulbs, thus seems to owe its unusual overall shape to selective forces acting primarily on the masticatory system.

Brain Size and Proportions

Although the brains of the strepsirhines are larger relative to body size than are those of nearly all other mammals, they are distinguished most strikingly from those of anthropoid primates by their relative smallness and lack of sulcal complexity. Among all primates the most remarkable specialization of the brain lies in the expansion of the neocortex; but Stephan and Andy (1969) note that while in the average "higher" primate the neocortex is enlarged 45.5 times over that of their "basal insectivores" (shrews, hedgehogs, and tenrecs), in the strepsirhines the equivalent enlargement is only 14.5. These authors have calculated what they call the "Index of Progression of the Neocortex," which expresses the degree of expansion of the neocortex over what would be expected in a "basal insectivore" of equivalent body weight. Among the lemurs, *Lepilemur* and *Hapalemur* show the lowest values of this index, at around 7.7. Next are *Microcebus* and *Cheirogaleus*, at 9.0–12.0. *Avahi* shows the smallest valve for an indriid, at about 11, and *Indri*, at around 14, is marginally higher than *Propithecus*. *Varecia* is higher yet, at 17.5, and *Lemur fulvus* scores 23.3. The highest value of all for a lemur belongs to *Daubentonia*, at 26.5. These numbers bring the aye-aye and the brown lemur just within the lower end of the range for anthropoids as exemplified by the howler monkey. However, it appears that in the strepsirhines cortical expansion has mostly involved the primary projection areas, while as a general rule in anthropoids it is the association cortex which has been emphasized (Noback and Moskowitz 1963). It should be noted that the figures just given require a cautious interpretation, since they are for the most part based on single individuals and are standardized by the notoriously variable characteristic of body weight.

According to Stephan and Andy (1969), the only brain structures to have undergone regression relative to body size among the primates are the olfactory bulbs. In almost all strepsirhines some such relative reduction may be detected, but in *Daubentonia* the index of progression of the olfactory bulb, which stands at 1.1, indicates some relative increase. Indices of 0.8–0.9 for the mouse and dwarf lemurs testify to a modest reduction in these animals, but quite handsomely exceed those of the lemurids (0.5–0.7). The lepilemurids and the indriines show the most marked reduction of all, with values in the range of 0.2–0.4, except in the case of *Indri*, which has a remarkably low index of 0.13, comparable with that of the callithricids, and even on a par with a specimen of *Cercopithecus ascan-*

ius. Again, these figures should be treated with some caution for the reasons already given.

Table 4.2 gives the volumes of various brain structures, as proportions of the volume of the brain as a whole, in a variety of Malagasy primates. These proportions are calculated from figures provided by Stephan, Bauchot, and Andy (1970), and are based on single individuals of each species. A tenrec and an Old World monkey are included for comparison.

Vision

All strepsirhines possess a postorbital bar, but lack postorbital closure. Among all extant lemurs the bony orbits are more or less well convergent and frontated; there are, however, powerful size factors operative here (see Cartmill 1972, for a discussion of allometric effects upon orbital size and orientation). It should be noted that there is no direct correspondence between the angle of divergence of the orbital and either the optic or the visual axes; in primates orbital divergence is always greater than that of the eyes themselves. Pariente (1979) states that the optic axes (axes of lens symmetry) of the eyes of the various strepsirhine species diverge at angles of 10–25°. Clark (1931) and Hill (1953) state that a considerable proportion of the optic fibers remain uncrossed in the optic chiasm of *Microcebus*; and although I know of no study of the pro-

Table 4.2. Brain volume (B.V. in cm³) of several Malagasy primates, plus a tenrec and a monkey, with the volumes of various brain structures expressed as a percentage of the total volume of the brain. MD: Medulla oblongata; CB: Cerebellum; MS: Mesencephalon; DC: Diencephalon; TC: Telencephalon; OB: Olfactory bulb; NC: Neocortex.

	B.V.	MD	CB	MS	DC	TC	OB	NC
Tenrec ecaudatus	2.3	13.2	14.5	5.9	7.2	59.1	12.5	12.0
M. murinus	1.7	5.0	13.6	4.2	7.6	69.6	2.4	46.5
C. medius	2.9	7.0	13.6	4.9	8.3	66.1	3.4	42.6
C. major	6.3	5.8	14.8	3.5	7.6	68.1	2.4	46.9
L. mustelinus	7.2	6.0	15.4	3.9	8.6	66.0	1.8	45.9
H. simus	8.9	5.5	16.0	3.5	8.8	66.2	0.9	48.8
L. fulvus	22.1	3.6	15.5	2.4	7.2	71.3	1.0	56.3
V. variegata	29.7	4.8	14.4	3.4	8.5	68.9	1.3	51.5
A. laniger	9.1	6.1	15.7	3.8	9.6	64.8	0.9	47.8
P. verreauxi	25.1	4.8	15.6	2.9	8.3	68.4	0.7	53.4
I. indri	36.2	3.8	15.0	2.7	8.6	69.9	0.4	55.2
D. madagascariensis	42.6	3.6	15.2	2.1	8.3	70.9	1.6	51.9
Macaca mulatta	87.9	2.3	10.2	1.6	5.1	80.9	0.1	72.2

SOURCE: Calculated from data given in Stephan, Bauchot, and Andy (1970).

portions of decussating and nondecussating retinofugal fibers in any other lemur, there can be no question that all Malagasy primates have binocular vision. Visual fields of each eye are not easy to measure, especially in nocturnal forms, but Pariente (1979) estimates for *Microcebus* a combined visual field of 230° or more. He remarks that the overlap between the visual fields of each eye in nocturnal lemurs is less than in diurnal forms, so the total visual field in the latter is smaller. Within the total field, Prince (1956) quotes for lemurs (species unspecified) a binocular field of 114–130°. These figures compare with a total visual field in catarrhines of 180–190°, and a binocular field of 140–160° (Prince 1956).

Almost all strepsirhines possess retinas of a type usually associated with nocturnality. Thus, for instance, most possess a tapetum lucidum, a highly reflective layer lying between the choroid and the retina, and composed of riboflavin crystals. Among the Malagasy primates, only *Varecia* and *Lemur* (with the notable exception of *Lemur catta*) appear to lack a tapetum of classic type; indeed, in most species of *Lemur* there is some question as to whether a tapetum is represented at all (see, for instance, Wolin and Massopust 1970). Pariente (1976) records that in the cheirogaleids the limits of the tapetum are very sharply defined, while in *L. catta*, *Hapalemur*, *Lepilemur*, and *Propithecus*, the tapetum fades out laterally as "digitations" of dark pigment invade its periphery. This difference is ascribed to varying arrangements of the granules in the adjacent pigment epithelium, which appears to be reflected in the structure of the tapetum itself; Pariente suggests that the distinctive structure of the tapetum in the cheirogaleids is directly related to their strict nocturnality, while its aspect in the other lemurs indicates some degree at least of accommodation to a diurnal way of life. The pigment epithelium itself is in most cases largely devoid of pigment, particularly in its central areas (hence the ability of the underlying tapetum to reflect light); in *L. catta* and *Hapalemur*, however, black pigmentation encroaches somewhat on the tapetum (Pariente 1975).

A more or less distinct fovea has been identified by Pariente (1975) in the retinas of *Lemur catta* and *Hapalemur griseus*. In both species receptors include cones as well as rods, and in the former, in particular, the ratio of cones to rods is relatively high: 1 cone to 5 rods (Kollmer 1930). Pariente (1975) has counted 1 cone to 17 rods in the macula of *Hapalemur*. In the cheirogaleids, in contrast, cones are exceedingly sparse: Kollmer (1930) recorded that the small cones he found in the retina of *Cheirogaleus* sp. existed in the proportion of 1 to 1,000 rods, while Rohen and Castenholz (1967) described the retinas of *Cheirogaleus medius* and of *Microcebus murinus* as being of a pure rod type.

There is some uncertainty as to the presence or relative abundance of cones in species of *Lemur* other than *L. catta*. Both Kollmer (1930) and Rohen (1961) describe the retina of *L. macaco* (which appears definitely to lack a tapetum) as being of mixed rod-and-cone type, while Rohen and Castenholz (1967) found some conelike receptors in *L. fulvus*, but doubted that they were true cones. In *Varecia variegata* these same authors were unable with certainty to demonstrate the presence of any cones at all. On the other hand, Pariente (1979) suggests that cones are indeed present, if only in low densities, in the retinas of all diurnal and "crepuscular" lemurs, although he does note that color discrimination at low light intensities must be exceedingly poor, and substantially less than perfect at higher levels. On the behavioral level, however, Bierens de Haan and Frima demonstrated as early as 1930 that some degree of color vision exists in *Lemur mongoz* as well as in *Lemur catta*.

The indriid retina was described in 1967 by Rohen and Castenholz, who found that *Avahi* has essentially a pure rod retina, but that *Indri* and *Propithecus*, while showing primarily rods, also exhibit cone-like receptors. In the two diurnal forms a clear central area is present, but while the entire fundus of *Indri* is only lightly pigmented, the peripheral parts of that of *Propithecus* show heavy concentrations of pigment.

In myosis (maximum opening) the pupil of the cheirogaleids takes the form of a vertical slit, while that of the other lemurs is generally round; in mydriasis (maximum closing) the pupil is round in all. Pariente (1979) quotes the surface ratio (between the areas of the constricted and dilated pupil) as 40 in *Phaner*, 22 in *Propithecus*, and 21 in *L. catta*, and remarks that the extreme value in the first emphasizes the importance among the strictly nocturnal forms of protecting the retina from intense light. All lemurs passess a well-developed nictitating membrane or "third eyelid."

As regards more central components of the visual system, it is worth noting that whereas in all living primates the lateral geniculate nucleus is distinctly laminated, with folded laminae, the strepsirhines are sharply distinguished in the way in which this folding is achieved. Among the lemurs the laminae are inverted, or folded inward. The fibers of the optic radiation, which project to the visual cortex, emerge from the medial aspect while the incoming fibers of the optic tract enter the nucleus laterally. In the higher primates, on the other hand, the laminae are everted, receiving the fibers of the optic tract from below and emitting those of the optic radiation dorsally from the convex folded surface (Clark 1971).

Olfaction

The olfactory sense is generally believed to play a more substantial role in the lives of strepsirhine than of anthropoid primates, and certainly olfactory communication is a much more conspicuous part of the behavioral repertoire of most strepsirhines than of most higher primates, certainly any catarrhine.

The external noses of strepsirhines, all of them apparently relatively macrosmatic, are characterized by the possession of a moist, naked rhinarium and a naked philtrum with a distinct median sulcus, this latter feature reflecting the medial tethering of the cleft upper lip to the alveolar process by a superior labial frenulum. The nostrils, more or less crescentic in shape, are convex laterally. Hill (1948) has described the "rhinoglyphics," or papillations of the rhinarium, of several strepsirhine genera; he records that in genus *Lemur* there is a regular papillary mosaic throughout, whereas in *Microcebus* and *Cheirogaleus* distinct papillary ridges are differentiated in the infranarial region and upon the philtrum. *Phaner* shows an intermediate condition. This elaboration of the sensory skin emphasizes the role of the rhinarial complex as a tactile organ, and parallels the often copious representation of vibrissae ("sinus hairs") on the muzzle (see chapter 3 for the vibrissae groups found in the various taxa). Further, the moist rhinarial complex serves to pick up particles and to transmit them via the superior labial tract and the incisive papillae to the bilateral vomeronasal (Jacobson's) organs. These lie forward at the base of the nasal septum and communicate with the palate via patent nasopalatine ducts. Vomeronasal organs are present in all lemurs; the structure, gross and fine, of that of *Microcebus* has been described in detail by Schilling (1970).

The olfactory muzzle of all lemurs is relatively capacious; the axons of the receptor cells of the olfactory mucosa pass to the olfactory bulbs via a large and multiply perforated cribriform plate. The internal bony structure of the nasal fossae of several strepsirhines has been studied by Kollman and Papin (1925), who point out a greater elaboration of the ethmoturbinals, particularly the first endoturbinal (E1), in the lorisiforms when compared to the Malagasy genera. These authors did not, however, include any cheirogaleids in their survey, and detailed study of the latter in this respect is still required. *Microcebus* and *Cheirogaleus*, however, do appear to show some of the expansion of E1 in its inferior portion which characterizes lorisiforms; certainly enough to cause some overlapping of the maxilloturbinal. *Lepilemur* is also said to share this trait (Kollman and Papin 1925).

The maxilloturbinal of *Lemur*, unlike that of the lorisiforms, is doubly rolled; in *Lepilemur* the element consists only of a single scroll. Among the indriines, *Indri* and *Avahi* both display only an inferior maxilloturbinal scroll, while in *Propithecus* a superior scroll is also found. The basal lamina of the maxilloturbinal is reduced in the indriines and *Lepilemur*, but is more extensive in *Lemur*. Not unexpectedly, the maxilloturbinal of *Daubentonia* is morphologically divergent from that of all other lemurs; it is somewhat stirrup-shaped, with a single inferior scroll.

In *Lemur* the rolled portion of the nasoturbinal is well developed, and extensions of the element penetrate both maxillary and frontal sinuses. The scroll within the former usually fuses with its wall in *Lemur*, but not in *Lepilemur*, where the element is also present. There is no penetration of the sinuses in *Daubentonia*. The uncinate process is normally well developed, but in *Lepilemur*, as in the lorisiforms, it is rolled only in its posterior part. Only *Lemur* lacks an anterior extension of this structure. In *Propithecus* and *Avahi* the maxillary sinus is closed off by fusion of the anterior border of the anterior lamina of the nasoturbinal with the lateral nasal wall; this fails to occur in *Indri*.

Lemurids, lepilemurids, cheirogaleids, *Avahi* and *Propithecus* uniformly exhibit four endoturbinals, which tend in most cases to diminish in size posteriorly. However, *Indri*, like *Daubentonia*, also

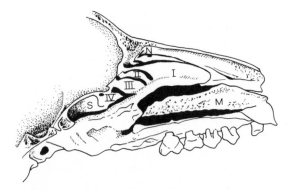

Figure 4.10. Sagittally sectioned cranium of a subadult *Lemur* (whence the lack of a frontal sinus) to show the turbinal bones. M: maxilloturbinal; N: nasoturbinal: S: sphenoid sinus; I–IV: ethmoturbinals I—IV.

possess a fifth. In *Indri* this element closes the orifice of the sphenoid sinus, as does E4 in the other indriines. Ectoturbinals (accessory ethmoturbinal scrolls) occur beneath the ethmoturbinals, including the nasoturbinal; in most cases these are two in number, but in *Daubentonia* there are three (Kollman and Papin 1925).

Very little is known of the olfactory physiology of the Malagasy primates, and it may be unwise to assume that olfactory acuity can be inferred directly from the surface area of the olfactory mucosa (absolute or relative), or indeed from any other single measure. In any event, at least some of the Malagasy primates have been shown empirically to display high levels of olfactory sensitivity; Schilling (1979) records, for instance, that *Cheirogaleus major* is capable of detecting which of five glass jars contains a cricket when the insect has been inside for only 45 seconds.

Audition

Vocal signals form an important part of the communication repertoire of all Malagasy primates, but almost nothing is known of the auditory physiology of these animals. All surviving lemurs (but not all extinct ones, see chapter 5) possess an inflated auditory bulla of petrosal origin, lack a bony external meatus, and show a "free" tympanic annulus which makes bony contact with the skull only at its superior margin. The relative size of both the annulus and of the bulla tends to increase with decreasing body size; evidently the volume of the hypotympanic sinus has to remain sufficiently large to prevent overdamping of the tympanic membrane regardless of the size of the animal. The bony structure of the ear region of the Malagasy primates is described under the rubric of cranial morphology; here I merely provide a few details of the morphology of the external ear.

As might be expected, the external ear tends in general to be relatively bigger among nocturnal than diurnal lemurs. Only in *Daubentonia*, however, is the external ear enlarged to an extent comparable with that seen among some mainland strepsirhines, especially galagids. In this genus the substantial membranous scapha is enormously elongate, but lacks the ribbing and other structural complications seen in galagid ears. The tragus is a simple ridge, the antitragus a low fold, the intertragal notch large.

Only for a short distance along its anterior border is the scapha of any lemur folded to produce a helix; the antitragus is invariably larger than the rather lobular tragus, producing a deep intertragal notch whose base is at the level of the meatus. A depression is usually formed posterior to the antitragus; the pinna is in most cases clothed in hair, at least on its medial aspect; this hair occasionally forms long, elegant tufts. Ear length measurements for the various taxa are given in chapter 3.

Mitchell and his colleagues (Mitchell et al. 1970, 1971; Gillette et al. 1973) have investigated the auditory thresholds of *Lemur catta*, *L. fulvus*, and *L. macaco* using a variety of behavioral techniques. These studies revealed auditory sensitivities of at least 60 dB at sound pressures below 1 dyne/cm² from 1 kHz to 32 kHz in *Lemur catta*, the species most intensively studied. Comparable results were obtained for the two other lemur species tested, as they had been by other investigators (e.g., Heffner and Masterton 1970) for lorises, pottos, and bushbabies. These findings indicate that the lemurs studied exhibit substantially greater auditory sensitivity at high frequencies than do anthropoids (catarrhines and platyrrhines appear to exhibit essentially similar sensitivities), but that this appears to be achieved at the expense of relatively diminished low-frequency sensitivity. Masterton et al. (1969) suggest that emphasis on sensitivity at high frequencies is a primitive characteristic, largely related to the need for accurate sound localization.

A study by Niaussat and Molin (1978) of audition in *Phaner furcifer* yielded a hearing range of 0.15 to 120 kHz, with a maximum sensitivity at around 16 kHz. This contrasts with a best frequency for *Lemur* spp. of about 8 kHz, below which these animals have lower auditory thresholds than *Phaner*.

LEARNING AND INTELLIGENCE

Two recent contributions have quite exhaustively reviewed the literature of laboratory testing of learning and intelligence in strepsirhines (Ehrlich et al 1976; Wilkerson and Rumbaugh 1979), and it is clear that virtually all such studies of Malagasy primates have been undertaken on members of genus *Lemur*. It is in this area of investigation that the concept of the order Primates as a *scala naturae* has perhaps been taken most literally, and many if not most workers in the field seem to have assumed, explicitly or otherwise, that the lemurs comprise the lowest rung in a progressive series which reflects itself in increasing learning ability. This distorting perspective, however, has been abandoned in the refreshing review by Ehrlich et al. (1976), and the brief remarks below are based largely on the commentary of these authors, who are at pains to point out that the persistent myth that strepsirhines cannot learn is based upon observations which are at the very least questionable.

Numerous studies have been reported in which lemurs and other strepsirhines have been trained to perform a wide range of simple tasks, including both unfamiliar responses and conditioned vocalization. In addition, their ability to acquire both conditioned suppression and passive avoidance responses has been demonstrated (e.g., Heffner and Masterton 1970). The fact that these animals often use their noses rather than their hands to displace test objects means that test equipment designed for anthropoids is often inappropriate for them, and that its use may thus vitiate comparisons between the two primate groups. Standard reinforcement schedules, using rewards such as food, appear to work well in training strepsirhines to perform tasks.

Lemurs have been shown to form object discrimination learning sets (e.g., Cooper 1974), with peak performances broadly comparable to those of some platyrrhine monkeys, although considerable differences in performance are reported to exist between individuals. In solving discrimination reversal problems (e.g., Rumbaugh and Arnold 1971) black and ringtailed lemurs showed a level of performance close to that of vervets; in a later study (Rumbaugh and Gill 1973), using a modified measure, lemurs fared rather less well than vervets, but better than talapoin monkeys. Similarly, in other complex tasks (those which involve the mastery of a general principle), such as delayed response, instrumentation, oddity, and relational problems, there appears to be overlap between the performances of anthropoids and strepsirhines (see references in Ehrlich et al. 1976).

Ehrlich et al. conclude that at least on the basis of the limited range of studies available, there appears to be no quantum distinction between "lower" and "higher" primates or even between primates and other mammals in the kind of problems mastered, although they note that "it is possible that prosimians and anthropoids would show qualitative differences in the type of learning transferred from one task to another where the tasks entailed different rules for correct solution" (1976:612). At present there are virtually no data which bear on this question; almost all available studies have used single tasks, which do not shed any light on this area of learning. In the limited area which has been investigated, however, the evidence suggests that lemurs cannot simply be written off as totally inferior in learning abilities to the "higher" primates, which lie monolithically above them in an all-embracing hierarchy of intelligence.

LOCOMOTION AND THE LOCOMOTOR APPARATUS

The locomotion of none of the Malagasy lemurs has been studied in detail, although general patterns are known for almost all genera. In particular, we almost totally lack quantitative data on locomotion and substrate preferences. One important distinction that has rarely been made in data collection by observers of free-ranging lemurs is that between the movements and postures involved in feeding and other such activities, and those adopted in group travel through the forest. Virtually all lemur species, whatever their basic locomotor morphology, are capable of adopting an astonishing variety of postures during feeding; specific locomotor adaptation does not in most cases appear to constitute a limiting factor in access to the various parts of the forest, although body size may have some influence in this respect, and at least in certain cases aspects of postcranial morphology may be related to feeding adaptation. On the other hand, in point-to-point travel through the arboreal environment, most lemur species display a clear-cut preference for particular types of substrate, hence parts of the canopy, and it is to such preferences that many of the grosser features of locomotor morphology appear to be related.

Locomotion and Posture

Although with the demise of *Palaeopropithecus* the Malagasy primate fauna no longer shows a complete spectrum from forelimb- to hindlimb-dominated locomotion, there is nonetheless very considerable locomotor variety among the surviving lemur genera. These have often been divided into quadrupedal types, such as the cheirogaleids and the lemurines, and vertical clinging and leaping types, such as *Lepilemur* and the indriines; but a simple division of this kind tends to obscure the very real differences to be found within each of these categories.

In general, the cheirogaleids exhibit a rapid, scurrying quadrupedal locomotion. The tiny *Microcebus* occupies what has been called the "fine-branch niche" and is at its most abundant in areas of dense foliage and tangled vegetation (Martin 1972a). Mouse lemurs run rapidly along networks of fine branches, readily leaping gaps in the substrate and landing on all fours. Travel of this kind usually occurs in bursts of several meters, interrupted by short periods of immobility. Walker (1979) reports that mouse lemurs may also cross gaps by reaching out for the next branch and grabbing it with the hands before letting go with the feet; this may correspond in part to the predatory pattern whereby an individual leaps out to snatch an insect without releasing its feet, then retracts (Martin 1972a). Mouse lemurs have also been noted to adopt a "flag posture" in which they cantilever themselves out from a vertical support, holding on only by the feet (Martin 1972a). Above-branch postures are the norm, but mouse lemurs will adopt a wide variety of postures while feeding, including quadrupedal suspension and suspension from the feet alone. These animals are not reluctant to come to the ground, which is normally visited to forage for beetles in the leaf litter of the forest floor, or to travel where gaps between trees of more than 3 meters or so have to be crossed (Martin 1973). Mouse lemurs usually travel quadrupedally while on the ground, but may progress in a series of short hops.

Being larger animals, the species of *Cheirogaleus* seek larger branches and run along the tops. According to Petter et al. (1977) the availability of large horizontal branches to some extent determines the levels of the forest which dwarf lemurs occupy. Generally, the body is held close to the support during quadrupedal running, and both species seem more reluctant to jump than are the mouse lemurs. Walker (1979) says that *C. medius* jumps more frequently than does *C. major*, but Petter et al. suggest that the reverse is the case, especially when the tail of the former is fat. Peyrieras (quoted in Petter et al. 1977) has reported regular jumps of several meters by a greater dwarf lemur.

Phaner exhibits considerable locomotor versatility. Fork-marked lemurs generally travel by rapid running along the tops of large horizontal and oblique branches, and unhesitatingly jump from one branch tip to the next. According to Petter et al. (1971), they can jump from four to five meters without losing much height. All cheirogaleids have somewhat keeled and pointed nails, but those of *Phaner* are the most claw-like of all, and aid their possessors in moving on vertical trunks, which they may descend head first. This ability to climb on the broad trunk surfaces of trees has been interpreted as an adaptation associated with gum-eating (Petter et al. 1971; Martin 1972b). Leaping between vertical supports is also common, and feeding postures include bipedal suspension.

The locomotion of *Mirza* appears to be closer to that of *Phaner* than to that of any other lemur, but is less rapid. Petter et al. (1971) have noted a tendency for these animals to progress at the lower levels of the forest by short hops between vertical supports of modest diameter.

Among members of genus *Lemur*, *L. catta* has long posed problems for categorizers of locomotion. While other *Lemur* species could be lumped conveniently as "arboreal quadrupeds," *catta* plainly could not. Sussman (1972, 1974) found that the ringtailed lemurs he observed spent more than 30% of their time on the ground, and that more significantly, over 65% of travel took place on the ground. Terrestrial locomotion is by quadrupedal ambling, running, or galloping, but occasionally bipedal postures or movement are seen. Sussman also notes that ringtails show a characteristic substrate preference in the trees: "*Lemur catta* progresses through the trees running up an oblique branch and leaping down to another" (Ward and Sussman 1979:578). Sussman states that contrary to other reports he never saw *catta* travel by leaping between vertical supports, but it seems likely that this was due to the structure of the forests in which he carried out his studies; he did not see it in sympatric *Lemur fulvus* either, but such travel was not unusual among *fulvus* in Mayotte, where it was the only practical means short of terrestrialism of movement through part of the study forest (Tattersall 1977c).

In contrast to the ringtails, the rufous lemur, which spent over 90% of its time in the upper levels of the forest, and which rarely descended to the ground (under 2% of time), tended during travel to seek horizontal supports within the canopy, often the fine terminal branches of large trees (Sussman 1972, 1974). At least as concerns support orientation, this accords well with my own observations on *Lemur fulvus* in Mayotte (Tattersall 1977c); horizontal supports were clearly preferred during travel, and branches of larger diameter were generally sought. Leaps of up to four meters or so were made where necessary between adjacent branch systems, although continuous passages were obviously preferred; and in young regrowth which conspicuously lacked horizontal supports Mayotte lemurs showed no reluctance to leap between relatively closely spaced (ca. 1–3 meters) vertical trunks of small to medium diameter. Overall, horizontal supports were used 61% of the time, oblique 33%, and vertical ones 6%. Supports of large diameter (> 10cm) were employed 14% of the time (a relatively low figure because branch orientation seemed to be more important in choice than diameter, and most large supports were oblique or vertical), medium ones (2.5–10 cm) 48%, small ones (1–2.5 cm) 33%, and fine ones (< 1 cm) 5% of time.

Other *Lemur* species (except, perhaps, for *Lemur coronatus*, which reportedly spends a substantial amount of time on the ground and which, indeed, appears in many ways to be ecological vicar in the dry north of *L. catta* in the dry south of Madagascar), in general exhibit locomotor patterns and substrate preferences reminiscent of those of *L. fulvus*. There is, however, some variation in the propensity for leaping shown by different species. Thus *L. mongoz* has been recorded to jump from one tree to another even where a continuous passage was available nearby (Tattersall and Sussman 1975), a choice which would be uncharacteristic of *L. fulvus*.

Despite its larger size, *Varecia* is no less agile in its arboreal locomotion than is *Lemur*. Ruffed lemurs will bound at considerable speed along large horizontal or oblique branches, and show considerable facility in climbing large vertical trunks. They will leap without hesitation between networks of fine branches at the periphery of the canopy, landing spreadeagled on a diffuse substrate, or forelimbs first where the objective is a single branch. They also show a great proclivity for bipedal suspension, and will readily feed hanging from their feet, as members of genus *Lemur* will; but in contrast to the latter they will equally readily descend from branch to branch by letting go with the feet while in this position, and falling head first. On the other hand, postures of bimanual suspension are exceedingly rare.

Daubentonia exhibits a slower and more deliberate quadrupedalism than that of *Lemur*, the essential characteristics of this slow climbing seeming to remain constant whether up and down vertical supports or along horizontal branches. Quadrumanous locomotion beneath horizontal supports and hindlimb suspension have also been noted in aye-ayes, as has the facility of movement on large trunks afforded them by their claw-like tegulae (Petter and Peyrieras 1970a; Walker 1979). Aye-ayes will jump quite readily, but may also proceed with great caution when passing from one branch system to another, not letting go with the feet until a sure grip by the hands is assured. Aye-ayes descend fairly frequently to the ground, where they progress quadrupedally, but where the great length of their manual digits causes them to dorsiflex their wrists and support their weight on the thenar and hypothenar pads (Walker 1979: Petter et al 1977); it is this that imparts the peculiar stiffness to their terrestrial gait.

The locomotor repertoire of *Hapalemur griseus* is not well known. Individuals have been recorded as walking and running along horizontal and oblique branches, as well as clinging to and leaping between vertical supports. On the ground gentle lemurs generally move quadrupedally rather than by hopping, but they can leap considerable distances between both horizontal and vertical supports.

The Alaotran subspecies is larger and heavier than those from the east and west coasts, and is reportedly less agile, hopping short distances between the reeds which provide its habitat, and climbing relatively slowly among them (Petter and Peyrieras 1970b; Walker 1979). A commonly adopted feeding posture in all subspecies is to sit with the trunk held erect. *Hapalemur simus* is said by Petter et al. (1977) to leap rapidly between vertical supports, but to flee quadrupedally on the ground if alarmed.

For *Lepilemur,* preferred supports are plainly vertical; even when an individual is at rest, with its long trunk flexed forward, its position is essentially upright. Sportive lemurs rarely come to the ground, but when there they get about almost invariably by hopping bipedally. In the trees they seem to travel in leaps almost exclusively, between vertical supports wherever possible. According to Charles-Dominique and Hladik (1971), who studied the small *L. m. leucopus* in the dry south, most leaps are from 1.5 to 2 meters, but distances of over 5 meters may be covered with some loss of height during the leap. Propulsion in leaping is provided solely by the hindlimbs, which leave the initial support last, but which then contact the landing support before the forelimbs.

All of the indriines are vertical clingers and leapers par excellence. No matter how the weight is supported, the trunk is usually held erect. On a horizontal substrate a resting indriid will seek ischial support; similar support is often found in the form of a branch when the animal is clinging, knees flexed, to a vertical trunk. Indriids rarely rest for long periods in a clinging position with the rump unsupported. Richard (1978) reports that the *Propithecus verreauxi* she studied spent 75% of the time when feeding in either a vertically clinging or vertical sitting posture, and that while all types of substrate except the ground were regularly used, such feeding occurred predominantly among the smaller branches. In leaping, the hindlimbs leave the departure support last and arrive at the destination support first, whatever the orientation of either. Leaps are made normally between vertical supports, and the leaps may exceed 10 meters in *Propithecus* and *Indri.* Travel is often by a rapid succession of leaps of this kind, the animals seeming to ricochet through the forest.

Although in a sense it seems paradoxical, much more suspensory locomotion involving the forelimbs is seen in the "hindlimb-dominated" indriines than among the quadrupeds. Armswinging is almost never exhibited, for instance, by members of genus *Lemur,* but it may be observed quite frequently in indriids (see fig. 4.11). On

Figure 4.11. *Indri* armswinging (left) and in mid-leap.

the other hand, quadrupedal progression is rare indeed among indriines. Bipedal or tripedal suspensory postures are also routinely adopted during feeding. These animals spend very little time on the ground, but when there they progress by hopping bipedally.

Morphology

Hill (1953) summarized the work which had been done on the postcranial anatomy of the Malagasy lemurs up to the time when his review was compiled. Since then, several investigators have continued research in this area; notable contributions include those of Jouffroy (e.g., 1962, 1975), Walker (1967a, 1974), Roberts and Davidson (1975), Ward and Sussman (1979), and Godfrey (1977). The following overview is intended only as a brief introduction to an area that is still imperfectly explored. Interested readers should consult the references given, particularly Jouffroy (1962), which is still the most complete account of the postcranial myology of the lemurs.

Perhaps the most immediately apparent morphological correlates of locomotor behavior among the lemurs lie in the linear proportions of the various body segments, and numerous indices have been devised to express such differences of proportion. One might note that there may be appreciable differences between the mean values of the same index for the same genus or species as measured and published by different authors (and even by the same author at different times!); but generally all show similar between-species relationships. The most exhaustive compilation of body proportions among the lemurs is that recently provided by Jouffroy and Lessertisseur (1979); most of the figures given in table 4.3 are derived from this source.

Among the extant lemurs as a whole the intermembral index (H + R/F + T) is remarkably low. This appears to be due almost entirely to the relative shortness of the forelimb in these animals, since compared to trunk length the hindlimb of none of the lemurs apart from the indriines is notably long for a primate. However, *Lepilemur* and the indriines are nonetheless clearly distinguished from all other lemurs by their unusually low values for this index. The elongation of the hindlimbs relative to the forelimbs which this reflects is the most obvious of the adaptations shown by these animals to their vertical clinging and leaping form of locomotion. However, while in *Lepilemur* the forelimbs and to a lesser extent the hindlimbs are short relative to trunk length, in the indriines the forelimbs as well as the hindlimbs are long in comparison to the precaudal vertebral column. A further distinction between *Lepilemur* and the indriines resides in the proportions of the foot. In the latter the tarsus is short relative to the length of the foot as a whole; in *Lepilemur* it is long, although less so than that of some cheirogaleids. The small relative size of the tarsus in the indriines, superficially surprising in

Table 4.3. Mean values of various postcranial indices, all expressed as percentages, for eleven Malagasy primate genera. H: length of humerus; R: of radius; F: of femur; T: of tibia; FL: of forelimb; VC: of precaudal vertebral column; HL: of hindlimb; TR: of tarsus; FT: of foot; HD: of hand.

	$\dfrac{H+R}{F+T}$	$\dfrac{R}{H}$	$\dfrac{FL}{VC}$	$\dfrac{HL}{VC}$	$\dfrac{TR}{FT}$	$\dfrac{HD}{FT}$
Lemur	69.7	108	81.0	117.2	37.3	68.0
Varecia	69.2	103	73.2	100.2	35.8	73.8
Hapalemur	67.1	103	84.0	126.7	35.6	65.1
Lepilemur	60.3	109	84.8	135.6	38.3	68.1
Cheirogaleus	71.4	100	71.7	109.3	39.9	63.7
Phaner	67.1	—	83.3	127.7	47.6	61.2
Microcebus	71.5	106	75.6	112.8	44.1	57.0
Indri	62.0	120	113.2	165.5	29.5	85.8
Propithecus	59.7	108	98.2	151.2	30.6	78.5
Avahi	56.3	120	100.9	162.8	31.0	77.1
Daubentonia	70.9	101	114.8	141.8	29.3	105.7

SOURCE: Data from Jouffroy (1975) and Jouffroy and Lessertisseur (1979).

a leaping form, results at least partly from the fact that these animals exhibit substantial elongation of both the metatarsals and the pedal phalanges. Martin (1972b) has pointed out that of all the indriines only the largest, *Indri,* exhibits any notable modification of the calcaneus for vertical clinging and leaping (as gauged from a shift in its calcaneal index/calcaneal lever arm ratio in the direction of the African galagids, a group of accomplished erect saltators, albeit with a biomechanically different leap), while Cartmill (1974b) has reasonably suggested that the long cheiridia of the indriines represent an adaptation for foraging in the peripheral forest canopy, as Van Horn (1972) had concluded for gibbons. One notable peculiarity of the indriines however, especially *Avahi,* is that they possess an extremely long femur relative to the other segments of the hindlimb.

It is also the proportions of the foot that most sharply set off quadrupeds of the "cheirogaleid" type from those of the "lemurid" type. While *Lemur* and *Varecia* possess tarsi of moderate length, those of *Phaner* and to a lesser extent *Microcebus* are greatly elongated, in correlation with their rapid springing form of locomotion. The slower, somewhat less saltatory *Cheirogaleus* has a lower ratio of tarsus to foot length, but one which nonetheless exceeds that of any lemurid. The cheirogaleines are also distinguished from the lemurids by their higher intermembral indices, while having even shorter forelimbs compared to trunk length. Further, with *Daubentonia,* the cheirogaleids are the only Malagasy primates in which the three hindlimb segments are subequal in length.

Daubentonia is most strikingly characterized in its body proportions by the length of its extremities. Its hand, in particular, is extraordinarily long. But while the elongation of the aye-aye hand is due largely to the great length of the digits, in the foot it is the metatarsals which show particular lengthening.

Among the Malagasy primates there is a good deal of variation in the robusticity of the skeletal elements. The long, slender femur of the indriines, for example, is rather gracile, while conversely the robusticity index of the humerus (minimum circumference as a percentage of maximum length) of *Lemur* and *Lepilemur* is, at 28–30, the highest among living primates (Jouffroy 1975). However, despite the wide disparities found among the lemurs in size, robusticity, body proportions, and so forth, the general plan of the muscular system is similar throughout Strepsirhini, and conforms to the overall primate pattern (Jouffroy 1962).

The structure of the forelimb and shoulder girdle among the

indriines clearly reveals the extent to which forelimb suspension is important in the repertoire of movement of these animals. The scapula in all three genera is broad, with a long vertebral border and a particularly extensive infraspinous fossa (fig. 4.12). This conformation reflects the strong development of the muscles of the rotator cuff, which function not only in lateral rotation and abduction of the humerus but in stabilizing and securing the shoulder joint. Roberts and Davidson (1975) have pointed out that such development is hardly necessary for a full time vertical clinger and leaper (although it is true that clinging to a vertical support does place the forelimb in tension), and they have related it to the propensity that the indriines show for climbing and hanging. In this context, comparison with the scapula of *Lepilemur* is particularly instructive. This latter bone is much narrower than that of any indriine, with much more restricted fossae (fig. 4.12); this may with some plausibility be attrib-

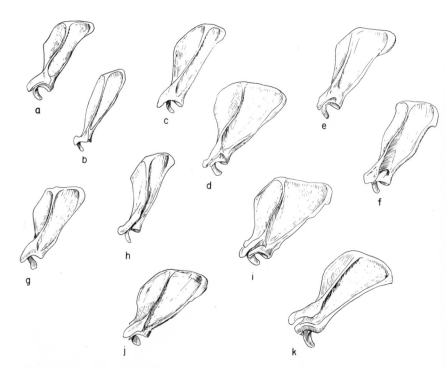

Figure 4.12. Scapulae of a variety of lemur species. a: *Cheirogaleus major*; b: *Microcebus murinus*; c: *Hapalemur griseus*; d: *Indri indri*; e: *Avahi laniger*; f: *Daubentonia madagascariensis*; g: *Varecia variegata*; h: *Lepilemur mustelinus*; i: *Propithecus verreauxi*; j: *Phaner furcifer*; k: *Lemur macaco*. Not to scale.

uted to the deemphasis of forelimb suspension in the sportive lemur's repertoire of movements.

Forelimb mobility is also emphasized in the structure of the indriine humerus. Among these forms the humeral head surpasses
the tuberosities in height to a much greater extent than it does in
any of the other lemurs (fig. 4.13). This has the effect of expanding
the articular surface of the humeral head, particularly in its upward-
facing part, thereby substantially increasing the mobility of the forelimb on the shoulder joint. In this respect the humerus of *Lepilemur*
closely resembles that of the more quadrupedal forms, such as *Le-*
mur, Varecia, and *Hapalemur,* where the greater tuberosity is subequal in height with the head. In these forms structural (as opposed
to muscular) stability is desirable, and is emphasized at the expense
of mobility. On the other hand, the humeral attachment of latissi-
mus dorsi in *Lepilemur* is rather distally placed, as it is in the indriines. The muscle is well positioned to function in absorbing tensile stresses generated in the forelimb, and its low insertion in
Lepilemur may be attributable to its role in counteracting tensile
stresses placed upon the arms during vertical clinging. In the indriines, of course, it is also well placed to absorb hanging stresses.

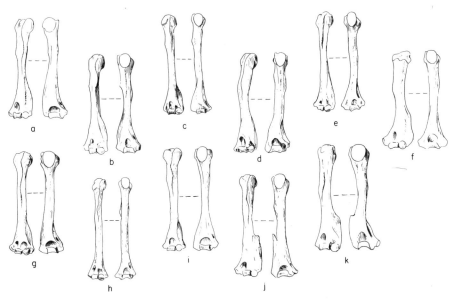

Figure 4.13. Anterior and posterior views of the humeri of: a: *Hapalemur gri-*
seus; b: *Lepilemur mustelinus*; c: *Microcebus murinus*; d: *Varecia variegata*; e:
Avahi laniger; f: *Phaner furcifer*; g: *Propithecus verreauxi*; h: *Indri indri*; i: *Lemur*
macaco; j: *Daubentonia madagascariensis*; k: *Cheirogaleus major.* Not to scale.

In the distal part of the humerus the indriines are particularly distinguished by the reduced size of the branchialis flange. This does not, however, appear to be due to any reduction in or distal migration of the origin of brachioradialis, the largest muscle of the forearm in all the Malagasy lemurs (Jouffroy 1962). According to Jouffroy's descriptions brachioradialis in *Propithecus* is greatly enlarged, and its origin proximally extended in relation to the condition of the muscle in Lemuridae. It seems probable that this extension of the origin of brachioradialis, which inconveniently restricts the amount of extension possible at the elbow, may be related to this muscle's fulfillment of a function similar to that of the rotator cuff more proximally, i.e., to stabilize the elbow joint under conditions of high tensile stress such as occur during manual suspension. *Lepilemur* conserves the muscular disposition associated with the development of a large brachialis flange, in this resembling the lemurids, where brachioradialis aids in maintaining the elbow joint in flexion under conditions of compression. The relatively small amounts of tensile force generated by *Lepilemur* in clinging to vertical supports apparently do not require specialization of brachioradialis, and it thus seems plausible to view the indriine condition purely as a suspensory adaptation.

In all Malagasy primates the deltoid crest of the humerus is long, sharp and prominent, and slightly overhangs the bicipital groove. An entepicondylar foramen is invariably present. The shaft of the ulna is relatively straight, while that of the radius is usually quite strongly bowed, with a long neck. The olecranon process of the ulna is long, with a large surface for triceps insertion. Jouffroy (1962, 1975) has noted a tendency throughout the group for many of the forelimb muscles to attach at points relatively remote from the articulations they cross.

Adaptations to vertical sitting, clinging, and leaping in the hind limbs of the indriines are not confined to linear proportions. A suite of adaptations can be identified which relates both to a habitual legs-flexed resting posture and to powerful leaping.

Features indicative of the posture of vertical sitting or clinging in the indriines and *Lepilemur* include the prolongation of the articular surface of the head of the femur on to its neck; this surface matches with the lunate articular surface of the acetabulum when the femur is strongly flexed and externally rotated. Also in this category is the posterior position of the tibial condyles vis-à-vis the long axis of the tibial shaft, and the backward tilt of their articular surfaces. This is associated with the backward-facing articular condyles of the femur. Another characteristic shared between the indriines and *Lepilemur*, and presumably related to habitual feet-first landing on vertical supports, is the great development of tibialis anterior, which effects dorsiflexion at the ankle with inversion of the foot. Tibialis anterior is a large muscle in all lemurs, but seems to be particularly large in these leapers. It is double in its central and distal thirds in all lemurs but the cheirogaleids (Jouffroy 1962). Apparently related to the absorption by the indriine hindlimb of substantial landing forces is the large size of the head of the femur, which fits into a correspondingly broad acetabulum. This transmits landing shock from the femur to the pelvis through the largest possible area, reducing unit forces.

One of the most striking differences between the indriids and the quadrupedal lemurs lies in the shape of the pelvis (fig. 4.14). In the cheirogaleids the ilium is long and narrow, and it is only marginally broader in the lemurids and *Daubentonia*. Both *Lepilemur* and *Hapalemur* are close to the lemurids in this regard. Among indriines, on the other hand, the ilium is considerably broader and somewhat blade-like, and the effective area available for muscle attachment is increased by the presence of a ligament joining the upper and lower anterior iliac spines. Gluteus medius, sartorius, and iliacus all surpass the ilium laterally to take additional origin from this ligament, while ventrally iliopsoas has an extensive origin over the iliac blade (Jouffroy 1962). Gluteus medius is given additional mechanical advantage by the raising of the greater trochanter of the femur, which is broad and amply surpasses the femoral head (fig. 4.15).

The peripheral vasti, especially vastus lateralis, are extremely large in the indriines, and the forward displacement of the articular surface of the distal end of the femur maximizes the mechanical advantage of rectus femoris on the tibial tuberosity. On the other hand, the hamstrings of the indriines are poorly developed relative to those of the quadrupedal lemurs (Jouffroy 1962).

Among the most remarkably developed muscle groups in Indriinae are the extensors of the back, the erector spinae. These are largest caudally (Hill 1953), and the sacrum shows an ossified median dorsal crest associated with erector spinae attachment. In *Lepilemur* this bony feature is less well developed than in the indriines, but is more so than among the quadrupedal lemurids. Together with the similarly substantial size in the sifaka of quadratus lumborum, a lateral flexor of the vertebral column, the great development of the erector spinae emphasizes the importance of hyperextension and control of the indriine vertebral column during leaping, and is matched by strong development of the abdominal musculature.

Lepilemur shares with the indriids a rather long lumbar region

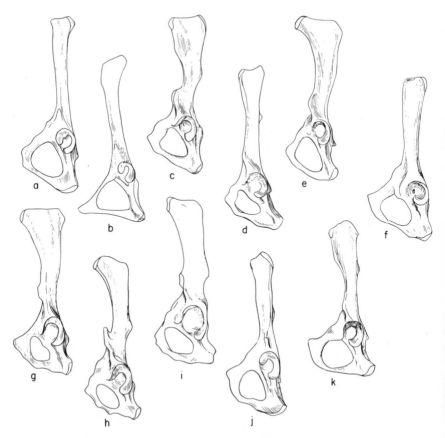

Figure 4.14. Innominate bones of: a: *Cheirogaleus major*; b: *Microcebus murinus*; c: *Varecia variegata*; d: *Lepilemur mustelinus*; e: *Hapalemur griseus*; f: *Phaner furcifer*, g: *Avahi laniger*; h: *Propithecus verreauxi*; i: *Indri indri*; j: *Daubentonia madagascariensis*; k: *Lemur macaco*. Not to scale.

with eight or nine vertebrae, although in the case of the latter the vertebral column is short in overall length compared to the limbs. In general, *Lepilemur* shows less strongly marked adaptations for bipedal leaping than do the indriines, mostly resembling the latter in postural adaptations; in its intermembral index *Lepilemur* clearly falls in with this group, but morphologically the sportive lemur fails to show some of the adaptations of the hind limb and pelvis which characterize the indriines. On the other hand, *Lepilemur* does exhibit some elongation of the tarsus, although less than what is found in some cheirogaleines. The indriines, then, exhibit a leaping pattern in which all the joints are subequally involved and where, for in-

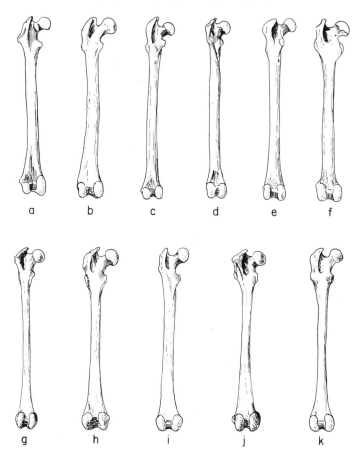

Figure 4.15. Posterior views of the left femora of: a: *Microcebus murinus*; b: *Varecia variegata*; c: *Hapalemur griseus*; d: *Lepilemur mustelinus*; e: *Indri indri*; f: *Phaner furcifer*; g: *Avahi laniger*; h: *Propithecus verreauxi*; i: *Daubentonia madagascariensis*; j: *Lemur macaco*; k: *Cheirogaleus major*. Not to scale.

stance, no short sartorius tends to limit mobility at the hip and knee joints as it does in quadrupedal springers, which tend to emphasize the ankle joint. This highlights the differences between the vertical clinging and leaping practiced by the indriines from that typical of *Tarsius* or some galagids, in which the tarsus may account for up to 50% or more of total foot length. *Lepilemur* exhibits a curious mixture of adaptations.

Ward and Sussman (1979) have recently investigated the extent to which morphological correlates of locomotion are found in the hindlimbs of two very closely related lemurs: *Lemur catta* and *Lemur fulvus rufus*. These authors found, not unexpectedly, that the anatomical patterns of the two were extremely similar. They did, however, note among other things that the foot of the rufous lemur showed greater mobility in all directions than that of the ringtail, the difference being on the order of that which separates more arboreal from more terrestrial primates in general. Further, the dominant component of hindlimb musculature in *catta* was associated with plantarflexion, while in *fulvus rufus* the greater part of the muscle mass was associated with pedal mobility and prehensility. These differences agree well with what we know of the locomotor and substrate preferences of the two lemurs, but the behavioral differences seem intuitively to be of greater degree than the anatomical.

Among the lemurs only the indriines and *Daubentonia* lack a caudofemoralis muscle, rare elsewhere among primates (Jouffroy 1962). In lemurids and lepilemurids this muscle shows a coccygeal origin, but among cheirogaleids it originates on the ischium. A variety of sesamoids are found in the hindlimbs of lemurs, notably the "superior patella" in the tendon of insertion of vastus intermedius. Two "fabellae" are found bilaterally in the tendon of origin of gastrocnemius, and a "cyamella" in the popliteus tendon (Jouffroy 1975). A character of the femur common to all extant lemurs, but otherwise found among primates today only in *Perodicticus*, *Galago* and the callitrichids is the third trochanter, associated with the insertion of gluteus maximus.

The hands of all lemurs except *Daubentonia* show "true" nails on all digits, although among the cheirogaleids, and *Phaner* in particular, the nails are quite strongly carinated and pointed. In *Daubentonia* the nails are greatly compressed laterally, and according to Clark (1936) show a distinct deep stratum, a characteristic of "true" claws. The nails are backed by apical tactile pads, which are complemented by three interdigital pads at the base of the digits, and by the thenar and hypothenar pads at the base of the palm. These pads are covered with papillary crests in all lemurs, but the degree of their differentiation differs substantially between genera. Among the cheirogaleines the pads are prominent and strongly differentiated, whereas among the indriines they are relatively poorly set off from the palmar skin, and indeed in *Indri* are distinguished only by their papillary cresting. The lemurids and lepilemurids are somewhat intermediate in this respect, with *Lepilemur* tending towards the indriid condition. Cartmill (1979) has suggested that a size factor is involved here, the larger forms having less elevated and less well-differentiated pads. This certainly holds good for the extremes, *Mi-*

crocebus having the best-differentiated pads and *Indri* the least, but there are certain exceptions to the trend, and as a rule the generalization applies best within closely related groups.

The indriines display interdigital webbing between the third, fourth, and fifth digits; in *Indri* this webbing surpasses the distal extremity of the first phalanx (Jouffroy 1975). The hand of none of the Malagasy forms except *Daubentonia* is extremely long by general primate standards, but the hands of the indriines are distinctly longer relative to the forelimb, and particularly to the vertebral column, than are those of the other lemurs. As noted, the length of the indriid hand is attributable to metacarpal and phalangeal elongation, and seems to be associated with feeding in small peripheral branches.

All strepsirhines, in sharp distinction to the rest of the primates, possess hands in which the morphological axis passes through the fourth, longest, digit, a condition sometimes known as "ectaxony." This condition is most strongly marked in those genera which display the greatest degree of manual elongation: in the indriines and *Daubentonia* the digital formula is 4>3>5>2, whereas in the other lemurs it is 4≥3>2>5 (Jouffroy 1975). The disposition of the long flexors and extensors of the digits seems to be unaffected by ectaxony, but the contrahentes and interossei do show some differences from the general primate pattern (Jouffroy and Lessertisseur 1959).

The pollex among the lemurs is long (only in *Avahi* does the thumb represent less than 24% of total hand length, whereas most higher primates fall below this figure), and is divergent from the other digits. The trapezo-metacarpal joint is angled away from the long axis of the hand, but permits no rotation, making the lemur thumb at best "pseudo-opposable," since it can move in only one plane. Further, the Malagasy primates exhibit only "whole-hand control"; all digits flex simultaneously, allowing little variation in prehensive pattern (Jouffroy and Lessertisseur 1959; Bishop 1964). Lemurs will grip a small object between the apical pads of the parallel digits and the interdigital palmar pads; at the same time the tip of the flexed thumb will contact the side of the object. The lemur hand conforms to the shape of a branch or other object when grasping it, so that the basic grasping gesture may result in many different prehensive grips (Bishop 1964).

All Malagasy primates exhibit a "toilet claw" on the second pedal digit; otherwise flattened nails (generally flatter than those of the hand) are present on all toes. Again, *Daubentonia* provides the exception, with compressed nails or pseudo-claws on all digits except the hallux, which does bear a flat nail. All pedal digits possess apical tactile pads with papillary crests; on the sole the pads are generally less well defined than those of the palm. Interdigital pads are in most cases relatively distinct, but proximally there is often less differentiation. As in the hand, the cheirogaleids and *Indri* represent the opposite ends of the spectrum; in the latter the entire sole area is covered with papillary crests, and in the other indriines and *Lepilemur,* crests are lacking on only a small central part of the sole.

Lessertisseur and Jouffroy (1974) have pointed out that when the foot is considered in relation to hindlimb length, the Malagasy primates display a complete spectrum. The indriines have relatively short feet (26–28% of limb length), the cheirogaleids and *Lepilemur* relatively long ones (33–35%). *Lemur, Hapalemur,* and *Daubentonia* are intermediate. Similarly, as noted earlier, the indriines (and *Daubentonia*) have short tarsi and long metatarsals, while the reverse is the case for *Lepilemur* and the cheirogaleids.

Like the hand, the lemur foot is ectaxonic, its morphological axis passing through the fourth digit, the longest. The long flexors and extensors reach each digit and conform to the general primate pattern despite the ectaxony of the foot; the intrinsic muscles, however, are affected by the shift in morphological axis (Jouffroy and Lessertisseur 1959).

The hallux of all lemurs is long, equivalent to 41–60% of the length of the foot. Relatively few higher primates exceed the lower figure (Jouffroy 1975). *Daubentonia* and *Microcebus* have the shortest hallux (40–43%), *Lemur* and *Hapalemur* follow (50–52%), while the indriines (53–60%) have the longest hallux relative to foot length found among all the primates (Jouffroy 1975). The hallux among all lemurs is quite strongly divergent (although never to the extent seen in some lorisids), and the second pedal digit is unreduced. All lemurs except the cheirogaleines possess an abductor hallucis longus derived from part of tibialis anterior (Jouffroy 1962). This muscle inserts at the base of the first metatarsal, and is lacking in the lorisids.

THE DIGESTIVE SYSTEM

The anatomy of the digestive tract among the Malagasy primates encompasses both relatively primitive conformations and a variety of divergent specializations. The best available summaries of the gross anatomy of the digestive organs in primates are those of

Hill (1958) and Hill and Rewell (1948); Hladik (1968) has published a comparative investigation of the microscopic anatomy of the intestinal mucosa which includes details of *Microcebus, Cheirogaleus* and *Lepilemur.* Hladik et al. (1971) and Charles-Dominique and Hladik (1971) have discussed the digestive system of *Lepilemur* in relation to its dietary habits. The notes which follow are intended as no more than a brief overview of the major features of the macroscopic anatomy of the digestive tract among the lemurs.

Parotid, submandibular, and both greater and lesser sublingual salivary glands are found among the lemurs; the parotid is well developed in *Lemur,* but the submandibular is the primary salivary gland in *Lemur, Microcebus, Indri,* and *Daubentonia,* all the Malagasy forms studied so far (Fahrenholz 1937).

The esophagus is quite uniform among the lemurs. The striated muscle at the upper end of the tube extends only a limited distance distally. The larynx is generally quite well developed and can be guided by the substantial epiglottis into the nasopharynx, permitting almost simultaneous breathing and swallowing.

In the cheirogaleids, as in the lorisoids, the stomach is a simple sac of globular form, with the esophageal and pyloric openings quite closely approximated. The area of the fundus is marked by rugae which radiate from the esophageal opening; elsewhere the lining is smooth. *Hapalemur* appears to be variable in the form of its stomach; according to Hill (1958) the organ is globular in some specimens, while in others it more closely resembles those of *Lemur* and *Lepilemur,* which are more elongated and pyriform, with long pyloric segments and well-separated pyloric and esophageal openings. In all three genera, long, well-marked rugae radiate from the cardia. However, while muscular taeniae are present in the pyloric antrum of *Lepilemur,* they are absent in *Hapalemur.* The stomach in all indriids is large and capacious, and Hill (1958) reports that in *Propithecus* the presence of a sort of raphe along the two primary surfaces indicates a tendency towards sacculation. He distinguishes a similar tendency in *Avahi* by virtue of a somewhat sacculated bulge in the ventral gastic wall. In *Indri* the stomach is less elongate than in other indriids, with a more expanded pyloric portion. The stomach of *Daubentonia* is relatively smaller and more globular, its esophageal and pyloric openings more closely approximated. Its cardiac and pyloric segments are differentiated by a transverse mucous fold.

In the small intestine there is relatively little macroscopic differentiation between the jejunal and ileal segments, although the duodenum is distinguished morphologically as well as topographically by the presence of a ligament of Treitz. In *Lemur* the capacity of the duodenum is great, rivalling that of the stomach (Hill 1958); the initial portion of the duodenum is also very wide in indriids. In contrast, Charles-Dominique and Hladik (1971) remark upon the shortness of the small intestine in *Lepilemur*. They note, however, that the very long villi of the ileum of this animal depart markedly from the general primate condition, and are matched elsewhere in the order only by those of the howler monkey.

The large intestine shows more variation among the lemurs than does the small intestine, the caecum perhaps most of all. This latter is well developed in all Malagasy primates studied. In *Microcebus* and *Cheirogaleus*, where the ileum and colon form a continuous tube, the caecum lies at a sharp angle to the ileo-caecal junction. In both genera it is of simple form, and is equal in diameter to the colon; but in *Microcebus* it is substantially longer than in *Cheirogaleus*. In *Phaner* it is considerably larger yet. Of the other lemurs, *Hapalemur* possesses a rather short, globular caecum which lacks apical narrowing; the rest all show at least some degree of caecal coiling and elaboration. In *Lemur* the caecum is long and voluminous, somewhat spiralled in its distal part. In *Propithecus* and *Avahi* the appendage is yet more capacious; the caecum of the former is spirally coiled in the newborn, but straightens out in adult life, while the caecum of the latter is much more convoluted. In *Indri* the caecum is yet more greatly developed, with complex convolutions, and is said by Hill (1958) to attain three times body length. This degree of elaboration is matched among the lemurs only in *Lepilemur*, where the enormously elongated caecum, irregularly convoluted and coiled, occupies almost half of the total volume of the abdominal cavity (Davies and Hill 1954). Hladik and Charles-Dominique (1971) note that in several respects the fine structure of the mucosa of the caecum and part of the colon in *Lepilemur* resembles that of the small intestine, and suggest that this is what accounts for the relatively small size of the latter. In *Daubentonia* the caecum is of more modest dimensions but is clearly differentiated into a globular basal sac and a cylindrical terminal segment.

The colon in *Microcebus* and *Cheirogaleus* is simple; there is no properly identifiable ascending colon, but rather an oblique transverse colon which connects the ileum to the terminal colon in a wide curve. In all the non-cheirogaleid lemurs the transverse colon is longer and is elaborated by the development of a looping ansa coli. In *Lemur* and *Daubentonia* the ansa is long but takes the form of a single loop; in the indriids it is greatly elongated and rolled into a compact spiral *(Propithecus)* or folded into multiple loops *(Indri)*, recalling in the former the condition seen in certain ruminants. In *Lepilemur* only the right half of the transverse colon participates in the formation of the ansa, which is kinked in both limbs; in *Hapalemur* the condition of the ansa is closer to that seen in *Lemur*.

Only one study has as yet been undertaken of the comparative efficiency of digestion among lemurs (Sheine 1979), in which attention was focused on the ability to digest cellulose. The species involved were *Lemur catta*, *L. fulvus*, and *Varecia variegata*. *Lemur catta* and *L. fulvus* were able to digest a significant proportion of ingested cellulose (29.9% and 21.2%, respectively) when food particle size was small, while *V. variegata* was totally unable to digest cellulose. None of the species showed any ability to digest cellulose at larger food particle size. It might be reasonable to suppose that certain other lemurs, such as the indriids and particularly *Lepilemur* with its highly specialized digestive tract and strongly folivorous diet (Charles-Dominique and Hladik 1971), may digest cellulose somewhat more effectively than do the species studied by Sheine.

THERMOREGULATION

Although the observation of "sunning" behavior in *Lemur catta*, *Propithecus verreauxi*, and various other lemurs has long provoked the speculation that internal mechanisms of thermoregulation among the Malagasy primates may be somewhat imperfect, few experimental data are available which bear on the question. Virtually all of those that are, moreover, concern cheirogaleids, particularly *Microcebus*.

In 1953 Bourlière and Petter-Rousseaux published the observation that the rectal temperature of *Cheirogaleus* tended to follow fluctuations in the ambient temperature. Further, these authors found that activity diminished somewhat with temperature, although even when the ambient temperature dropped to its minimum of 14° C, their subject animals failed to exhibit lethargy. Bourlière et al. (1956b) later broadened their study to include certain other lemurs, and found that the rectal temperature of *Hapalemur griseus* varied between 32° and 36° C, of *Cheirogaleus major* between 25° and 32° C, of *C. medius* between 21° and 32° C, and of *Microcebus murinus* between 25° and 34° C. They noted that rectal temperatures tended to be both higher and more stable among larger-bodied forms than among smaller ones, and that their data, although preliminary, did not suggest seasonal fluctuation in body temperature among their subjects.

This last observation was of particular interest in the context of Kaudern's (1914) suggestion, repeated by several subsequent authors, that *Microcebus* and *Cheirogaleus* undergo an annual period of torpor, equivalent to the hibernation of some Northern mammals. Starmühlner (1960) gave this period as July–September. During a field study in 1968, however, Martin (1972a) was unable to find any evidence to support this statement in the case of *Microcebus*, although he never observed *Cheirogaleus* between mid-July and mid-September, which did suggest a period of dormancy for the dwarf lemur. In the light of these observations, Martin (1972a) remarked that the capacity of these small lemurs to store fat in the tail is probably indeed related in the case of *Cheirogaleus* to the need to tide the animals over a period of dormancy, while in *Microcebus* it serves merely to supplement the limited food available during periods of scarcity.

Pursuing the question of possible seasonal torpor in *Microcebus*, Andriantsiferana and Rahandraha (1974) exposed mouse lemurs to temperatures of 7° C. Among females, rectal temperatures fell gradually to 28° C in from three to eight days, while the animals retained normal activity. It then fell rapidly to 15° C and then to ambient temperature where it stabilized, at which point the animals became totally inactive. The two males in the study never achieved such hypothermia, losing only a few degrees in rectal temperature during exposures of five and eleven days. Maximal survival time in hypothermia was five days, and although following the restoration of normal temperatures hypothermic individuals rapidly regained normal activity, again with a "step" at 28° C, they were unable to awaken spontaneously from such cold-induced dormancy. Further, their loss of body weight per day of hypothermia was far greater than that observed in true hibernators. On the basis of these observations, Andriantsiferana and Rahandraha concluded that *Microcebus* could not be regarded as hibernator, although females, at least, easily enter into and recover from hypothermia, which appears to have no deleterious effects if not too prolonged.

In 1973 Andriantsiferana and Rahandraha (1973b) reported that central temperatures in *Microcebus* held captive in Madagascar varied from 30°–38° C, and that they did indeed show some seasonal fluctuation roughly parallel to changing ambient temperatures. Decrease in central temperature, however, anticipated falling ambient temperature by a couple of months. Further, these authors noted that from March to September, central temperature varied greatly both from day to day and between cage mates, while on average remaining low. From September to March on the other hand, the higher central temperatures of individual *Microcebus* remained stable, and differences in temperature between cage mates were insignificant. Andriantsiferana and Rahandraha (1973a) also found that in the two months (February–March) when ambient temperatures were still high, but when the central temperatures of their mouse lemurs dropped, both food consumption and body weight increased. They suggested that at this period the drop in body temperature and its short-term fluctuations were due to the redirection of available energy to a build-up of fatty reserves, a hypothesis in accord with the fact that weight gain ceased with the drop in ambient temperature at the end of March.

Russell (1975) has approached the question of seasonal body temperature fluctuation in *Microcebus* and *Cheirogaleus* from another angle, monitoring body temperature while maintaining captive *M. murinus* and *C. medius* under conditions of constant temperature, humidity, diet, and day length. He found that all the animals studied showed correlated annual change in body weight, tail fat storage, sexual activity, and body temperature, even where no cues were provided by variations in the stimuli already noted. In March and April, all animals were active during the twelve hours of darkness provided, and at rest during the twelve hours of light. During the dark phase, body temperatures were consistently higher than during the light phase (*M. murinus:* 37.0–39.2° C, dark phase; 29.0–38.4° C, light phase; *C. medius:* 37.0–38.0° C, dark phase; 34.5–37.0° C, light phase). In no case did body temperatures fall below 6° C above ambient temperature. From September through December, body temperatures were lower during the light phase, approaching or reaching ambient temperature, at which time the animals were clearly lethargic, i.e., unresponsive to stimuli. This condition was reached by *M. murinus* below 29° C. At this time of year the animals showed a degree of variability during the dark phase; at relatively high body temperatures many remained lethargic while others became active. It was during this period that the cheirogaleids became sexually quiescent, and achieved their highest body weights for the year.

The fact that annual fluctuations in body temperature occurred in the absence of the external temperature changes present in the experiments of Andriantsiferana and Rahandraha, as well as the observation by the latter of a two-month lag between decrease in body and ambient temperatures in *Microcebus*, seem to indicate that the annual body temperature cycle in these animals is at least in part under endogenous control; this may also be true for the daily cycle observed (Russell 1975).

Clearly, there remains much to be learned about the complexities of thermoregulation in the cheirogaleids; and equally clearly, it would be unwise to extrapolate from the little we know of the cheirogaleids to other Malagasy primates. The one suggestive observation available comes from Bourlière et al. (1956), who noted that larger forms tend to have both higher and more constant body temperatures; the variation in rectal temperature of *Hapalemur griseus*, not a very large animal, is reported to be in the region of 4° C, not vastly more than that of the rhesus monkey (2° C: Myers 1971). It seems at least a good guess that the larger diurnal lemurs will eventually prove not to show very large variations in body temperature, daily or seasonal; but at the moment the data are lacking.

REPRODUCTION

External Genitalia and Genital Tract

Hill (1953) and Petter-Rousseaux (1962, 1964) have described these structures among the lemurs at some length, and only a few notes are provided here. In all male lemurs the penis is pendulous, terminiating in a glans protected by a prepuce, and is provided with a baculum. This bone is bifid at its tip in cheirogaleids, *Lepilemur*, and the indrids; simple in form in the lemurids and *Daubentonia*. In length the baculum varies from 9 mm *(Lemur fulvus)* to 28 mm *(Daubentonia);* according to Fiedler (1959) the length of the baculum of a *Microcebus* of 13 cm body length is 11 mm. The urethra opens transversely and dorsally near the tip of the penis in cheirogaleids and lemurids; terminally in the indriids. The glans is scaly in all lemur species, and bears differentiated spines in several genera (e.g., *Phaner, Lemur, Avahi*). In all species the scrotum is postpenial and permanently extra-abdominal, although its size may vary markedly from season to season. Petter-Rousseaux (1962, 1964) notes that the scrotum of *Microcebus* doubles in linear dimensions between the sexually quiescent part of the year (7.5 x 6 mm) and the period of sexual activity (15 x 12 mm). The scrotum is generally hairy, the notable exception being *Lemur catta*, and in many species bears glandular skin (see chapter 3). The ductus deferens and vesicular ducts unite to form a common ejaculatory duct in the lemurids; this last structure is very short in *Microcebus,* and in *Hapalemur* the former ducts share a common opening while remaining distinct. In the indriids they are completely separate. The seminal vesicles are well developed in all genera but *Daubentonia,* where they are greatly reduced; in most cases the prostate encircles the urethra and is not distinct from it. Cowper's glands are large.

In all female lemurs the urethral opening lies at the tip of a well developed clitoris, often provided with a small, sometimes terminally bifid, os clitoridis. The uterus is invariably bicornuate, with a short median body and a pair of conical horns which are sometimes elongated (particularly in the cheirogaleids), but which may, as in *Lepilemur, Lemur,* and the indriids, be very short (Petter-Rousseaux 1962, 1964). The vagina is long, and is separated from the uterus by a cervix. In *Microcebus* and *Cheirogaleus* gestation of (normally) two or three embryos takes place simultaneously in the two well-differentiated uterine horns; in *Lepilemur* one horn only is involved in gestation, the other remaining small and well defined, while in the indriids gestation takes place in the uterine body itself and the two horns become indistinct (Petter-Rousseaux 1962, 1964).

Placentation

The mode and development of placentation in strepsirhines has received much attention in view of its potential importance in phylogenetic reconstruction (see, for instance, Luckett [1974, 1975], to which reference may be made for detailed descriptions of placentation and placental development). Amnion formation in lemurs is by folding (not by cavitation as in the anthropoids), and the polar trophoblast is lost before implantation occurs. In the early stages, the vascular mesoderm of the yolk sac joins with the chorion to produce a choriovitelline placenta which is subsequently supplanted by the definitive chorioallantoic placenta. In all lemurs—and strepsirhines—studied, the definitive placenta is epitheliochorial and diffuse, i.e., the chorionic epithelium, via its outer trophoblastic layer, invades the whole of the endometrium instead of being restricted to small areas of the chorion. There is a substantial separation of the maternal and fetal blood supplies by the vascular endothelia and surface epithelia of both chorion and endometrium; exchange between the two is accompanied by the development of vascularized trophoblastic villi which are received by crypts of the uterine epithelium. This contrasts with the situation in anthropoids, where maternal tissue is entirely lacking in the placental barrier; even the maternal capillary walls break down, allowing the chorionic trophoblast to be bathed directly by the maternal blood. The epitheliochoral placenta is nondeciduate, i.e., following parturition the maternal and fetal tissues separate cleanly at the junction between the chorionic and uterine epithelia, with no loss of blood.

Sexual Cycles

All Malagasy primates exhibit distinct seasonality in breeding, although breeding periods may be of considerably varying length, and females of almost all species appear to be polyestrous. In Madagascar, *Microcebus* breeds from September to March, and Martin (1972a) has suggested, on the basis of indirect evidence, that at least some females bear two litters during this breeding period, some 75% of the females conceiving early in the season, and a second pregnancy peak occurring after January. There is some question as to whether second pregnancies within a single breeding season are due to the occurence of a postpartum estrus (Van Horn and Eaton 1979), although these authors suggest that the observation of Andriantsiferana et al. (1974) of conception by two lactating females during such postpartum estrus reinforces this possibility. Petter-Rousseaux (1964) studied the length of estrous cycles in *Microcebus* by examining vaginal smears, and found that, although they were exceedingly variable, most cycles fell in the range of 45–55 days, the periodicity apparently being unaffected by the isolation or sexual behavior of the females involved. Petter-Rousseaux (1962, 1964) and Martin (1972c) have shown that estrous cycles are affected in captive *Microcebus* by photoperiod, with long photoperiods inducing and shorter ones inhibiting such cycles. Female mouse lemurs kept captive in the northern hemisphere show breeding seasons almost six months out of phase with those in Madagascar (Petter-Rousseaux 1970).

As has been noted, male mouse lemurs show considerable changes in testicular size; the testes attain their maximal volume and spermatogenic activity during the long photoperiods of the breeding season, when they attain eight times their quiescent volume (Spühler 1935). They begin to increase in size during August, and probably begin to decline in April or May (Martin 1972a). Female mouse lemurs display a sealed vulva during the times of sexual quiescence; during a "proestrous" phase of four to six days the skin around the vulva becomes swollen and pink, and the vaginal orifice begins to open. Full opening usually occurs in around 3 days, and may last for a further 2 to 3 days, at which time brightly colored folds of skin surround the vagina. Gradual closure of the vagina then takes place over a period of 5 to 10 days. (Petter-Rousseaux 1962, 1964).

Less detailed observations have been made on *Cheirogaleus major* by Petter-Rousseaux (1962, 1964), from which it appears that reproductive cycles in the dwarf lemur resemble those of *Microcebus*, with mating occurring from October onwards, and births spanning the period November to February. The vulva is sealed until the estrous period: swelling of the external genitalia takes place over 3 to 5 days, and vaginal opening lasts for a similar time during which copulation occurs; then the orifice closes, although more slowly than in the case of *Microcebus*. Estrous cycles are repeated at intervals of about 30 days for the duration of the breeding season or until conception is achieved. Postpartum estrus has been found in dwarf lemurs at DUPC following the loss of an infant; conception has been noted as soon as one to two weeks thereafter (E. L. Simons, pers. comm.). Testicular changes in *Cheirogaleus* males apparently mirror those in *Microcebus*.

Among the other lemurs so far studied, only *Varecia*, possibly *Hapalemur* (see Petter et al. 1977, contra Petter-Rousseaux, 1968), and some individuals of *L. fulvus* and *L. coronatus* (M. Stuart, pers. comm.) show total closure of the vagina outside the period of sexual activity. In *Varecia* the closed vulval area pinkens and begins to open at the beginning of proestrus, achieving a wide orifice after about seven days; ovulation occurs four to six days after maximum opening is attained (Foerg 1978). If conception is not achieved, a second ovulation may occur about 40 days after the first (and possibly a third, some 40 days thereafter); but male testicular size and spermatogenesis, which appear to peak in synchrony with the initial female oestrus (Foerg, in prep.), have by this time so greatly diminished that impregnation is highly improbable. In captivity at DUPC births of *Varecia* are very highly synchronized, occurring over a short period, and reflecting the extremely brief breeding season. Females housed together cycle within a few days of each other, and the entire DUPC colony cycles within a period of two to three weeks. In Madagascar, captive *Varecia* give birth in October–November (Petter et al. 1977), suggesting a breeding season around mid-July.

Lemur catta is the best known species of its genus from the point of view of reproductive cycles as from many others. Evans and Goy (1968) reported a mean cycling period, on the basis of vaginal smears, of 39.3 days, with a range of 33–45 days. Full vaginal estrus lasts on average 4.7 days, and is accompanied by swelling and pinkening of the external genitalia, which deflate rapidly following mating. Postpartum estrus is known to be stimulated by separation from, or loss of, an infant; females will normally experience a postpartum estrus if such separation occurs within two to three weeks following parturition, although the percentage of females doing so declines as the infants age (Van Horn and Eaton 1979). Males display a testicular cycle which parallels the estrous cycle of the females (Andriamiandra and Rumpler 1968).

In captive *Lemur catta* it has been demonstrated by Van Horn (1975) that estrous cycles are inhibited by long photoperiods and reactivated by a return to short photoperiods. As might be expected, the magnitude of photoperiod change need not be high to produce these results. Matings in the wild occur during a period of decreasing day length. On the basis of field observations, A. Jolly (1966) believed that mating was restricted to a period of two weeks in April, during which individual females were receptive for only a day or two, although this was preceded by a nonreceptive period of "pseudoestrus" a month in advance. Such pseudoestrus has also been noted in *L. fulvus* by L. Vick (unpubl.). Budnitz and Dainis (1975) have, however, noted that the major *L. catta* birth peak in August–September is succeeded by a period of some births in October–November, which suggests that females failing to conceive in April may be impregnated during later cycles. Petter et al. (1977) state that among ringtails captive in Tananarive the sexually active period, as determined from vaginal smears, lasts from May to August, with two or three cycles exhibited during these months.

Sexual cyclicity in other *Lemur* species is less well known, although observations at DUPC suggest that in *Lemur mongoz* unimpregnated females will cycle only two to three times during a relatively restricted breeding season, at intervals of 29–36 days (Evans and Goy [1968] quote 36–37 days), while in those species, such as *Lemur fulvus* and *L. macaco,* which habitually live in larger social groupings, such females may cycle six or seven times (at intervals of 30 and 33 days respectively) over a much more diffuse period of breeding activity (E. L. Simons, pers. comm.). At DUPC (northern hemisphere) pregnant female *L. macaco* and *L. fulvus* may be found at any time from October through March; Petter et al. (1977) state that the breeding season of these species in Madagascar lasts from April through June, with births occurring from August through November. Male testicular cycles clearly exist in these animals, but precise data are lacking.

The breeding season of *Lepilemur* is said by Petter et al. (1977) to fall between May and July, with births occurring from September to November. Information for the various populations of *Hapalemur* is too imprecise to be worth quoting.

Data on the timing of births in *Propithecus verreauxi* suggest that only a single ovulation occurs per breeding season in this form, although a "pseudoestrus" comparable to that of *Lemur catta,* and also involving flushing of the vulva, may occur a month or so prior to ovulation (Richard 1974, 1978). Female receptivity is extremely limited in duration; Richard (1978) quotes maxima of 42 and 36 hours for two females studied in southern Madagascar. Breeding appears to take place between January and March, and to be quite highly synchronized; the birth season of *P. v. coquereli* has been reported to last 21 days (Richard 1974, 1978), and of *P. v. verreauxi* only 10 days (A. Jolly 1966).

Data are lacking for *Indri,* but Pollock (1975a) noted sexual behavior only in January and February, and a single birth in mid-May; he suggests that birth intervals of two to three years are the norm for this lemur. *Avahi* appears to give birth in or around August (Petter-Rousseaux 1962), which would place the breeding season in late March or April.

Martin (1972b) has discussed the significance of breeding synchronization and seasonality of birth among the lemurs. He regards them as adaptations to the seasonality of food availability, and suggests that the timing of birth is related to the rate of maturation of the infants of each species: infants must have time to grow and accumulate adequate tissue reserves during the wet season to allow them to survive during the nutritionally more marginal conditions imposed by the dry season.

Gestation Periods and Litter Sizes

The length of gestation may vary substantially, both within and between the various lemur species, the larger ones tending to longer gestation periods. Thus *Microcebus* exhibits an average gestation period of 63 days in contrast to the 162 days of *Propithecus* (see table 4.4), and a range of 54–68 days. In all cases gestation lengths are long compared to those of non-primate mammals of comparable size. In many cases our knowledge of gestation periods is based on only one or two observations, while in others (for instance that of *Indri,* whose large body size suggests that it may have the longest gestation period of all) no reliable information at all is available. In

Table 4.4. Gestation periods, litter sizes, and sexual maturation in various Malagasy primates.

	Average Gestation Period (days)	Modal Litter Size	Approx. Age at Sexual Maturity (months)
Lemur catta	136	1	21–24
Lemur fulvus	120	1	20–24
Lemur macaco	128	1	20–24
Lemur mongoz	128	1	
Varecia variegata	102	2	20
Hapalemur griseus	140	1	
Lepilemur mustelinus	135	1	18
Microcebus murinus	63	2–3	8–12
Mizra coquereli	86	2	
Cheirogaleus major	70	2–3	
Cheirogaleus medius	61	2–3	12–24
Phaner furcifer		1	
Avahi laniger	155(?)	1	
Propithecus verreauxi	162	1	30
Indri indri		1	36(?)

SOURCE: Data from Petter-Rousseaux (1962). Petter et al. (1977), and DUPC.

similar fashion, there is some within-species variation in litter size (*Varecia variegata*, for example, may have from one to five infants); and in some species which most commonly display single births twinning is frequent enough to be regarded as a normal event. A number of lists of such rates has been published, but these necessarily depend on data from captive colonies based on restricted founding stocks, and may be biased in a number of ways. Figures on births at DUPC over the period 1970–1978 yield the following for rates of twinning in several *Lemur* species and subspecies: *Lemur catta*, 16.1% of births; *L. f. macaco*, 30%; *L. f. fulvus*, 3.9%; *L. f. rufus*, 4.1%; *L. f. albifrons*, 29%. Table 4.5 gives modal litter sizes for several lemur taxa.

Infant Development and Sexual Maturation

Infant lemurs, in correlation with their relatively long gestation lengths, are born at a relatively advanced stage of development; only in the cheirogaleids are the eyes not open at birth, and even here they open within a couple of days of that event. Among the cheirogaleids infants are born in tree hollows or in nests constructed of leaves, and are normally transported in the mother's mouth; only in *Phaner* do infants regularly cling to the mother's fur following their emergence from the natal nest. Initially *Phaner* infants ride ventrally, later dorsally. *Microcebus* infants are able to climb and to grip branches shortly after birth, and Petter et al. (1977) indicate that *Cheirogaleus* infants may be slightly more advanced yet at birth. Weaning takes place at about 40 days in *Microcebus*, and in *Cheirogaleus* at about 45 days, although consumption of fruit may begin as early as around 25 days.

In *Lepilemur* the infant is able to move around fairly well immediately after birth. It is transported by the mother in her mouth for the first few weeks of life, but later moves to her fur; the mother generally leaves the infant on a branch or in a tree hollow while foraging (Petter-Rousseaux 1962, 1964). *Hapalemur* infants are also carried initially in the mouth of the mother (although they are permitted to grasp the mother's fur at the same time), but within a day or so they move to the fur of her sides or back (Petter et al. 1977). Lactation in *Lepilemur* appears to last about four months (Petter-Rousseaux 1962, 1964); infants begin to feed on solids at about one and a half months, and by two and a half months are sampling the entire range of the adult diet. Lactation in *Hapalemur* is said by Petter-Rousseaux to last about six months.

The two lemurid genera differ strikingly in the development of the infant at birth, as might be expected from the disparity in their gestation lengths. Immediately following birth in *Lemur* infants are able to grip the mother's fur strongly; initially they ride ventrally (in a transverse position in *L. macaco* and *L. fulvus;* longitudinally in *L. catta*), but after about two *(L. catta)* or four *(L. fulvus)* weeks switch to the mother's back (Klopfer and Klopfer 1970; Sussman 1977b). In *Varecia,* on the other hand, the mother prepares a nest in advance and then gives birth to a relatively undeveloped offspring, which cannot cling to her fur or move around. Transport of the infant is in the mother's mouth, and most of the time the infant is left in the nest or in dense foliage high in the canopy (Petter 1962). Development appears to be relatively rapid, and by the end of a month the infant moves around independently, if inelegantly. *Lemur* infants may suckle up to the age of five months, but begin sampling fruit after a little more than a month (Petter-Rousseaux 1962, 1964), and *L. catta* definitely ingest solid food by the end of the second month (Klopfer and Klopfer 1970). Vick and Conley (1976) note that dorsal riding is no longer tolerated by *L. fulvus* mothers after about three months.

The infant *Propithecus* rides on the mother's belly, in a transverse position, from the time of birth until switching to her back at between about three and six weeks (Eaglen and Boskoff 1978). It is

three to four months before the offspring is able to follow the
mother independently, and it may still be carried after six or seven
months (Petter-Rousseaux 1962, 1964). Nursing probably continues
throughout this period. In *Indri* Pollock (1975a,b) observed dorsal
riding up to five months. Eaglen and Boskoff (1978) state that consumption of solid food by infant *Propithecus* probably begins in most
cases only at the end of three months, although Richard (1976)
noted "tasting" of solid food at between one and two weeks of age.
Pollock (1975a,b) has seen *Indri* infants gnaw at fruits by two months
of age, but, again, it is not known whether they were ingesting at
this stage. Eaglen and Boskoff conclude that infant development in
Propithecus, and probably also in *Indri*, is considerably slower than in
lemurids; Richard (1976), on the other hand, notes that the devel-
opmental pattern of *Propithecus verreauxi*, as expressed in the various
stages of infant dependence on the mother, more closely resembles
that of the equally arboreal *Lemur fulvus* than that of the more terrestrial *Lemur catta*.

Age at sexual maturity is difficult to generalize about both because it can be measured in a number of different ways, and because it is variable both within and between sexes of the same species. For example, male *Varecia* are normally unable to impregnate
females until about 32 months of age, although some at DUPC have
sired offspring in their second year while others have failed to show
even testicular enlargement in their fourth breeding season (Foerg,
in prep.). Female *Varecia* mature somewhat earlier, usually exhibiting first estrus at about 20–22 months. Some local vulval pinkening,
characteristic of females of all ages during the breeding season, may
be seen in nine-month-old females, although in these young individuals it does not precede cycling. Possible hormonal correlates of this
phenomenon are under investigation (R. Foerg, in prep. and pers.
comm.). Table 4.4 lists the ages at which the females of several lemur species generally begin normal cycling; these ages are, as the
example of the ruffed lemur shows, subject to some variation.

BLOOD PROTEINS AND GENERAL HEMATOLOGY

The work of Buettner-Janusch and his colleagues has shown
that one major hemoglobin is normally found in *Lemur fulvus, L. macaco, L. mongoz, Cheirogaleus major, C. medius, Microcebus murinus,* and
Propithecus verreauxi, while two major hemoglobins are present in *Lemur catta, Varecia variegata, Hapalemur griseus, Lepilemur mustelinus,*

and *Phaner furcifer* (Buettner-Janusch et al. 1971). In the case of *Propithecus verreauxi,* however, a polymorphism exists; the anodal mobility of the hemoglobin of nine *P. v. verreauxi* was greater than that of six *P. v. coquereli,* probably because of α-chain differences. Buettner-Janusch et al. believe that the existence of two major hemoglobins in *L. catta, Varecia,* and *Hapalemur* is attributable to gene duplication at one hemoglobin locus.

One general characteristic of strepsirhine hemoglobin is the large proportion of alkali-resistant hemoglobin present in adults (Buettner-Janusch and Twichell 1961); among anthropoids, only fetal hemoglobin contains a substantial alkali-resistant component. Moreover, lemur hemoglobin β-chains appear to have a number of residues which are homologous to residues in human γ-chains (Hill and Buettner-Janusch 1964). Again, the major hemoglobins of newborn lemurs, unlike those of newborn higher primates, cannot be distinguished from those of adults (Buettner-Janusch et al. 1972). The possibility of a distinct embryonic hemoglobin in lemurs cannot yet be eliminated, however, although if, as seems possible, the functional properties of anthropoid fetal hemoglobin are dictated by the possession of a hemochorial placenta, there would be no reason to expect this.

In general, the α-chains of primate hemoglobins are stable compared to the β-chains (Buettner-Janusch and Hill 1965), i.e., they vary much less between species. Table 4.5 gives tentative values for the numbers of amino acid replacements in the α- and β-chains of *Homo sapiens, Lemur fulvus,* and *Propithecus verreauxi* hemoglobins, and shows, for instance, that between 12 and 20 amino acid residues differ in the α-chains of the Malagasy forms and man, while up to 38 differences are found between the β-chains. As noteworthy as the

Table 4.5. Differences among the hemoglobins of *Homo sapiens, Lemur fulvus,* and *Propithecus verreauxi.*

Species	Amino Acid Replacements	One-Step Mutations	Two-Step Mutations
α-chains			
Lemur fulvus vs. *Homo sapiens*	15	12	3
Propithecus verreauxi vs. *Homo sapiens*	19	15	4
Lemur fulvus vs. *Propithecus verreauxi*	13	7	6
β-chains			
Lemur fulvus vs. *Homo sapiens*	31	21	10
Propithecus verreauxi vs. *Homo sapiens*	33	23	10
Lemur fulvus vs. *Propithecus verreauxi*	16	11	5

SOURCE: The data are tentative, and are taken from Buettner-Janusch (1967).

differences between the lemurs and man, however, are the substantial differences between *Lemur* and *Propithecus*; these are particularly striking in the case of the normally relatively stable β-chain. Buettner-Janusch (1967) remarks, and the table shows, that most of the mutations between the hemoglobins of *Lemur fulvus, Propithecus verreauxi,* and man are one-step, i.e., require the substitution of only one base in the DNA triplet.

In stark contrast to the hemoglobins, the transferrins in genus *Lemur* show a high degree of polymorphism (Nute and Buettner-Janusch 1969). In 47 individuals of *Lemur fulvus* 23 electrophoretically distinct phenotypes were demonstrated, and in·11 *Lemur catta* 5 phenotypes were found. Other species examined included *Lemur macaco* (10 individuals, 3 phenotypes), *L. mongoz* (4,4), and *Varecia variegata* (3,2). Inheritance in *Lemur fulvus,* and presumably also the others, appears to be via non-dominant autosomal alleles, with at least 11 alleles involved in the case of the brown lemur. Polymorphism on this scale rivals anything reported in other mammalian species.

Among other erythrocytic and serum proteins investigated in *Lemur fulvus* (these also include haptoglobin, glucose-6-phosphate dehydrogenase, 6-phosphogluconate dehydrogenase, and carbonic anhydrases I and II), only red cell acid phosphatase shows any polymorphism (Mason and Buettner-Janusch 1977); 3 phenotypes were observed. Thus, out of a total of 8 loci so far investigated in this species, 2, or 25%, have proved to be polymorphic, a figure comparable to those reported for other mammalian species (Mason and Buettner-Janusch 1977). *Lemur catta, Lemur mongoz, Varecia variegata,* and *Propithecus verreauxi* were found to show only one red cell acid phosphatase phenotype, while *Hapalemur griseus,* like *Lemur fulvus,* showed three.

Turning to the cellular, as opposed to the molecular, aspects of hematology among the Malagasy primates, Bergeron and Buettner-Janusch (1970a,b) have reported on erythrocyte and leucocyte counts and morphology in various lemur species held in captivity both in the United States and in Madagascar. Erythrocyte morphology of all the lemurs examined (various *Lemur* species and subspecies, *Varecia, Lepilemur, Cheirogaleus, Phaner,* and *Propithecus*) was generally similar to that of other mammals, but with a mean diameter of 5.4μ, these cells were rather smaller than those of man (7.2μ), and comparable to those of galagos. Only *Cheirogaleus major* showed a significant difference in erythrocyte diameter from other Malagasy primates; the low mean diameter of 4.2μ is consistent with a high total erythrocyte count of about 13–14 x 10^6/mm^3. This latter value ranged in other lemurs from about 8–10 × 10^6/mm^3, i.e., somewhat higher than that of man (about 5 × 10^6/mm^3) and other nonhuman primates which have been studied so far. Average hemoglobin concentrations in most of the lemur species studied ranged from about 16–18 g%, a little higher than the average for man (15 g%); but *Lepilemur* (12 g%) and to a lesser extent *Cheirogaleus* (13.6–14.8 g%) showed lower values.

Total leucocyte counts of lemurs were found to depend upon the conditions of captivity of the animals; those which had been isolated from the vagaries of the natural environment for relatively long periods yielded substantially lower values. However, average total leucocytes even of well-isolated lemurs (ranging approximately from 10–16 × 10^3/mm^3), were rather higher than that of man (7 × 10^3/mm^3), although comparable to the values yielded by other nonhuman primates tested. Differential leucocyte counts of lemurs are largely noteworthy for the low neutrophil-to-lymphocyte ratio found; in man this ratio is considerably greater than one, while in all lemurs tested it was substantially less. In this respect the lemurs resemble galagos and most other nonhuman primates studied.

The various classes of lemur lymphocytes correspond quite well to the categories observed in man and other mammals. The lymphocytes are very variable, but fall broadly into two groups, large and small, in all lemurs except *Phaner*, in which the one individual examined in this respect displayed almost entirely small lymphocytes, each with a prominent nucleolus.

CHAPTER FIVE

The Subfossil Lemurs

Every species of lemur living today had become known to science before fossil evidence began to accumulate showing that the extant Malagasy primates, for all their variety, are in fact merely the survivors of a more extensive primate fauna of extraordinary adaptive diversity: one which, moreover, has become impoverished only in recent times. It is unfortunate that the inevitable difference in the nature of our information on the living and extinct lemurs compels for the most part separate discussion of the two groups; but it should not be forgotten that when we discuss the living lemurs we are speaking only of a partial fauna, and that many primate niches in the Malagasy ecotope were until very recently occupied by animals phylogenetically close to, but in many cases adaptively very distinct from, their surviving relatives. To what extent the living forms have expanded the ecological space which they originally occupied to fill the niches vacated by those now extinct we shall never know for certain; but it seems probable that, at least in most cases, the answer to this is "not very much." In this chapter I summarize what is known of the recently extinct lemurs, and of the reasons for their disappearance; one can only regret, bitterly, that these fascinating animals have escaped fuller scientific knowledge by so short a time.

Although the remarkable Flacourt missed being the first European to describe a living lemur, it may well be that he was the first to describe one of the lemur species now extinct. For in the list of "animaux terrestres" in chapter 38 of his *Histoire* occurs the following:

> *Tretretretre* ou *tratratratra,* c'est un animal grand comme vn veau de deux ans, qui a la teste rond, & un face d'homme: les pieds de deuant comme vn singe, & les pieds de derriere aussi. Il a poil frisoté, la queuë courte & les oreilles comme celles d'vn homme. Il ressemble au *Tanacht* descrit par Ambroise Paré. Il s'en est veu vn proche l'estang de Lipomami aux enuirons duquel est son repaire. C'est vn animal fort solitaire, les gens du païs en ont grand peur & s'enfuient de luy comme luy aussi d'eux (1661:154).

Since Flacourt's descriptions appear uniformly to be sober accounts of animals most of which are readily identifiable with species existing today, it is quite possible that the "tretretretre" indeed existed in the area of Fort-Dauphin[1] during the time Flacourt was there, even though his choice of words seems to make it clear that Flacourt never actually saw one himself. In any event, the description is clearly of a large primate which, allowing for some exaggeration in the reporting of size, might have been any of several of the larger lemur species now known only from their subfossil remains.

Until 1893 the tretretretre, however graphic Flacourt's description of it, could have been dismissed as fantasy. But on June 15 of that year, C. I. Forsyth Major presented to the Royal Society a description of the skull of *Megaladapis madagascariensis,* a Malagasy primate vastly larger than any lemur then known from the island (Major 1894). Since that time, a succession of excavators has discovered the subfossil remains of six genera and up to fifteen species of extinct primates in recent deposits scattered over the center, south, and west of Madagascar. These extinct lemurs (denoted by daggers) are listed together with the living ones in the classification presented in chapter 3; all are more or less closely related to forms still extant today. A discussion of their affinities will be found in chapter 6.

THE SUBFOSSIL SITES

The sites from which the subfossil remains of extinct lemurs have been recovered are all small, local deposits of recent origin.

[1]Lake Lipomami is not shown on Flacourt's map (1661, facing p. 43) of the Fort Dauphin region, but from several references in both the *Histoire* (1658) and the *Relation* (1661), it appears that a village named "Lypoumami" lay on the route from Fort-Dauphin to the larger village of "Fanshere." This latter settlement is shown on Flacourt's map in the approximate location of the modern village of Fanjahira, about twenty km northwest of Fort-Dauphin.

Lying within modern, active, drainage systems, they are isolated from major sedimentary sequences and thus cannot be stratigraphically correlated (Walker 1967a). Radiocarbon dates are, however, available in a number of cases. Mahé (1965) and Mahé and Sourdat (1972) have classified the majority of the sites into three categories: caves or solution cavities in cliffs or karstic formations (e.g., Andrahomana, Ankazoabo); marshes in the central highlands resulting from the blockage by volcanic activity of streams (e.g., Ampasambazimba); and coastal marshes on the western and southern margins of the island (e.g., Ambolisatra). Early investigators were far more interested in the fossils than in the deposits from which they were recovered, and in general failed to record much information about the sites themselves, although a notable exception to this is the classic central highland site of Ampasambazimba (e.g., Raybaud 1902; Standing 1908a). More recent discussions of various of the sites include those of Walker (1967a), Raison and Vérin (1968), Mahé and Sourdat (1972), and Tattersall (1973a); an overall review of the sites is provided by Tattersall (1973b). Figure 5.1 gives the locations of the principal sites, with radiocarbon dates where available.

Despite their isolated occurrence and limited extent, the subfossil sites have yielded an extraordinary abundance of fossils, both in the variety of species represented and in the sheer quantity of material recovered. Primates, however, represent only a tiny proportion of the total number of specimens, the most commonly occurring fossils, depending on the location, being those of various chelonians, crocodiles, *Hippopotamus lemerlei*, and the giant ratite *Aepyornis*. Among the primate fossils, apart from the extinct forms discussed below, occur (although rarely in abundance or in good condition) the remains of individuals belonging to genera which still survive today. Apart from those of *Daubentonia*, such remains from the southern and southwestern sites are readily attributable to species which still inhabit those areas; Lamberton's (1939) *"Propithecus verreauxioides"* from Ankazoabo, supposedly distinguished from *P. verreauxi* by its slightly larger size and various minor characters, cannot in fact be sustained as separate from the latter.

In contrast, bones allocable to extant genera which have come from the sites in the central highlands of Madagascar, a region now almost totally denuded of forest and aboriginal fauna, provide us with a unique glimpse of the earlier distributions of the forms they represent. Thus postcranial remains of the genera *Lepilemur, Hapalemur, Lemur, Propithecus, Avahi, Indri* and *Cheirogaleus* have been been recovered from Ampasambazimba and the sites around Antsirabé (Lamberton 1939). *Hapalemur* is represented by a distinct species

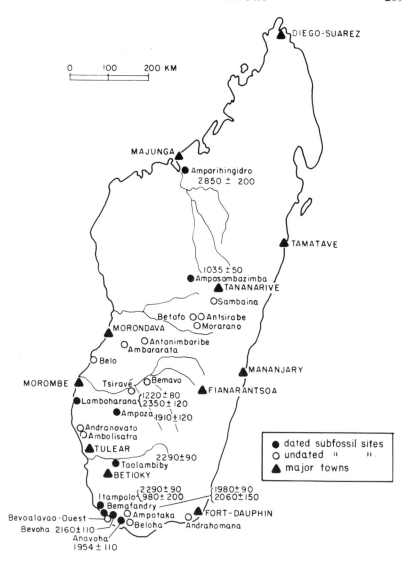

Figure 5.1. Subfossil sites of Madagascar.

(but a large one, and the larger forms of *Hapalemur* are limited to eastern Madagascar); among the other extant genera the monotypic *Indri* is restricted now to the eastern humid forests, while *Cheirogaleus* is apparently present at Ampasambazimba in the form of the larger eastern species *C. major:* I have not seen the cranium involved, but Standing (1905) gives its maximum length as ca. 61 mm, which falls at the high end of the range of *C. major*. None of the other fossils can be allocated to particular species within their respective genera, all of which are represented today by populations in both the eastern and western regions of Madagascar, although the area of distribution of western *Avahi* is of limited extent and lies far to the north of the plateau sites.

On the basis of the extant lemur taxa represented in the subfossil assemblage, then, it might be concluded that on balance the fauna of the central plateau is most likely to have been essentially of eastern, rather than western, aspect. The flora, however, is distinctive: a study by Perrier de la Bathie (1928) of fruits and seeds collected from the peat surrounding the fossil beds at Ampasambazimba led this competent scholar to the conclusion that "Toutes ces plantes sont des espèces de bois des pentes occidentales, formation qui devait couvrir, avant son déboisement, le Massif d'Itasy tout entier et dont on retrouve encore les vestiges aux alentours du lac" (p. 25).

The Itasy Massif, on which Ampasambazimba is situated, lies within the Central phytogeographical Domain of the Eastern Region (see fig. 2.2), but Perrier de la Bathie's description of the flora of the site (as well as Humbert's [1955] phytogeographical map) suggests that it lies within the Western division of this Domain, where the vegetal climax is (or was) a low and relatively sclerophyllous forest (Koechlin 1972), in contrast to the taller, more humid forest which characterizes those parts of the Central Domain further to the east. The presence at Ampasambazimba of *Indri*,[2] in particular, whose current distribution is limited to the northern half of the eastern high rain forest, thus presents us with something of a puzzle. Lamberton (1939) perceived a slightly greater robusticity in the fossil long bones he studied than in his comparative material of *I. indri*, and on this basis suggested that the subfossils might represent a different species of the genus. This speculation cannot, however, be

----

[2]Although the overwhelming preponderance of evidence suggests that the *Indri* material was recovered at Ampasambazimba, it is just possible that it may have come from Antsirabé. This latter site, of which little is known, clearly lies within the Eastern division of the Central Domain, as defined in Humbert's map of 1955. According to Koechlin (1972) the climax vegetation of this division is closer in nature to the eastern rainforest in which *Indri* is found today than is that of the Ampasambazimba region. A provenance of Antsirabé for the *Indri* bones might thus seem somewhat more likely for ecological reasons, although still explicable only with difficulty in view of the localization of *Indri* even within the eastern forest. In any event, it may be unwise to lump together all of the plateau sites for discussion, as has traditionally been done, although admittedly the extinct lemur faunas of Ampasambazimba and Antsirabé appear to be closely comparable.

sustained on the basis of the inadequate material available at present.

THE EXTINCT LEMURS

Subfossil material belonging to genera or species now wholly extinct can be attributed to seven subfamilies, some of which survive today: Archaeolemurinae, Palaeopropithecinae, Indriinae, Lemurinae, Hapalemurinae, Megaladapinae and Daubentoniinae. Each of these groups is discussed separately in the following notes.

Subfamily Archaeolemurinae G. Grandidier, 1905

This subfamily comprises two genera, *Archaeolemur* and *Hadropithecus,* the former being known from two species, the more gracile *A. majori* and the more robust *A. edwardsi.*

Although as we have seen the first extinct lemur to be brought to scientific attention and the first to be named was *Megaladapis madagascariensis,* the earliest to be described in print (Major 1893) was an individual of *Archaeolemur majori.* The material consisted of an incomplete subadult calvaria from "a marsh on the southwest coast of Madagascar," which Major, in the amply fulfilled expectation that better material would shortly be available, refrained at the time from naming. The genus and species thus owes its name to Filhol (1895), who based it upon a humerus and the proximal portions of a radius and ulna from Belo-sur-Mer. Later in the same paper, Filhol proposed the name *Lophiolemur edwardsi* for "two mandibles and several postcranial bones" from the same site. In the following few years discoveries of *Archaeolemur* remains at virtually all of the principal subfossil sites in Madagascar spawned a host of new names. The history of study of the genus has been reviewed in detail by Tattersall (1973b); suffice it to say here that by 1902 Guillaume Grandidier was able to synonymize *Lophiolemur* Filhol, 1895, *Dinolemur* Filhol, 1895 (in part), *Nesopithecus* Major, 1896, *Globilemur* Major, 1897, and *Protoindris* Lorenz, 1900, with *Archaeolemur* (G. Grandidier 1902). Within the genus Grandidier recognized two species: *A. majori,* which included *N. australis* Major, 1896, *P. globiceps* Lorenz, 1900, and *Bradylemur bastardi* G. Grandidier, 1900; and *A. edwardsi,* containing *L. edwardsi* Filhol, 1895, and *N. roberti* Major, 1896. But he retained his genus *Bradylemur* G. Grandidier, 1899, for the species *B. robustus.* It remained for Standing (1908a) to demonstrate the complete synonymy of *Bradylemur* with *Archaeolemur*, *B. robustus* now being subsumed under *A. majori*. At the same time, however, Standing named a third species of *Archaeolemur*, *A. platyrrhinus*, on the basis of one of the series of *Archaeolemur* skulls recovered from Ampasambazimba; This specimen should be placed in *A. edwardsi* (Tattersall 1973b).

At present, then, only two valid species of *Archaeolemur* can be recognized. If a third species is in fact known, it is known only in the form of a single cranium from the sole northwestern site, Amparihingidro, excavated recently by Mahé (1965). This individual is intermediate in morphology between the relatively robust *A. edwardsi* of the central highlands and the more gracile *A. majori* of the south and southwest, although it falls within the size range of the former; but since these two species are separated by differences of degree rather than kind, and since both are variable, it seems at the very least premature to make a taxonomic judgement on the basis of a single specimen.

A potential problem also exists in the naming of the species conventionally referred to as *A. edwardsi*, the species name of which derives from the *"Lophiolemur"* material from Belo described by Filhol in 1895. Of the two valid species of *Archaeolemur*, the gracile form in all diagnostic material is limited to the lowland sites of the south and southwest, while the robust is in equivalent material known only from the sites of the central highlands. The unsatisfactory type series of both *majori* and *edwardsi* were recovered from the same site, Belo, the most northerly of the southwestern coastal subfossil deposits, and it is at the very least highly unlikely that two different species of *Archaeolemur* are in fact represented in this fossil assemblage. Since definite attribution to species can only be made where good cranial specimens exist, neither suite of type material (see above) would be completely diagnostic even if it could be definitely identified today, something which I have been unable to do. But rather than regard *edwardsi* as a junior synonym of *majori*, and look to the next available species name, *roberti* Major, 1896, for the robust form, it seems reasonable for the sake of stability to give the *edwardsi* type material the benefit of the doubt, and to continue to use Filhol's original name for the robust species, particularly since Grandidier drew the distinction between gracile *(majori)* and robust *(edwardsi)* species as early as 1902.

Hadropithecus, a much rarer and apparently monotypic form, has a simpler history. In 1899 Lorenz von Liburnau described a right dentary (apparently VNHM 1934 XV 1/1) from the cave site of

Andrahomana as the holotype of a new genus and species, *H. sten-ognathus*. Shortly thereafter, the same author (Lorenz 1900a) described a cranium (VNHM 1934 IV 1) of a much younger individual from the same site as the type of another new genus and species, *Pithecodon sikorae*, and in 1902 he effected the synonymy of the two. The most outstanding subsequent discovery of *Hadropithecus* remains, now known from Ampasambazimba as well as from five of the southern and southwestern sites, was by Lamberton, who in 1931 recovered a well-preserved adult skull (of a considerably older individual than Lorenz's) from the site of Tsiravé (Lamberton 1938). In his publication of 1902 Lorenz declared his belief in a close relationship between the genera *Hadropithecus* and *Archaeolemur*, but it was left to Grandidier (1905) to unite them in the subfamily Archaeolemurinae, later formally diagnosed and placed within Indriidae by Standing (1908a).

It is largely on the basis of the archaeolemurines that the myth of the former presence in Madagascar of "advanced," "monkeylike," primates arose. As early as 1896, Major placed his rather fragmentary "*Nesopithecus*" material from Antsirabé in its own family of Anthropoidea, "intermediate in some respects between the South American Cebidae and the Old World Cercopithecidae, besides presenting characters of its own" (1896:436). This daring assessment was based on molar morphology and on the erroneous supposition that the single facial skeleton which Major possessed at the time had originally exhibited full post-orbital closure. The accumulation of more material in the following years failed to sway Major from this line of thought; in 1900 he concluded that "*Nesopithecus*" and its allies formed a "side-branch of the evolving line from lemurs to monkeys, branching off close below the Cercopithecidae" (1900a:499). Although G. Grandidier (1905), in the earliest comprehensive monograph on the subfossil lemurs, came down firmly for the opposite viewpoint, stating that "il est d'abord hors de doute qu'il faut les ranger parmi les Lémuriens" (1905:135–36), and that the attributes which had led Major to his view were "caractères acquis par une similitude de vie" (1905:136), Standing (1908a) returned to an interpretation not dissimilar to Major's, claiming that "there is much more reason for regarding these simian features as general ancestral characters and the condition of the recent Malagasy lemurs as specialized" (1908a:103).

The conception of the subfossil Malagasy lemurs as somehow "monkeylike" has shown remarkable tenacity in surviving even into the recent literature. There can be no doubt, however, that Grandidier's assertions of 1905 were fully justified, and that the subfossil lemurs, highly derived in many characters though some of them undoubtedly are, in no way represent any "advance" in "level of organization" over their living relatives (Tattersall 1973b).

For all practical purposes, the two species of *Archaeolemur* differ only in size (mean maximum cranial length of *A. majori* 130.6 mm, *o.r.* 122.3–135.0 mm, *n* = 17; of *A. edwardsi* 147.0 mm, *o.r.* 139.6–153.5, *n* = 17) and in robusticity, the latter displaying in particular, varying degrees of sagittal and nuchal cresting, while the former almost invariably does not (fig. 5.2). But although there is little or no overlap between the two species in cranial dimensions, the ranges of dental dimensions do overlap considerably, rendering uncertain the attribution to species of jaw or dental remains alone (Tattersall 1973b). The postcranial skeletons of the two species are likewise extremely similar, again differing only in size (about 15%) and in the greater robusticity of *A. edwardsi* (Walker, 1967a).

Hadropithecus stenognathus falls within the size range of *Archaeole-*

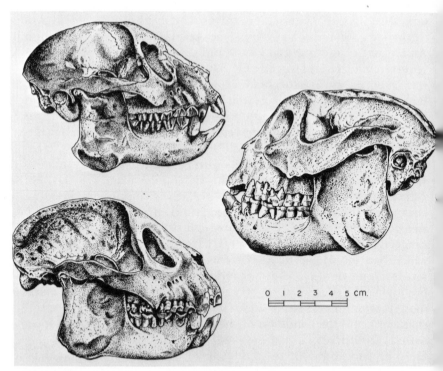

Figure 5.2. Lateral views of skulls of the three archaeolemurine species. Top left: *Archaeolemur majori*; bottom left: *A. edwardsi*; right: *Hadropithecus stenognathus*.

mur, the two known crania measuring 128.2 mm (Andrahomana) and 141.8 mm (Tsiravé) in maximum length. The postcranial skeleton of this species is only poorly known.

Detailed descriptions of the crania and postcrania of the archaeolemurines have been provided recently by Tattersall (1973b) and Walker (1967a), respectively, and it seems most useful here merely to emphasize those features which depart most markedly from the ancestral indriid condition, most closely approximated today, at least in the skull and dentition, by *Propithecus* (see chapter 6). Apart from the considerably larger size of the archaeolemurines and the consequent much heavier build of their skeletons, these lemurs are most sharply distinguished in the skull from the living members of their family by the components of the masticatory apparatus.

The archaeolemurines are distinct from the indriines in both the number and the morphology of their teeth (fig. 5.3). The archaeolemurine dental formula is $\frac{2.1.3.3}{1.1.3.3}$, i.e., a premolar has been retained in each quadrant of the jaw which has been lost in the indriines. The upper central incisor of *Archaeolemur* is enlarged and spatulate, while I^2 is much smaller and closely approximated to it. A small diastema usually separates these teeth from the upper canine, which is robust but rarely projects much beyond the occlusal plane. The premolars embody what is perhaps the most striking dental adaptation of the genus: closely approximated one to another, these teeth form a single longitudinal shearing blade. The posterior upper premolar is incorporated into this blade in its mesial moiety, but its distal part functions with the molar row; these latter teeth are quadricuspid, as are those of the indriines, but are subsquare and bilophodont, the paracone-protocone and metacone-hypocone pairs being united by transverse crests. While the incisor and canine are clearly derived from a procumbent condition in the lower jaw of *Archaeolemur*, they are inclined forward at only about 45 degrees to the tooth row. The anterior lower premolar is robust and caniniform, but is not high-crowned; its distal moiety forms the mesial portion of the continuous shearing blade in which P_3 and P_4 participate. The lower molars closely resemble the uppers, except that the lingual cusps are higher than the buccal ones.

The dentition of *Hadropithecus* (fig. 5.3) is yet more derived. Its incisor teeth, upper and lower, are greatly reduced and quite orthally implanted, as is the associated lower canine. The upper canine and the caniniform anterior lower premolar are similarly much reduced, the latter being continuous with the longitudinal shearing blade of P_3. The posterior lower premolar, however, is molariform,

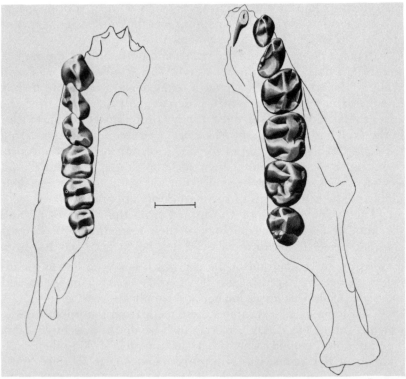

Figure 5.3. Lower cheek dentitions of *Archaeolemur edwardsi* (left) and *Hadro-pithecus stenognathus* (right). Scale is 10 mm; drawing by Carl Wester.

with a cruciform arrangement of rounded ridges radiating from a central protoconid. The upper premolars increase in size and complexity posteriorly; P⁴ is molariform. The anterior molars, upper and lower, are large and subsquare, with high, rounded enamel folds replacing the transverse crests of *Archaeolemur*. The last molars, upper and lower, are reduced. But perhaps as unusual as the morphology of the teeth of *Hadropithecus* is the way in which they wear: the entire crowded battery of cheek teeth rapidly becomes worn flat,. with thick enamel ridges enclosing shallow basins in the softer dentine, as in a variety of ungulates.

The great depth of the mandibular corpus in *Archaeolemur* and *Hadropithecus* (fig. 5.2), the expansion of the gonial region, and the elevation of the temporomandibular joint are all characteristically indriid traits, but the fossil forms are distinguished from the living ones by the much greater robustness of their dentaries and their possession of fused and less oblique mandibular symphyses. In view of its much larger size, the cranium of *A. majori* in particular, is extraordinarily *Propithecus*-like, although its long dental arcade is somewhat posteriorly shifted, slightly shortening the prezygomatic face and, posteriorly, crowding the structures of the cranial base. *Archaeolemur edwardsi* and *Hadropithecus* display sagittal cresting, and all the fossil species display some degree of nuchal cresting, which reflects the proportionately much larger musculature to be expected in bigger animals; other evidence of this relative muscular hypertrophy shows up, for instance, in the highly developed attachments of both superficial and deep masseters, and of the digastric (both bellies). The ear region of the archaeolemurines is similar in essentials to that of the indriines, and has been exhaustively described by Saban (1956, 1963).

Although Clark (1945) concluded that the endocranial cast of *Archaeolemur* showed similarities to the brains of anthropoids, and Piveteau (1950) considered that the external morphology of its frontal lobe displayed an evolutionary level reminiscent of ceboids and cercopithecoids, Smith (1908) and, more recently, Radinsky (1970) have emphasized that the brain of *Archaeolemur* is indriid in external morphology. The endocranial cast of *Hadropithecus* appears to differ little from that of *Archaeolemur* (Radinsky 1970).

In its cranial construction *Hadropithecus* differs from the species of *Archaeolemur* chiefly in its remarkable orthognathy, a correlate of the diminution of its anterior dentition. This shortness of the face appears to be responsible for the foreshortening of the braincase relative to *Archaeolemur* (Tattersall 1974; Roberts and Tattersall 1974). A further major distinction lies in the elevation of the neurocranium relative to the facial skeleton in *Hadropithecus*, leading to the forward displacement of the frontal sinus and the raising of the zygomatic process of the temporal bone; at the same time the face is deepened by the downward extension of the maxilla. In the mandible the angle of the fused symphysis of *Hadropithecus* is more vertical than in *Archaeolemur*, again in response to the abbreviation of the face, while the mandible itself is more robust.

Although any attempt to link the archaeolemurines phylogenetically with the Old World monkeys is clearly futile, it is nonetheless true that adaptively, these lemurs do bear comparison with certain of the more terrestrially adapted cercopithecoids. The postcranial skeleton of *Archaeolemur* is well known, and has been analyzed recently by Jouffroy (1963) and Walker (1967, 1974), who have emphasized the adaptations it shares with those of terrestrial cercopithecoid monkeys, and by Godfrey (1977), who has pointed to primitive retentions from an indriid-like (and, at a further remove, *Lepilemur*-like) ancestor, as well as resemblances to a wider variety of monkeys. Such primitive retentions appear to be in features not associated with vertical clinging and leaping, but rather with quadrupedal locomotion and a posture of vertical sitting (Godfrey 1977).

In life *Archaeolemur* appears to have been a powerfully build and rather short-legged quadruped, with reduced powers of leaping. Walker (1967, 1974) has described characters of the forelimb, femur, and extremities, in particular, which parallel the condition found in baboons,[3] and while Godfrey makes a broader range of comparisons (most of them to monkeys which are at least semiterrestrial), both authors would presumably agree that *Archaeolemur* possessed many of the adaptations of a terrestrial quadruped which nonetheless retained the ability to exploit arboreal food resources, emphasizing above-branch stabilization (cf. Rose 1973) while in the trees. Such locomotor and postural behavior would, in fact, closely parallel that of baboons of the genus *Papio*, which, in contrast to the closely related but exclusively terrestrial *Theropithecus*, show considerable facility in the arboreal milieu.

The postcranial skeleton of *Hadropithecus* is very poorly represented, but those elements which are known resemble those of *Archaeolemur* except in being more gracile (Walker 1967, 1974). This led Walker to suggest that *Hadropithecus* may have been a more fully committed terrestrial cursor.

Such analogies between the archaeolemurines and the large cercopithecids assume added significance when cranial adaptations are considered, for in many ways the distinctions between the skull and dentition of *Archaeolemur* and *Hadropithecus* echo those that exist between *Papio* and fossil and living *Theropithecus* (C. J. Jolly 1970a, Tattersall 1973b).

Most of the cranial and especially dental peculiarities of *Theropithecus* relative to the closely related but much less derived *Papio* may be ascribed to its specialized diet and feeding behavior (C. J. Jolly 1970b, 1972). The highly adaptable *Papio*, although often found today in open habitats, appears essentially to be a woodland-

---

[3]Walker's list of such features includes, in the humerus: high and prominent greater tuberosity, small brachioradialis flange, small and backwardly displaced medial epicondyle, poorly rounded capitulum, and wide, well-excavated olecranon fossa; in the ulna: posteriorly and transversely compressed styloid process; in the femur: anterior convexity of the shaft, large size of the greater trochanter, shallow patellar groove.

or deciduous forest-dwelling type. On the other hand, *Theropithecus* is adapted to life in open, treeless areas. Living in a more varied habitat, *Papio* displays a more opportunistic type of feeding behavior, which involves the exploitation of resources both in the trees and on the ground, than does *Theropithecus*, whose sustenance is derived entirely from terrestrial sources. Numerous studies of geladas (e.g., Crook 1966, 1967) have shown that these animals exist on a diet of the blades, seeds, and rhizomes of grasses, supplemented with the leaves of low shrubs, the bulbs of small plants, and ground-living arthropods. All of these food items are small, most of them are relatively tough, and as they are obtained at or near ground level, they are generally gritty.

In correlation with this unusual diet, the gelada possesses a reduced anterior dentition, an adaptation particularly evident in some of the large Pleistocene members of the genus. This reduction is due to the fact that the incisors are virtually unused in the preparation of food, morsels of which are instead picked up and conveyed directly to the cheek teeth by the fingers. Additionally, the molars are large and high-crowned, and become rapidly worn flat, in which condition they show alternating ridges and valleys of enamel and dentine, respectively; this reflects the tough, abrasive nature of the food consumed (Jolly 1970b, 1972).

These specializations, and others discussed by C. J. Jolly 1970a,b) and Tattersall (1973b) are precisely analogous to those distinguishing the highly derived *Hadropithecus* from the more generalized (for the clade) *Archaeolemur*. The conformation of the archaeolemurine masticatory apparatus indicates the adoption in these animals of a more powerful mode of chewing than that of the more plesiomorph indriines (Tattersall 1973b, 1974); but *Hadropithecus* quite evidently shows adaptations to a powerful grinding mode of mastication which are lacking in *Archaeolemur*. Dental wear in the latter is most pronounced on the incisors and on the premolar shearing blade, while in *Hadropithecus* the entire premolar-molar series becomes worn flat in the same way as does that of the gelada, and also shows considerable interproximal attrition, another concomitant of powerful grinding.

The evidence thus suggests that *Archaeolemur* used its premolars rather in the manner of incisors, while its molars evidently processed relatively unabrasive materials. Bilophodont dentitions characterize, besides *Archaeolemur*, both the leaf-eating and the primarily frugivorous cercopithecids, and are also characteristic of certain other browsing mammals. Colobine monkeys, however, exhibit higher, sharper cusps and transverse lophs on their molars than do the more frugivorous cercopithecines (Pilbeam in Pilbeam and Walker 1968), which share with *Archaeolemur* a relative bunodonty. Further, the molar teeth of primate frugivores tend to be small relative to body size, as in *Archaeolemur*, when compared to those of leaf-eaters (Kay 1975). The presumptive evidence is thus strong that *Archaeolemur* was primarily frugivorous, or at least dietarily generalized, consuming relatively large food items, such as hard-skinned fruits, which required incisal/premolar preparation but relatively little molar processing. *Hadropithecus*, on the other hand, was highly specialized dentally for the consumption of smaller, tougher food items which required no incisal preparation but heavy grinding by the cheek teeth, and which must have been conveyed to the mouth by the hands (C. J. Jolly 1970a, Tattersall 1973b,c). The gelada also habitually forages from an upright sitting position, and C. J. Jolly (1970a) believes that certain features of *Hadropithecus* may indicate a similar proclivity. The evidence to which Jolly points is weak; but the adaptations to vertical sitting to which Godfrey (1977) has pointed in the postcranial skeleton of *Archaeolemur* might suggest that the ancestor of *Hadropithecus* which first committed itself to the ground was preadapted to foraging from this position.

It seems reasonable, then, to hypothesize that *Archaeolemur* and *Hadropithecus* were the Malagasy equivalents of *Papio* and *Theropithecus*, respectively, in Africa: the one a relatively generalized digitigrade quadruped, perhaps primarily terrestrial, but also at home in the trees, and with a more varied, if chiefly frugivorous, diet; the other a more specialized, terrestrially-committed, "manual grazer."

Subfamily Palaeopropithecinae Tattersall, 1973

Two genera belong to this indriid subfamily: *Palaeopropithecus* and *Archaeoindris*.

Among the fossils recovered by Guillaume Grandidier at Belo in 1898 was a single right mandibular fragment of large size, containing the posterior premolar and the two adjacent molars. Impressed by the resemblance of these teeth to those of *Propithecus*, Grandidier named the new species he created to receive the specimen *Palaeopropithecus ingens* (G. Grandidier 1899). Subsequently the same author ascribed to his new species further mandibular fragments, found at Ambolisatra (G. Grandidier 1900), and shortly thereafter Standing (1903) was able to report the discovery at Ampasambazimba of two skulls of *Palaeopropithecus*. Each of these

Standing attributed to a new species: the smaller to *P. maximus*, the larger to *P. raybaudii.* By the time he came to prepare his great monograph of 1908, however, Standing had available 13 crania of *Palaeopropithecus*, all of which, including the type of *raybaudii*, he allocated to *P. maximus* (Standing 1908a). In the same work Standing also described the postcranial remains of *Palaeopropithecus*, but under the impression that they belonged to *Megaladapis grandidieri*, because in the absence of associated skeletons he allocated the longest bones to the species with the longest skull. The history of the attribution of postcranial bones to *Palaeopropithecus* is a complex one, admirably summarized by Lamberton (1947), who also suggested the synonymy with *Palaeopropithecus* of *Bradytherium madagascariense*, proposed by G. Grandidier (1901) to accommodate a "sloth-like" femur from Ambolisatra.

Since Standing's time, almost all authors have concurred in recognizing two species of *Palaeopropithecus*: the supposedly smaller *P. ingens,* which is unknown from complete cranial material and which occurs in the sites of the south and southwest (Belo, Ambolisatra, Ampoza, Lamboharana, Taolambiby, and the sites of the lower Menarandra Valley); and *P. maximus*, from the central plateau (Ampasambazimba, Morarano). This specific distinction is, however, hard to sustain (Walker 1967; Tattersall 1973b), particularly since *P. ingens,* which is unknown from complete cranial material and which impossible to discriminate absolutely between congeneric species of living lemurs on the basis of osteology alone and that there is a strong tendency for the subfossil lemur genera to be divisible into plateau and coastal species; but in dealing with fossil forms one is obliged to resort to the concept of morphospecies, and in the case of *Palaeopropithecus* the available evidence of morphology and size is inadequate to support polytypic status for the genus.

In the case of *Archaeoindris* problems of this sort fail to arise: the genus is known only from Ampasambazimba. Founded by Standing (1908b) on the basis of right and left maxillary fragments, *Archaeoindris fontoynonti* remained otherwise unknown or unrecognized, and separable only with difficulty from *Palaeopropithecus*, for well over a decade. But in 1924 Lamberton recovered an unassociated cranium and mandible, both in good condition, which amply justified Standing's creation of a new genus and species (Lamberton 1929, 1934). In his monograph of 1934 Lamberton assigned to *A. fontoynonti* a few immature long bones, and both he and Walker (1967, 1974) also allocated to this species a femur described by Standing (1910) under the name of *Lemuridotherium*. These attributions cannot be defini-

tive, but are reasonable in terms of what is known of the fauna at
Ampasambazimba.

In striking contrast to the archaeolemurines, the palaeopropithecines are clearly united to the indriines by their dental morphology, while diverging from the living forms in their cranial architecture, albeit to a lesser degree than has generally been supposed
(Tattersall 1973c).

Palaeopropithecus ingens possesses a long (mean maximum cranial
length 194.5 mm, *o.r.* 181.0–211.0 mm, $n = 8$), relatively low, and robustly built skull (fig. 5.4). A small degree of nuchal cresting is invariably present; the superior temporal lines are strongly marked
and occasionally coalesce posteriorly into a sagittal crest. The rela-

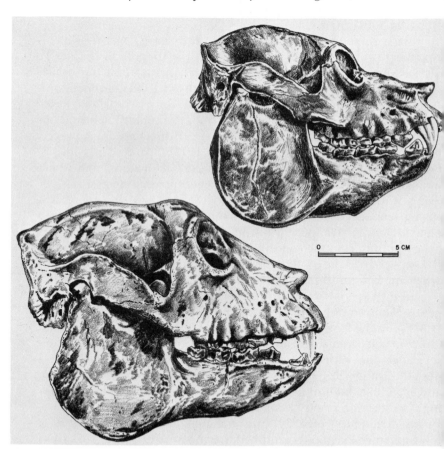

Figure 5.4. Lateral views of the skulls of *Palaeopropithecus ingens* (above) and
Archaeoindris fontoynonti. Cranium and mandible of *Archaeoindris* not associated.

tively small and well-frontated orbits are heavily ringed by bone, and are markedly oriented upward relative to the plane of the tooth-row—although not relative to the plane of the horizontal semicircular canal (Mahé 1968); the post-orbital constriction is pronounced, and the posterior root of the massive zygomatic arch is somewhat raised. The occipital plane is at right angles to that of the cranial base and the occipital condyles are prominent, their articular surfaces being strongly curved but oriented chiefly downward, indicating that the head was habitually held at a considerable angle to the neck. Lateral to these is a massive structure consisting of the partially coalesced and highly developed styloid, mastoid/postglenoid, and paroccipital processes (fig. 5.5). The mandible is also massive: the long, oblique fused symphysis displays a deep genial fossa and

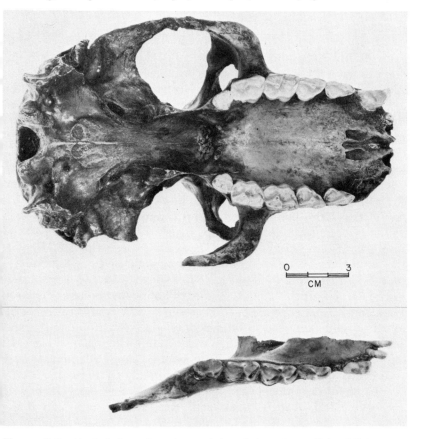

Figure 5.5. Ventral view of cranium and dorsal view of mandible of *Palaeopropithecus ingens.*

a pronounced inferior torus. The rugose gonial region vastly accentuates the indriid tendency to expansion, recalling the condition in the howler monkey, and presumably indicating similarly the development of an enlarged hyoid apparatus.

The skull of *Archaeoindris* (fig. 5.4) is larger and more robust yet (length of single known cranium 269.0 mm). The neurocranium is raised relative to the face when compared to *Palaeopropithecus*, while the face itself is shortened anteroposteriorly and deepened from top to bottom. A broad rather than high sagittal crest is present, as is a smaller degree of nuchal cresting. In correlation with the relative abbreviation of the face there is some foreshortening of the cranial base. The orbits are relatively somewhat larger than those of *Palaeopropithecus* (but smaller than those of the diurnal indriids), and less upwardly oriented but also slightly less frontated; the postorbital bar is strongly built.

Both *Palaeopropithecus* and *Archaeoindris* share unusual conditions of the nasal and auditory regions. The premaxillae are enlarged bilaterally in their superior portions to produce paired protuberances projecting from the superolateral margins of the nasal aperture. In the ear region, prominent auditory bullae, as seen in the indriines and archaeolemurines—and indeed, in all the living lemurs—are entirely lacking, much of the area they would otherwise occupy being taken up with the enormous paroccipital/mastoid/styloid structure referred to earlier. Presumably because of the presence of this massive development, the tympanic ring lies deep within the petrosal and communicates with the exterior via a bony tube. Saban (1956, 1963) has interpreted this conduit as a lateral extension of the ectotympanic itself, while Szalay (1972) regards it as petrosal in origin. The tube may, in fact, be complex in formation. Despite the absence of an inflated bulla, the tympanic cavity is nonetheless of considerable volume (Saban 1956, 1963, 1975), and I have suggested, in view of the similar absence of an inflated bulla in the large *Megaladapis*, that such absence represents what is effectively an allometric reduction: that in the large, relatively small-brained forms, sufficient space is already available *within* the structure of the cranial base to accommodate a tympanic cavity of a volume adequate to prevent overdamping of the tympanic membrane (Tattersall 1973c).

In their dentitions the palaeopropithecines closely resemble each other and, in their cheek teeth, the indriines, especially *Propithecus* (fig. 5.5). All share the same dental formula, with a single lower incisor and only two premolars in each quadrant of the dentition. In *Palaeopropithecus* the medial upper incisor is large and subcylindrical, while the lateral is greatly reduced. Only upper incisor alveolae are known in *Archaeoindris*; these indicate a similar conformation. In the lower jaw, the incisor and canine are clearly derived from a procumbent, indriinelike condition, but are short and stubby, not closely approximated, and quite orthally implanted. This apparently also applied in *Archaeoindris*. The upper canine in *Palaeopropithecus* is stout and sharp but quite short, resembling that of *Propithecus; Archaeoindris* seems to have been similar in this respect.

The wrinkled upper premolars of the palaeopropithecines are more strongly developed posterolingually than those of the indriines, which they otherwise resemble. Both lower premolars are mesiodistally elongate, and the caniniform is low crowned. The wrinkled upper molars of the palaeopropithecines are close morphologically to those of *Propithecus*, which they resemble particularly in their buccal elaboration. In contrast to the indriines, however, the area of the hypocone in M^{1-2} of the palaeopropithecines is developed as a shelf, rather than as a cusp. The lower molars are more elongate than those of the indriines; concordantly with this, M_{1-2} possess distinct paraconids.

This close correspondence between the dentition of the palaeopropithecines and *Propithecus* has generally been viewed as providing the starkest possible contrast with the divergence in cranial structure between the indriines and palaeopropithecines. But although the cranial differences separating the two groups are in many respects substantial, one cannot but feel that they traditionally have been overstated (Tattersall 1973c, 1975a). *Palaeopropithecus*, with its long and low cranial contour, is not unreminiscent of *Indri:* the relatively smaller brain and orbits of the subfossil form, presumably at least largely allometric manifestations, account for many of the major apparent distinctions between the two genera in overall cranial appearance. Thus, for instance, the more depressed profile of *Palaeopropithecus* in the area of the junction of the orbital and nasal capsules is probably to be traced to the relative reduction of its orbits. In *Indri*, conversely, the interorbital area, because it is flanked by the relatively large orbits, is filled out by the development of considerable frontal sinuses.

Propithecus and *Archaeoindris* both possess crania which are relatively abbreviated anteroposteriorly. Structurally, *Propithecus* is extraordinarily similar to *Indri* (indeed, in the collections of the BMNH there is a *Propithecus diadema* skull [BMNH 1533.6] which is distinguishable from *Indri* by almost nothing except its dentition); virtually all the dissimilarities between the crania of the two genera are directly due to their differences in facial length. The shorter face of *Propithecus* is functionally linked to the abbreviation of its neurocranium and the forward shifting of the origin of temporalis. Thus, although relative to skull length the cranium of *Propithecus* is rather high when compared with that of *Indri,* this is not a real difference in functional terms. In *Archaeoindris,* however, the deepening of the face and neurocranium does appear to be biologically significant, although the relative longitudinal proportions of the splanchnocranium are determined by the requirements of the masticatory apparatus in a pattern very similar to that evinced by *Propithecus.* It seems plausible that the functionally real facial deepening in *Archaeoindris* reflects the need to resolve greater occlusal forces than those generated by *Palaeopropithecus.*

Suggestions as to the mode of life of *Palaeopropithecus* have been varied. Standing (1908a) and, following him, Sera (1935), proposed that the animal had been aquatic,[4] but largely on the basis of wrongly attributed postcranial remains. Carleton (1936) came closer in proposing a sloth-like mode of existence, but it was not until 1947 that Lamberton concluded that they had been "des animaux de grande taille se déplaçant dans les arbres en se suspendant par les mains à la manière des Orangs-outans" (1947:41). The recent investigations of Walker (1967, 1974) have fully confirmed Lamberton's hypothesis.

Walker has calculated an intermembral index for *Palaeopropithecus* of 147.04, indicating extreme lengthening of the arms relative to the hindlimbs; this figure falls between the mean values calculated by Schultz (1930) for the siamang (148.5) and orangutan (142.8). At 95.33, the brachial index of *Palaeopropithecus* falls slightly below that of the orangutan (100.3). These indices already suggest strongly that *Palaeopropithecus* practiced a form of suspensory locomotion, and this is borne out by the osteology of its postcranial skeleton. The following brief discussion is based on Walker's analysis (1967a, 1974).

The incompletely known hand and foot of *Palaeopropithecus* are long, hook-like structures, the phalanges being thin, elongated, dorsally-curved and strongly grooved by the flexor tendons. The absence of malleoli on the thin tibia and fibula indicate the sacrifice of stability for mobility in the ankle joint. These features would seem to indicate an exclusively arboreal existence for *Palaeopropithecus,* since neither the ankle nor the foot is in the least suited for weight-

---

[4]The first suggestion of this nature relative to any of the subfossil lemurs appears to have been that of Major (1900b): "The remarkable shortness and flattening of the *Megaladapis* femur calls to mind the same bone of aquatic Mammalia" (1900b:494).

bearing. The proximal end of the femur shows adaptations for extreme mobility of the hip joint; again, stability is sacrificed. The shallow patellar groove and widely spaced condyles at the distal end of this bone indicate a considerable potential range of rotation at the knee joint. In the pelvis an erect posture is suggested by the strong anterior curvature of the ilia, since this implies the presence of a strong abdominal musculature of the kind required to support the viscera in this position, as well as by other features, including a long ilium but relatively short ischium, anterolateral and inferior orientation of the acetabula, and a broad, flat pubis.

In the forelimb, the humerus has a large, globular head which greatly surpasses both tuberosities, permitting a great range of movement at the shoulder joint. The deltoid crest extends well down the shaft, as in living arm-swingers. At its distal extremity, the humerus shows clear separation of the capitulum and trochlea by a pronounced keel: adaptations permitting a high degree of both pronation and supination. The ulna shows a poorly developed olecranon process, and both this bone and the radius are bowed strongly, the former posteriorly, the other laterally, again as in living arm-swingers.

All the available evidence, then, points to the conclusion that *Palaeopropithecus* was adapted to a primarily suspensory form of locomotion which emphasized mobility and below-branch stabilization, and in which all four long, curved extremities were involved. This is probably the only way in which an animal of such size could have exploited the outer canopy of the trees, and as both Lamberton (1947) and Walker (1967, 1974) have proposed, the nearest living primate analogue of *Palaeopropithecus* appears to be the quadrumanous orangutan. The retention by *Palaeopropithecus* of a relatively primitive (for an indriid) cheek dentition suggests that its dietary regime may not have been greatly different from that of the living indriines. This diet is seasonally highly variable (see chapter 7), but is exclusively vegetarian.

The few known postcranial bones of *Archaeoindris* are in some ways reminiscent of *Megaladapis*, and pending the discovery of more complete remains it seems reasonable to follow Walker (1974) in assuming a similar locomotor type.

Subfamily Indriinae Burnett, 1828

In 1905 Standing described a new indriine genus and species, *Mesopropithecus pithecoides* (fig. 5.6a), on the basis of four crania from

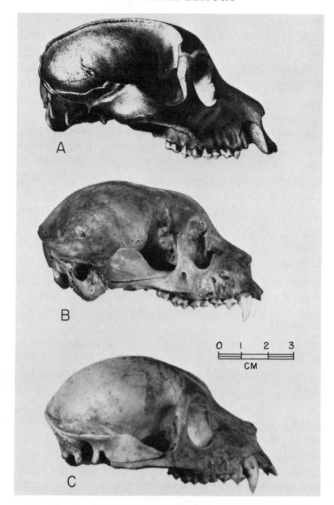

Figure 5.6. Crania in lateral view of A: *Mesopropithecus pithecoides*; B: *Mesopropithecus globiceps*; C: *Mesopropithecus "platyfrons."*

Ampasambazimba. These crania resemble quite closely those of *Propithecus*, but differ in being slightly larger and considerably more robustly built, adults possessing sagittal and nuchal crests, massive zygomatic arches, broader snouts and hence more parallel toothrows, and (as inferred from their alveolae) larger upper incisors and canines.

Thirty years later, Lamberton (1937) reported the recovery of four more *Propithecus*-like crania from Tsiravé and the sites of the lower Menarandra Valley. One of these was the cranium of *"Propithecus verreauxioides,"* to which reference has already been made, and which does not deserve distinction from *P. verreauxi.* For the others Lamberton created a new genus, *Neopropithecus,* intermediate between *Mesopropithecus* and *Propithecus,* and two new species: *N. globiceps* (fig. 5.6b) for the single individual from Tsiravé, in recognition of a marked convexity in its frontal profile, and *N. platyfrons* (fig. 5.6c) for the two crania from the lower Menarandra area. Subsequently, Lamberton (1939) redescribed this material in detail, and reported the discovery of further specimens, including a new skull of *N. globiceps* from the cave site near Ankazoabo. It should be noted that Lamberton's Ankazoabo is not the subprefectoral center of that name, but is instead a small village on the coast a few miles north and west of Itampolo. New material of *N. platyfrons* included a calvarial fragment, some mandibular pieces, and four humeri.

Carleton (1936) tentatively assigned certain postcranial elements from Ampasambazimba to *Mesopropithecus,* while Lamberton (1948) referred a further femur from this site to *M. pithecoides* and a group of femora from Tsiravé and Beloha and several tibiae from Beloha, Tsiravé and Belo to *Neopropithecus.* Walker (1967) restudied these postcranial remains and accepted Lamberton's attributions while rejecting those of Carleton.

Some years ago I reviewed the crania of the subfossil indriines (Tattersall 1971), and concluded that the morphological evidence is inadequate to justify the separation of these extinct indriines into two genera. Further, among the southern and southwestern fossils there is inadequate variation to maintain a specific distinction. I thus recognize two species of *Mesopropithecus: M. pithecoides* Standing for the more robust plateau species (mean maximum cranial length 97.9 mm, *o.r.* 94.0–103.5 mm, $n=4$), and *N. globiceps* (Lamberton) for the more gracile lowland form (mean maximum cranial length 91.7 mm, *o.r.* 83.3–94.8 mm, $n=4$). Clearly, the species of *Mesopropithecus* are closely related to *Propithecus,* which they closely resemble in both their dentitions and their cranial proportions, although the genus is extremely derived in its presumed postcranial architecture. Of the two species, *M. pithecoides* is the more apomorphic cranially, showing sagittal and nuchal cresting and broadening of the anterior part of the muzzle; *M. globiceps* remains more primitive, and more *Propithecus*-like, with its more gracile build and anteriorly-narrowing snout.

Walker (1974) has described the few postcranial elements attributed to *Mesopropithecus* as in some respects resembling those of a small *Megaladapis,* but regards them as being inadequate for the reconstruction of locomotor habit.

Subfamily Daubentoniinae Gray, 1870

Among the subfossils mentioned by Guillaume Grandidier in his monograph of 1905 were a few anterior teeth from Lamboharana, referable to a form of the aye-aye, *Daubentonia* (his *Chiromys*). Some of these were pierced for stringing. In a later contribution Grandidier (1929) noted that these teeth were much larger than those of the living aye-aye. Subsequently, Lamberton (1934) described an almost complete postcranial skeleton from Tsiravé, unhappily unaccompanied by cranial or dental remains, and two humeri from Anavoha. These bones were placed by Lamberton in a new species, *Chiromys robustus*, now known as *Daubentonia robusta*. Essentially, these bones differ from those of their living congener only in being some 30 percent larger in their linear dimensions.

Subfamily Megaladapinae Major, 1894

The genus and species *Megaladapis madagascariensis* was founded by Major (1894) to accommodate a large cranium and partial left and right dentaries (BMNH M4848) recovered by J. T. Last at Ambolisatra. In his discussion of these specimens, Major noted that "a *superficial* examination of the skull will certainly not suggest its classification among the Lemuroidea" (1894:19), but his choice of name reflects his conclusion as to its affinities, "although no *close* approach to *Adapis* is implied" (1894:27). Major dwelt at length on the morphology of *Megaladapis*, concluding that the animal could not be regarded as conserving a primitive condition, but rather that it showed the results of "retrogressive evolution"; he classified the form in its own family of the *"sub-order Lemuroidea."*

In a later paper Major described a brain cast of his specimen (1897). He was chiefly impressed by the small size of the brain, its lack of convolutions and the almost total reduction of the forebrain. This "reptilian-like conformation" confirmed him in his belief that *Megaladapis* was the product of a retrogressive evolutionary trend "for I trust that no anatomist will maintain that this was the primitive condition in Lemuroids." (1897:49).

Two years later, Guillaume Grandidier (1899), reporting on his own excavations at Ambolisatra, described and figured two femora, a partial maxilla and an isolated tooth as belonging to subfossil lemurs. The larger of the two femora was found "with" the *Megaladapis* mandible described by Major, and was ascribed by Grandidier to the same species. The second femur, smaller but similar in morphology to the first, was assigned to *Megaladapis* (?), pending more thorough study. The isolated tooth, a lower third molar, closely resembled that of *M. madagascariensis*, but was of much larger size. Grandidier named this specimen *Peloriadapis edwardsi*, and placed the new taxon "dans la même famille" as *M. madagascariensis*. Later in the same year, Grandidier (1899) described a maxillary fragment with two molars, assigning this specimen also to *P. edwardsi*. The association of the larger of the femora with the mandible of *Megaladapis madagascariensis* was spurious; this specimen should in fact have been assigned to *P. edwardsi*, and its smaller counterpart to *M. madagascariensis*.

In the next year Major (1900b) described a series of mandibular and maxillary fragments from Andrahomana (BMNH 7371–2) as representing a new species, *Megaladapis insignis*. Apart from their much greater size, these teeth very closely resembled those of *M. madagascariensis*, so much so that Major felt justified in extrapolating from the size of the teeth to the size of the skull which he predicted would prove to measure at least 330 mm in length when found.

Grandidier, meanwhile, was continuing work at Ambolisatra, and in the same year (1900) reported further finds of *Megaladapis* at this site, among them the posterior portion of a cranium. However, while Grandidier recognized the specimen as *Megaladapis*, he believed that "une différence assez sensible dans le hauteur de l'occipital empêche de la ranger parmi les *M. madagascariensis*" (p. 216), although he felt the material to be insufficient to warrant the erection of a new species.

Lorenz von Liburnau, always ready to propose new genera and species on the basis of photographs, then (Lorenz 1900a, b) described two new genera and species, *Palaeolemur destructus* and *Mesoadapis destructus*, and one new species, *Megaladapis brachycephalus*, from photographs sent him by the collector Sikora, who had recovered the material concerned at Andrahomana. The actual specimens, however, were sent to London, so that within a few months Major (1900a) was able to write that all were synonymous with his *Megaladapis insignis*, and that "*Mesoadapis*," which Lorenz had felt to be most strikingly distinct, had anyway been founded on a juvenile skull (BMNH 7369). At the same time, however, Major agreed with Grandidier's assessment of the generic distinctness of the two femora from Ambolisatra, although he did suggest their correct associations. Major also revised his earlier opinion as to the familial distinctness of *Megaladapis* from other Malagasy prosimians, but did not make unequivocally clear his new conclusions. From his discussion, however, it appears most likely that he would have regarded

Megaladapinae, Indriinae, Lemurinae and Daubentoniinae as subfamilies within the same family.

In 1902 Grandidier effected a more formal and more extensive synonymy of the genus *Megaladapis* than Major's of two years earlier. Grandidier synonymized *Peloriadapis* with *Megaladapis,* and included in the species *M. edwardsi* the expanded *M. insignis,* together with *M. dubius,* a species provisionally erected by Lorenz (1900b) to contain an ulna from Andrahomana.

Lorenz, who by then had accumulated several excellent specimens of *M. edwardsi* from Andrahomana and had accepted Grandidier's synonymy, published in 1905 a monograph on the species, in which he claimed to be able to distinguish two distinct types among the five crania of this species in his collection. These he named *M. edwardsi brachycephalus* and *M. edwardsi dolichocephalus.* He also attempted the first restoration of the full skeleton of *Megaladapis,* which he believed indicated that the animal was arboreal, though perhaps with "habits analgous to those of the cave-bear." Grandidier (1905) disagreed with Lorenz's subspecific division; in his view the differences between the two groups were attributable to sexual dimorphism.

Although as early as 1903 Standing had named a new species of *Megaladapis, M. grandidieri,* from Ampasambazimba, a full description and documentation of the species awaited the publication of his monograph five years later (Standing 1908a). In the same work, Standing described a massive femur under the name of *Megalindris gallienii,* and assigned the limb-bones of *Megaladapis* from Ampasambazimba to *Palaeopropithecus,* and vice-versa. From a study of the postcrania, Standing concluded that *Megaladapis* possessed an armswinging mode of locomotion, a conclusion in a sense correct, because *Palaeopropithecus,* to which the bones in question belonged, did in fact move very much in this way. On the other hand, faced with the problem of explaining why the skulls of *Palaepropithecus* from Ampasambazimba appeared to have been bitten by crocodiles, Standing discovered aquatic features in the postcranial skeleton of *Megaladapis* under the impression that the bones were those of *Palaeopropithecus.*

Standing followed Major in classifying Megadadapinae as a subfamily of Lemuridae (in which he also placed Galaginae and Lorisinae), as also did W. K. Gregory in 1915. Standing did remark, however, that *Megaladapis* was so close to Lemurinae as barely to deserve distinction. Anthony and Coupin (1931) took this view a step further and actually included *Megaladapis* in Lemurinae.

Lamberton's major work on *Megaladapis* appeared in 1934. In this work, Lamberton raised again the question of cranial variation in *Megaladapis edwardsi*, and concluded that three geographical races could be recognized in the species. He also suggested on the basis of the scapular morphology of *M. grandidieri* that *Megaladapis* might have been a powerful climber, a conclusion that he abandoned in later publications. Subsequently, Lamberton contributed numerous notes detailing various aspects of the cranial and postcranial morphology of *Megaladapis*. Principal among these were his study of the ear region (1941), which revealed the close similarity in the structure of the bony ear between *Megaladapis* and the other very large subfossil lemurs, *Palaeopropithecus* and *Archaeoindris*, to which it bears no other obvious resemblances; his analysis of the milk dentition (1938b); and his investigation of the relationship between hypophyseal development and body size (1952).

The first study of the brain of *Megaladapis* since Smith's of 1908 was produced by Hofer in 1953; this work also constituted the first detailed description of the internal anatomy of the skull of *M. edwardsi*, and the first approach towards a comparative and functional analysis of its form.

Saban (1963) briefly, but at greater length than had been attempted earlier, described the morphology of the temporal bone of *M. edwardsi*, and found that in specialized characters this element resembled only those of *Palaeopropithecus* and *Archaeoindris*. This conclusion was as troubling to Saban as it had been years earlier to Lamberton, since in other characters, in particular the dentition, *Megaladapis* appeared totally disparate from these two closely related genera. Also in 1963, Zapfe produced a flesh reconstruction of *Megaladapis*, in the course of which he pointed out that the large grasping extremities of the animal constitute unequivocal evidence of arboreal habits; this was the first such suggestion to have been made in almost thirty years. Precisely what form of locomotion *Megaladapis* possessed, however, remained obscure until Walker's study of 1967.

Three species of *Megaladapis*, then, are currently recognized (fig. 5.7). The largest, *M. edwardsi* (G. Grandidier) (mean maximum cranial length 296.0 mm, *o.r.* 277.0–317.0, $n = 10$), is found in the sites of the south and southwest, and is characterized by a large dentition (mean mesiodistal length of M^1: 18.8 mm, *o.r.* 16.0–21.3 mm, $n = 14$). Also in the south and southwest occurs the smallest, *M. madagascariensis* Major (mean maximum cranial length 240.5 mm, *o.r.* 235.0–244.0 mm, $n = 3$), with an expectedly smaller dentition (mean mesiodistal length of M^1: 14.0 mm, *o.r.* 13.2–15.7 mm, $n = 5$). Only known from the high plateau is *M. grandidieri* Standing, whose cranial length (mean maximum 288.5 mm, *o.r.* 273.0–300.0, $n = 3$) almost matches that of *M. edwardsi*, but whose dentition (mean mesiodistal length of M^1: 15.4 mm, *o.r.* 14.4–16.0 mm, $n = 9$) is closer in size to that of *M. madagascariensis*. Apart from these differences in craniodental proportions, the adaptive significance of which is still obscure, the three species of *Megaladapis* are sufficiently similar in skeletal morphology to be described together, at least at the general level employed here.

The cranium of *Megaladapis* is narrow and greatly elongated. Robustly built, it invariably possesses strong nuchal and sagittal crests. The orbits are quite divergent, and are heavily encircled by bone. The facial region is extremely long, and its axis sharply retroflexed on the plane of the cranial base; it is this flexion, termed "airorhynchy" by Hofer (1953), which produces the apparent elevation of the neurocranium relative to the facial skeleton that has been noted by several authors. The nasal bones are extremely long, projecting considerably beyond the anterior margin of the palate, and are downturned at their tips, forming an overhang above the nasal aperture (fig. 5.7). Together with the curious texture of the bone in this region and on the frontal, this may hint at the presence in life of a mobile snout.

The zygomatic arches are massive and of substantial vertical extent; the temporal fossa is capacious, with a pronounced postorbital constriction. The occipital plane is almost perpendicular to that of the cranial base, and the round foramen magnum opens almost directly to the rear. The elongated occipital condyles are almost vertical; their articular surfaces are expanded inferiorly, where they face laterally. The nuchal region is divided in the midline by a high, strong ridge, which runs from the nuchal crest to the foramen magnum and is bordered on either side by rugose shallow depressions, indicating the presence of a highly developed neck musculature. The paroccipital processes are elongated rearward into strong, projecting, gutterlike protuberances, suggesting strong digastric development as does the large attachment area on the mandible for the anterior belly of that muscle.

The braincase is remarkably small for a primate of such size, and is restricted to the posterior portion of the postfacial skeleton. Anterior to this tiny neurocranium proper are enormous frontal sinuses which extend forward to overlap the orbits. These great chambers are best interpreted as space-fillers: the muscular system required to operate the long jaws of *Megaladapis* simply could not

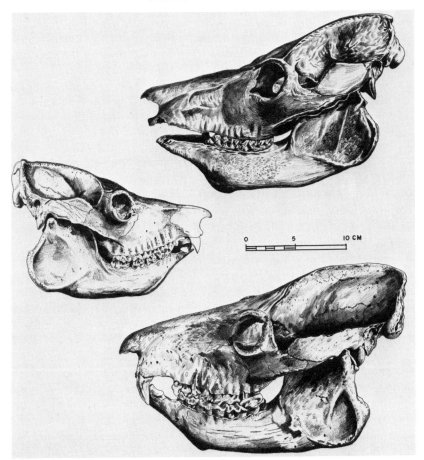

Figure 5.7. Crania in lateral view of the species of *Megaladapis*. Above right: *M. grandidieri*; left: *M. madagascariensis*; below right: *M. edwardsi*.

have functioned if the braincase were positioned immediately behind the face. Roberts and I (Roberts and Tattersall 1974) have described a model of mandibular operation in mammals in which the more anterior the bite-point, the more effort is required of the posterior (horizontal) fibers of temporalis; conversely, the more posterior the bite-point, the more emphasis is placed on the anterior (vertical) fibers of this muscle. The two ends of this adaptive spectrum are admirably illustrated among the lemurs by the short-faced, high-skulled *Hadropithecus*, and the long, low cranium of *Megaladapis*. In the latter, the anterior portion of temporalis was relatively small, while the posterior part of this muscle was hypertrophied, as shown in the development of pronounced nuchal and (posterior) sagittal cresting. The small size of the brain, coupled with the need to shift temporalis backward, necessitated considerable spacing between the braincase and the facial skeleton. The development of truly enormous frontal sinuses fulfilled this need. The failure of the brain itself to intrude into the immediate postorbital area is reflected in the lack noted by Radinsky (1970) of orbital impressions in the endocast, and in the unusual length of the olfactory tracts.

The attenuation, for mechanical reasons, of the upper post-facial skeleton of *Megaladapis* also impinges on the structure of the cranial base, which exhibits a similar drawing out. But perhaps the most atypical feature of the cranial base lies in the loss of an inflated auditory bulla, and the consequent rearrangement of the foramina of the region. Although the ventral part of the bony ear of *Megaladapis* does protrude (especially anterolaterally) more than does its homologue in the palaeopropithecines (this protrusion being somewhat masked by the development of a large postglenoid complex), there is no bullar inflation of the kind typical of the living lemurs. As I have noted in discussing the palaeopropithecines, this may well in effect represent an allometric reduction. Again as in the palaeopropithecines, the tympanic ring lies at the end of a long ossified tubular external auditory meatus, probably of ectotympanic formation.

The auditory area is confluent anteriorly with a robust and salient postglenoid process. This process bears a clearly defined articular surface on its anterior aspect, which is matched by a corresponding surface descending from the medial part of the mandibular condyle. The lower jaw of *Megaladapis* is large and robust, but the corpus is not notably wide relative to its depth. The large angle is not hooked, as it tends to be in other lemurids; but neither is it posteroventrally protruded, as it is in the indriines. The fused symphysis has a long, sloping planum alveolare, terminating in twin genial pits separated by a salient crest; there is no inferior transverse torus.

The dental formula of *Megaladapis* is identical with that of *Lepilemur*: no upper incisors are present in the adult. Instead, the presence of bilateral bony ridges in the upper incisal region suggests the replacement in life of those incisors by a horny pad such as is characteristic of some ruminants. The upper canine is sharp and laterally compressed, and is separated from the premolars by a long diastema. These latter teeth, broadening posteriorly, most closely resemble those of *Lepilemur*, as do the upper molars, which in both genera display buccal cingula and parastyles, lingual buttressing of the para- and metacones, and distal displacement of their lingual moieties. However, the upper molars of *Megaladapis* all exhibit mesostyles, and progressive enlargement towards the rear (fig. 5.8).

Megaladapis possesses a dental comb in which the incisors and canines are shorter and stubbier than are those of its living relatives. The lower anterior premolar is strongly caniniform and closely approximated to the toothcomb; it is separated from the simple posterior premolars by a substantial diastema. The lower molars very distinctly increase in size from front to back (fig. 5.8). In both M_1 and M_2 a small metastylid is closely approximated to the metaconid; this latter is isolated from the entoconid, leaving the talonid basin open lingually; and on all lower molars a distinct crest runs forward from the protoconid. The posterior molar is extraordinarily elongated, recalling the condition found in certain ungulates, with a large hypoconulid at the rear of a well-developed talonid heel.

Walker (1967a, 1974) has studied the postcranial skeleton of *Megaladapis,* and has concluded that the closest living locomotor an-

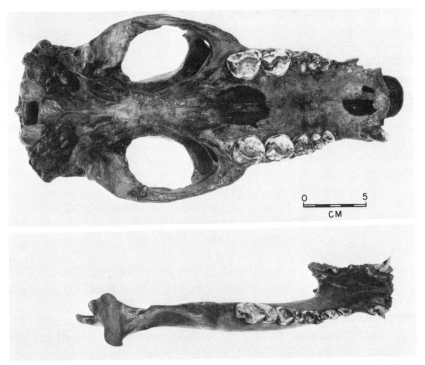

Figure 5.8.   Ventral view of cranium and dorsal view of mandible of *Megaladapis edwardsi.*

alogue to this animal is not a primate, but is instead the koala, *Phascolarctos*, a phalangerid marsupial. This conclusion has been reinforced by the recent study by Jungers (1977).

The long, somewhat curved extremities of *Megaladapis* are clearly powerful grasping organs, and in themselves are evidence of an arboreal habit for *Megaladapis* (Zapfe 1963). The limb proportions quoted by Walker and Jungers vary from 121.4 for *M. edwardsi* to 114.0 for *M. madagascariensis*, i.e., in all species the forelimbs are somewhat longer than the hindlimbs. As Jungers (1977) has pointed out, these proportions reflect a reduction in hindlimb length rather than lengthening of the forelimb. The intermembral indices of the species of *Megaladapis* are higher than those of the koala, but Walker (1974) believes that in view of the reduction of the intermembral index with body size in both species of the large *Megaladapis* and the much smaller koala, a *Megaladapis* of koala size would have koalalike limb proportions.

The pelves of *Megaladapis* and the koala share long ilia, with pronounced beaking of the anterior inferior iliac spines; and short, stout ischia and pubic symphyses. The femur in both cases prossesses a short neck, globular head, strong development of the greater trochanter, pronounced condylar retroflexion and other common features, while the tibiae are likewise closely comparable in the morphology of the condylar facets and in possessing short but strong malleoli (Walker 1967a, 1974). The hindlimb of *Megaladapis*, like that of the koala, was clearly adapted to habitual strong flexion; and Jungers (1977) has explained the unusual anteroposterior flattening in *Megaladapis* of the femoral and tibial diaphyses, and their pronounced cortical thickness, as responses to strong transverse bending forces.

In the forelimb, the head of the humerus of *Megaladapis* faces backward, and this bone also shows an expanded brachialis flange together with a distinct and rounded capitulum. The olecranon process of the ulna is long, and the sigmoid notch restricted; the coronoid and styloid processes are long. The scapula is wide, with a well-developed spine and acromial and coracoid processes (Walker 1976a, 1974). In all these features the forelimb of *Megaladapis*, adapted for powerful flexion, particularly at the elbow, is matched by that of the koala.

The locomotion of *Phascolarctos* is a modified version of vertical clinging and leaping (Walker 1974), in which the habitual resting posture is similar to that of the vertical clinging lemurs, with the ascending tree trunk grasped by the feet and the hands wrapped around the support. The adaptations to strong flexion at the knee and hip joints (e.g., rounded femoral head with the articular surface extended on to the neck; backward-facing femoral condyles) in *Megaladapis* suggest that such a resting posture was also typical of this primate: a posture which would have been facilitated by the varus orientation of the femoral bicondylar angle and the raising of the lateral condylar facet of the tibia, which would have produced an advantageous inversion at the knee during vertical clinging (Jungers 1977).

The koala progresses up vertical supports in a series of short hops, and also leaps short distances between adjacent vertical supports. In proposing that the locomotion of *Megaladapis* was similar to that of the koala, Walker (1967a, 1974) also noted that the powers of leaping were probably yet further reduced in the Malagasy form. In a large "vertical clinger and climber" this reduction would seem to produce a useful tradeoff: the shortening of the hindlimb, producing a relatively long (and powerful) forelimb, is particularly advantageous in a large-bodied animal moving on vertical substrates (Cartmill 1974; Jungers 1977), and this advantage appears to be reflected in the increase of the intermembral index with size in *Megaladapis*.

The koala analogy receives additional support from study of the cranial adaptations of *Megaladapis*. In his initial description of *M. madagascariensis,* Major (1894) noted numerous similarities between the skull of this animal and that of *Phascolarctos*. I have noted (Tattersall 1972) that two of these common features, i.e., the retroflexion of the facial skeleton on the plane of the cranial base, and the directly caudal orientation of the foramen magnum combined with the vertical positioning of the occipital condyles, appear to form part of a functional complex of particular significance. In both of these attributes *Megaladapis* departs markedly from the condition characteristic of the vertical clingers surviving in Madagascar today, and even from *Palaeopropithecus,* whose backward-opening foramen magnum is combined with primarily downward-facing occipital condyles.

If *Megaladapis* and *Phascolarctos* shared the same feeding habits, however, the otherwise inexplicable cranial morphology of *Megaladapis* falls into adaptive perspective. The koala feeds (as do many lemurs), and *Megaladapis* presumably fed, by cropping leaves from branches pulled by the hands within reach of the mouth. The combination of facial retroflexion with occipital condyles in line with the vertebral column in effect turns the head into a functional extension of the neck, maximizing the range within which cropping of leaves (or other foodstuffs) is possible from a single clinging position. The great elongation of the face of *Megaladapis,* which superficially appears maladaptive in a vertically clinging animal, may be explained as a further addition to this functional food-obtaining complex: a suite of adaptations advantageous for the relatively unagile koala, and presumably even more so for the vastly bulkier *Megaladapis.*

Other features of the skull also suggest a browsing habit, probably folivory, for *Megaladapis* (Tattersall 1972). The probability that *Megaladapis* cropped leaves from branches is reinforced by the loss in the adult of the upper incisors and their presumed replacement by a horny pad, as in bovids, and possibly also by the diastema between the canine (or caniniform) and the premolar posterior to it, whose function was presumably analagous to that found in browsing animals which use their tongues to transfer food from the anterior part of the mouth to the cheek dentition. If the highly vascularized nature of the bone covering the nasal region and the downward flexion of the tips of the nasals does indeed imply the possession in life of a mobile snout, this structure would also have formed part of the same complex.

The Malagasy primates, then, have solved the problem of combining arboreality with great body size in two ways (Walker 1974). In the suspensory locomotion of *Palaeopropithecus* a way was found which has also been adopted by other primate groups; but in the case of *Megaladapis* the solution is unique within Primates.

Subfamily Lemurinae Gray, 1821

Among the lemur subfossils described in 1895 by Filhol was a humerus, found at Belo, to which was given the name *Lemur insignis.* At the same time, Filhol described a mandible and a smaller humerus from the same site as *L. intermedius.* Subsequently, G. Grandidier (1899) gave the name *Palaeochirogalus jullyi* to two isolated molars from Antsirabé, and in 1900 Lorenz described a photograph of a partial immature cranium from Andrahomana which he thought might belong to this species. In his review of 1905, G. Grandidier, with more ample material to hand, assigned all these forms to *L. insignis.*

By 1908, the ever-faithful site of Ampasambazimba had yielded 11 crania which Standing (1908a) concluded were "evidently closely allied . . . to *L. insignis*" (1908a:119). Nine of these Standing assigned to the species *L. jullyi,* which he had created in 1904 to contain the posterior portion of a cranium from Ampasamabazimba; and on one small individual he based a new species, *L. majori,* while at the same time sinking his earlier (1904) species *L. maziensis,* founded on an incomplete cranium, into *L. jullyi.*

Except for a couple of comparative notes by Lamberton, and an analysis of the postcranial skeleton by Carleton (1936), study of these forms languished until 1948, when Lamberton reviewed the fossil material, by then abundant, and proposed that genus *Lemur* should be divided into two subgenera: *Lemur (Lemur)* for the living species, and *Lemur (Pachylemur)* for the extinct ones. I am inclined to believe that morphologically the subgeneric separation of the subfossil species from its closest living relative, *Varecia variegata,* is justifiable, but that in effect it serves primarily to complicate the nomenclature of the group. I have thus not retained it here.

Whether two species may in fact be recognized in the assemblages of material from the southern and southwestern sites and from Ampasambazimba is open to question. The lowland form is slightly smaller and more gracile (maximum cranial length according to G. Grandidier [1905]: ca. 115 mm), while that of the plateau is larger (Standing's [1908a] figures give a mean maximum cranial length of 121.9 mm, *o.r.* 114–5–126.0 mm, *n* = 11) and slightly more heavily built; but whether these small differences are specifically significant is unclear. Provisionally, however, it seems reasonable to maintain the distinction, although this does pose a problem of nomenclature. Standing (1904) described *L. jullyi* as a new species, but the specific epithet became preoccupied by G. Grandidier's *Palaeochirogalus jullyi* when that author later synonymized the latter with *L. insignis.* However, since *P. jullyi,* although based on nonspecifically diagnostic material, came from the plateau site of Antsirabé, it is at least a good guess that it is conspecific with *L. jullyi* from Ampasambazimba, in which case the problem fades, authorship of the species *jullyi* going to G. Grandidier.

Both cranially and postcranially, subfossil *Varecia* quite closely resembles its extant relative. Cranially, the fossil form is distinguished most clearly from the living one by its larger size, its more robust build, including possession by almost all individuals of sagittal and nuchal cresting (fig. 5.9), and by the more forward orientation of its orbits. Postcranially, Walker (1967a, 1974) has found greater robusticity and a slightly reduced intermembral index (98.2), the limbs being thus more nearly equal in length than those of *V. variegata.* This implies some de-emphasis of leaping, and might be

Figure 5.9. Lateral view of skull of *Varecia insignis.*

indicative of a greater degree of terrestrialism, although Walker (1974) notes that there are few other features of the skeleton which support this conclusion.

Subfamily Hapalemurinae

The new species *Hapalemur gallieni* was created by Standing (1905) to receive a pair of right dentaries from Ampasambazimba which differed from *H. griseus* in size and in a few minor characters (not specified) of the teeth. Standing's view did not change with the later discovery of a cranial fragment, but with somewhat more material at his disposal, Lamberton (1939) thought otherwise, and proposed the new genus *Prohapalemur* to accommodate the species. Among the distinguishing characters noted by Lamberton was the presence of but a single upper incisal alveolus, and molarization of the middle as well as the posterior upper premolar.

None of Lamberton's material was undamaged in the upper incisal region, and since I have been unable to examine it myself I am unable to confirm Lamberton's proposed dental formula. One might note, however, that the occasional absence of an upper incisor would not be an unexpected anomaly, particularly in a genus whose upper incisors are anyway reduced. Ampasambazimba *Hapalemur* is only marginally larger than *H. simus*, if at all, and appears in comparable features to resemble it quite closely: in particular, the two share lingual enlargement of both posterior premolars and large hypocones on the upper molars, both derived character states. On the other hand, the fossil form retains in known specimens the primitively large hypoconulid on M_3 which also characterizes *griseus*, but not *simus*. Sample sizes of both the fossil form and *simus* are, however, very small, and it is quite possible that the two are in fact conspecific, although pending better knowledge it seems appropriate provisionally to retain the specific distinction. Generic separation, however, at least if *simus* is to be retained within *Hapalemur*, is unwarranted. The genus is an arbitrary category, and among the lemurs there are large differences in the variation in both size and morphology found between species allocated to the same genus. In contradiction of an earlier opinion (Tattersall and Schwartz 1974), I now believe it to be most useful to include the closely related *griseus* and *simus* within the same genus despite their differences; and if this is done for the living forms, then must it also be done for *H. gallieni*.

EXTINCTION OF THE SUBFOSSIL LEMURS

From early in their study it has been apparent that the subfossil lemur remains are of quite recent date, although the notion that they are of "Pleistocene" age has survived, mainly in secondary accounts, to the present. Standing, however, was under no such illusion, and estimated the "age of the uppermost of the lemuroid remains . . . [at Ampasambazimba] as probably not more than five centuries" (1908a:71), so that "One may at any rate from a biological point of view regard all these subfossil Malagasy Lemuroids as the contemporaries of extant species in other parts of the island" (1908a:71).

More recently, radiocarbon dates (fig. 5.1), determined mostly on bone, chelonian carapace, or wood samples, have become available for some of the sites (Mahé 1965; Mahé and Sourdat 1972; Tattersall 1973a). These range from 2850 ± 200 yr. B.P. (Amparihingidro) to 980 ± 200 yr. B.P. (Itampolo). I have argued elsewhere (Tattersall 1973a) that the Itampolo determination (one of two; the other, of 2290 ± 90 yr. B.P., was made on a sample recovered from much lower down in the section) probably postdates substantially the last occurrence of fossil lemurs at this site; but it is closely matched by the date of 1035 ± 50 yr. B.P. obtained by Dr. J.C. Vogel on a tibial diaphysis of *Megaladapis* from Ampasambazimba (Tattersall 1973a).

Although it remains unknown exactly when man first arrived in Madagascar, the indications are that this event took place within the past two millennia, and certainly by about 1500 B.P. (Battistini and Vérin 1972); the younger radiocarbon dates thus establish that the subfossil fauna survived intact into the human period. Other evidence of the contemporaneity of man with the extinct lemurs is available from a variety of sites, either in the form of artifacts intermixed with the lemur remains (e.g., Ampasambazimba, Taolambiby), or of fossils cut, burned, modified or dispatched by human agency (e.g., Andrahomana, Lamboharana).

Interestingly, the condition of the subfossils from the various sites differs considerably. Bone from the sites of the south and southwest often gives the impression of being almost fresh, while that from Ampasambazimba, for instance, is quite heavily permineralized. These differences are, however, attributable to the conditions of preservation, rather than to age; complete diagenetic alteration of bone can take place extremely rapidly (perhaps in from one to ten years) in a wet and alkaline environment where drainage is poor (Berner 1968), as in the sites of the volcanic highlands. Moreover, the encasement of some fossils from Ampasambazimba (e.g., the cranium of *Archaeoindris*) in lava does not suggest, as Jully (1899) proposed, that a volcanic catastrophe overwhelmed the local fauna: the lava was undoubtedly cemented to the fossils post-mortem, after deposition and during the process of fossilization.

Apart from spectacular explanations such as Jully's of extinction events at particular sites, theories which have been advanced to explain the disappearance of much of the lemur fauna within the recent past have fallen into two categories: climatic change, and the intervention of man.

The climatic change model has emerged from studies of the subfossil sites of the dry south and southwest of Madagascar. Decary (1930), noting that many of these subfossil-bearing deposits represent the remnants of former lakes, proposed that an aridification of the climate caused the large lemurs to congregate around the ever-dwindling water sources of the region, in which their remains were eventually preserved. This theme has recently been taken up in modified form by Mahé and Sourdat (1972), who believe that man merely finished off the remnants of a southwestern lemur fauna already declining rapidly as a result of climatic drying. It seems more likely, however, that the drying up of the southwestern sites was due merely to changes in local drainage patterns: changes perhaps ultimately due to the deforestation of the highlands, with the consequent increase in the runoff of water from the interior. There appears to be no satisfactory independent evidence of any major climatic change in Madagascar within the past few thousand years.

Walker (1967b) and Battistini and Vérin (1967) have more plausibly argued for the role of man as the exclusive agent of lemur extinction in Madagascar: either directly, through hunting, or indirectly, through the destruction of the forest habitat. Clearly, it is the disappearance of the forest which will ultimately signal the extinction of the entire lemur fauna and which has already resulted in its extirpation throughout vast areas of the island. But although moderately extensive areas of many of the aboriginal forest types still exist today, they are populated by only a partial primate fauna. The genera now missing were all large, slow-moving or terrestrial, and diurnal (Walker 1967b): precisely those most vulnerable to direct predation by man. It is the smaller, quicker, more agile forms which have survived. For the present.

CHAPTER SIX

Phylogeny and Classification

Our understanding of the evolutionary relationships of the Malagasy primates has been bedeviled by their geographical uniqueness at least since the second decade of this century. The extant strepsirhines clearly form a monophyletic group; and this, taken together with the common occurrence of most of them on Madagascar, has led to the conventional view that the lemurs represent the endpoints of an adaptive radiation stemming from a single ancestral form which was by some means isolated on Madagascar at an early date. Furthermore, consideration of the interrelationships between the various lemur taxa has been handicapped by the total lack of an appropriate fossil record not only in Madagascar itself, but also in pre–Oligocene Africa, whence the precursors of the recent lemurs must have been derived. Conceptions of lemur interrelationships have thus tended until recently to be embodied in the dark viewing glass of classification, rather than in explicit phylogenetic hypotheses.

Early classifiers tended to give equal rank to the indriids, lemurids (in the broadest sense), and various mainland strepsirhine groups, while reserving a separate major category for the aye-aye; they showed the most uncertainty in the placement of the cheirogaleids. Thus, for instance, Mivart (1864) proposed a tripartite division of his suborder Lemuroidea into the families Cheiromyidae (the aye-aye), Tarsiidae, and Lemuridae. Within the last of these families, Mivart recognized four subfamilies: Indrisinae, Lemurinae (including the mouse lemur and, with a query, the dwarf lemur), Nycticebinae (= Lorisinae), and Galaginae. Thirty years later, Forbes (1894) retained the same family groupings, and also recognized four subfamilies within Lemuridae. The complements of Indrisinae and Lorisinae remained as they had been under Mivart's classification (and remain today), but Forbes transferred the mouse and dwarf lemurs from Lemurinae to Galaginae. Standing (1908a) was the first to produce a classification which reflected the proper affinities of the aye-aye. This author recognized three families of "lemurs": Indrisidae, Lemuridae, and Tarsiidae. His Indrisidae included the aye-aye (subfamily Chiromyinae) as well as the indriines. His Lemuridae contained the subfamilies Lemurinae, Galaginae, and Lorisinae, and within the galagines he included *Cheirogaleus* and *Microcebus*.

Students around the turn of the century thus did not feel themselves inhibited by zoogeographical considerations, and it was not until 1915 that Gregory proposed a classification of essentially "modern" aspect, in which the Malagasy primate fauna as a whole was separated from other strepsirhines. Gregory's "series" Lemuriformes contained three extant families: Lemuridae (embracing Cheirogaleinae and Lemurinae), Indrisidae, and Daubentoniidae. This has remained essentially the standard classification of the living lemurs ever since; it was, for example, largely adopted by Simpson (1945), who nonetheless mysteriously failed to follow Gregory's excellent classification of the subfossil lemurs. Simpson's only departure from Gregory's arrangement of the living forms was to raise the aye-aye to superfamilial status: a rather remarkable move in a classification which included Plesiadapidae as a mere family within the superfamily Lemuroidea.

In his 1915 contribution, Gregory surmised that while *Adapis* itself represented a specialized side branch among the adapids, the late Eocene *Pronycticebus* possessed the craniodental morphology appropriate in an ancestor of both the Malagasy lemurs and the lorisiforms. Several decades later, Simons (1962) suggested that *Pronycticebus* might bear some special relationship to the lorises alone. Apart from this, most attempts to assess the relationships of the strepsirhines on the basis of fossil evidence have amounted to pointing out certain resemblances between one or more adapids and one of the extent lemurs, normally *Lepilemur* (e.g., Gregory 1920) or *Hapalemur* (e.g., Gingerich 1975), the living form in question thus automatically becoming the most "primitive" of the extant lemurs.

In recent years, however, more emphasis has been laid upon comparative studies of the living forms in the attempt to understand evolutionary relationships among the lemurs. Most such investigations have focused upon the place of the cheirogaleids within Strepsirhini, following the initiative of Charles-Dominique and Martin (1970), who drew attention to similarities between the lorisids and cheirogaleids, and in particular to the behavioral resemblances between the mouse lemurs and Demidoff's galago. The purpose of this contribution, however, was not to establish the closeness of the relationship between lorises and cheirogaleids but rather to emphasize the retention in both groups of "ancestral primate" characteristics. Charles-Dominique and Martin's provisional phylogeny shows a "lemur-loris stock" giving rise separately to a "galagine stock," ancestral to mainland strepsirhines, and to a "cheirogaleine stock" from which independently sprung the indriids, lemurids, and cheirogaleids. It was thus left to Szalay and Katz (1973) to propose that the similarities shared between lorises and cheirogaleids are derived, not primitive, and that this indicates a closer relationship between these two groups than the latter possesses with the other Malagasy forms. Cartmill (1975) subsequently undertook a detailed investigation of carotid circulation, bullar formation, and medial orbital wall structure in the strepsirhines, and concluded that the morphological evidence pointed to the derivation of the lorisiforms from "a cheirogaleid near *Allocebus*." He felt it premature, however, to transfer Cheirogaleidae to Lorisiformes.

Schwartz and I felt less hesitation when we took this taxonomic step on the basis of a comparative investigation of craniodental morphology among the Malagasy strepsirhines, living and subfossil (Tattersall and Schwartz 1974, 1975). A more vexing problem, we found, lay in the determination of the relationships of the indriids with the other strepsirhines; provisionally, we placed the indriids, their subfossil relatives, and *Daubentonia* together as the sister group of the remainder of Strepsirhini, albeit with some misgivings. We have since found reason to conclude otherwise, and it is with our reappraisal that the next section of this chapter will be concerned.

In parallel with morphological investigation of this kind a certain amount of work has been done on the molecular systematics of the strepsirhines. The scope of such studies has been limited, however, at least partly because of the difficulty of obtaining the necessary samples from many Malagasy species. Goodman and his colleagues (e.g., Dene et al. 1976) have extended their investigations of immunodiffusion reactions in serum proteins of primates to include *Cheirogaleus, Microcebus, Propithecus, Lepilemur,* and several species of *Lemur* (*Daubentonia* was also tested, but with an inadequate sample). In this type of study, antigenic distances derived from immunodiffusion plates are converted into branching diagrams which are believed to represent sequences of phylogenetic events. Results imply that the Malagasy strepsirhine fauna as a whole is monophyletic, the mainland strephsirhines being equally related to all of the Malagasy forms. *Propithecus* falls closer to *Lemur* than does either to the cheirogaleids, while *Daubentonia* is the sister of the *Lemur-Propithecus* group.

In their major conclusion, these results are at variance with those derived by Cronin and Sarich (1978) on the basis of immunological comparisons of serum albumins. The work of these authors on lemur albumins has unfortunately so far been published in little more than summary. On its basis they divide Strepsirhini into four approximately coequal lineages which they say originated at around the same point in time: the lorisiforms, cheirogaleids, lemuroids (including indriids), and *Daubentonia*. Cronin and Sarich claim to find no special relationship between the cheirogaleids and the galagos (although their phylogenetic diagram identifies a taxon "Cheirogalagidae" [fig. 1, p. 288]), and find that *Hapalemur* is more closely related to *Lemur* than is *Varecia*. *Lepilemur* is placed as the sister of the *Lemur/Varecia/Hapalemur*/indriid group (Sarich and Cronin 1976).

It is difficult to evaluate the varying conclusions of these two studies both against each other and against the alternative conclusions derived from the consideration of morphology. One may, however, express certain reservations about the ways in which molecular change is measured in studies of this kind. For instance, albumins are highly complex molecules of which only a small part is immunologically active; the relationship of immunological distances to differences in the primary structure of these molecules in different species is far from clear. This is an important consideration, since the argument that immunological similarlity is in itself indicative of phylogenetic proximity is certainly not invariably reliable. Thus while the insulins of two (we hope) distantly related mammals, man and pig, are antigenically quite similar, there is often little cross-reactivity between different primate species in gonadotropin antisera. Additionally, as Uzzell and Pilbeam (1971) have pointed out, molecular comparisons are by their nature phenetic, and such comparisons may disguise the amounts of molecular (or other) change will have actually occurred in the course of evolution.

Another potential alternative to morphological analysis in the unraveling of evolutionary relationships is provided by karyology, an approach particularly favored in lemur studies by Rumpler (e.g., 1975; Petter et al. 1977). But although chromosomal information is indeed of great potential systematic importance (see the discussion by Hamilton and Buettner-Janusch [1977]), particularly at low taxonomic levels, it is at present inadequate to support general schemes of relationship. Of all available approaches then, comparative morphology remains for the moment our best avenue toward understanding of the evolutionary relationships among the lemurs, and it is on such evidence that the provisional theories of relationship proposed here are based.

EVOLUTIONARY RELATIONSHIPS

Some years ago Schwartz and I undertook a comparative analysis of craniodental morphology in the lemurs (Tattersall and Schwartz 1974, 1975). In reconstructing the evolutionary relationships of the lemurs on this basis we adhered to the principles of phylogenetic systematics, also known as the "cladistic" approach. The essence of cladism lies in the recognition of nested sets of organisms based on the common possession of evolutionary novelties, or "derived" character states. While acknowledging that the retention of "primitive" character states—those inherited from a remote ancestor—may contribute greatly to overall similarity between organisms, cladists reject the use of such states in forming sets. The hierarchical arrangement of taxa resulting from clustering by synapomorphies (shared derived states) is embodied in a branching diagram known as a "cladogram," which can be regarded as expressing the hypothesized recency of common ancestry amongst the taxa represented. It seems superfluous to discuss here the methods and assumptions of cladism, extensive exposition of which may be found in a variety of publications, e.g., Eldredge and Tattersall (1975), Eldredge and Cracraft (1980); these remarks are intended merely as preamble to the following summary of our recent reappraisal (Schwartz and Tattersall in prep.), using the same methodology, of the (primarily) craniodental evidence for lemur interrelationships.

Major Groups

Most recent attempts to understand the relationships of the cheirogaleids with other lemurs have focused on the structure of the auditory bulla and the condition of the carotid arterial system (see discussion in chapter 4). The possession of an ascending pharyngeal artery, entering the cranium in front of the bulla to join the cerebral arterial circle, is most convincingly interpreted as a synapomorphy linking the cheirogaleids with the lorisids (lorisines plus galagines), while the common possession of an intrabullar ectotympanic by the cheirogaleids and other lemurs is most plausibly viewed as a primitive retention. In view of the phylogenetic relationship that this implies, the results of certain recent investigations of soft tissue characteristics may be of interest. For instance, Reng (1977) has shown that the placenta of *Microcebus murinus* resembles that of *Galago demidovii* in being of mixed structure, with a syndesmochorial placental nucleus surrounded by a ring placenta of epitheliochorial type. Similarly, Straus (1978) has pointed to the fact that in its formation of an orthomesometrial "nidation plaque," and in the subsequent central ovoimplantation, the mouse lemur shows a clear similarity to *Loris*. In another area, Zilles (Zilles et al. 1970; Zilles and Schleicher 1980) found a remarkably close correspondence between the mouse lemur and Demidoff's galago in the areal patterns and cytoarchitectonics of the neural cortex.

In an earlier study, Schwartz and I (Tattersall and Schwartz 1974, 1975) tentatively expressed what might best be called a "feeling" that the "*Indri*-group" (the indriines, their subfossil relatives, and *Daubentonia*), formed the sister group of the rest of Strepsirhini. We later abandoned this view, and nested the *Indri*-group within a lemur clade forming the sister of Cheirogaleidae + Lorisidae (Schwartz et al. 1978). I retain this overall arrangement here, although we now reject the reasoning on which it was originally based and we nest the *Indri*-group differently (Schwartz and Tattersall in prep.; see also below).

Although the adapids have for many years been regarded by most authorities as "archaic lemurs," and as early as 1920 Gregory was emphasizing the close dental similarities between adapids and *Lepilemur*, few attempts have been made to pinpoint more specific relationships between adapids and any living lemur. After all, the known representatives of the two groups are found in different hemispheres and are separated in time by fifty-odd million years. Even Gingerich, with his proclivity for identifying ancestors, stated that "The molars of *A[dapis] parisiensis* and *H[apalemur] griseus* are so similar that if they were found together in the same fossil bed they would undoubtedly be placed together in the same primate family and possibly in the same genus" (1975, p. 75), but then concluded that this remarkable resemblance merely "indicated the close relationship of . . . *A. parisiensis* to the living lemurs and lorises" (p. 75). At the other extreme, Cartmill and Kay (1978) have claimed that "no synapomorphies linking adapids to living strepsirhines have so far been identified" (p. 210). These authors, however, limited their discussion to plesiomorphies—primitive character states.

I believe that perhaps the most basic point of resistance to linking adapids to a particular extant lemur taxon lies in their separation both in time and space. If *Adapis* were to be found in the Eocene of Madagascar it is improbable that most scholars would object to the linking of this form to *Hapalemur* among the living lemur genera; similarly, if *Adapis* survived in Europe today it is likely that this species would be fitted into the taxonomic framework that embraces the extant lemurs, and that elaborate scenarios would be devised to explain its geographical remoteness from its relatives. The isolation of the lemurs in *both* time and space, however, seems to have been too much for most scholars; and indeed, any step beyond a vague acknowledgement that adapids lie somehow at the "base" of Strepsirhini would necessitate a profound readjustment in our ingrained notions of primate phylogeny. Yet it has been argued powerfully (see, e.g., Schaeffer et al. 1972; Eldredge and Tattersall 1975; Eldredge and Cracraft 1980) that neither time nor geography has any *necessary* connection with evolutionary relationship. Although on average organisms existing earlier in time may tend to plesiomorphy, there is no guarantee that the state of a character in a species occurring early in the fossil record will be primitive compared to its condition in a relative known from later in time. Similarly, there is no demonstrable and consistent link between geographical distribution and the phylogenetic (as opposed to adaptive) history of taxa.

Of course, to accept a relationship between any particular lemur and an adapid (except possibly *Pelycodus*, which lacked symphyseal fusion and perhaps also orthal incisors [Simons, 1972]) requires that the toothcomb, a highly distinctive structure, be regarded as primitive for Strepsirhini as a whole, and as subsequently lost in known adapids. This is, admittedly, a lot to swallow; but other cases exist of the replacement of the toothcomb by an orthal anterior dentition (notably in *Hadropithecus*), and Gingerich (1975) has remarked that in *Adapis* the lower incisors functioned as a unit with the canines. Gingerich interprets this as a plausible precursor condition to the tooth scraper of the lorises and lemurs, but the argument could equally well apply in reverse to explain the transition away from an ancestral strepsirhine dental scraper (Schwartz and Tattersall 1979). In any event, even if available evidence is too weak to permit a high degree of certainty, it is certainly worth inquiring whether the striking resemblances between the cheek teeth of the lepilemurids and certain adapids represent synapomorphous states or simply a remarkable case of convergence. In view of their distributions, it is hard to see them as plesiomorphies. A further implication of an adapid-lepilemurid common ancestry would be the independent enlargement of the brain in several lemur lineages. This hardly represents a problem, however; an increase in the brain: body size ratio seems to have characterized virtually all primate groups during the Tertiary, as indeed mammals in general. Independent enlargement of this kind would further help to explain the great disparities in the relative sizes of the different components of the brain in different lemur genera documented by Stephan and his co-workers (e.g., Stephan and Andy 1969), and noted in chapter 4.

In the discussion which follows, I outline the synapomorphies on which the cladograms in figures 6.1–6.3 and 6.5 are based. I wish to emphasize that these cladograms merely represent what Schwartz and I consider the best hypotheses of relationship to account for the distributions among the Malagasy primates of the various states of the characters we examined. In many cases they are founded on less than entirely satisfactory data, and I make no claim that they are definitive; but they are testable propositions which may be compared to alternative cladograms derived from the analysis of other, or more, characters. It should be pointed out, however, that the falsification of one part of a hypothesis of relationships as expressed in a cladogram does not reflect on the validity of the relationships hypothesized in independent parts of the same cladogram.

It seems simplest to discuss the cladograms by characterizing the ancestral morphotypes represented at each node; in other words, by giving the synapomorphies which unite each clade and subclade.

Figure 6.1 shows our hypothesis of relationships among the major strepsirhine groups. The common ancestor of Strepsirhini as a whole possessed an inflated auditory bulla of petrosal origin which extended laterally beyond the inferior edge of the tympanic ring. Also present were a "toilet" claw on the second pedal digit and a tooth comb composed of three teeth bilaterally; the morphology of this structure was probably reminiscent of that seen in lorisines, with relatively short, robust, and orthal teeth. Among the primitive retentions of this form would have been been: promontory artery larger than stapedial; postorbital bar; bicornuate uterus; nails at least on

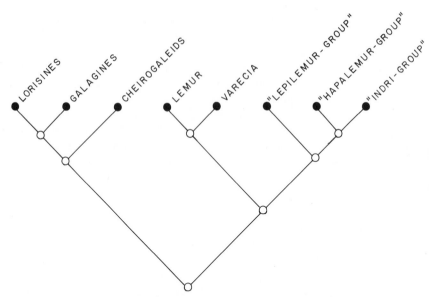

Figure 6.1. Hypothesis of evolutionary relationships among the major groups of Strepsirhini. For the purposes of this figure only, "*Lepilemur*-group" indicates *Lepilemur* + *Notharctus*, and "*Hapalemur*-group" indicates *Hapalemur* + *Adapis* + *Megaladapis*.

hallux and pollex; vomeronasal organ communicating with a moist, naked rhinarium; four premolars; slender mandibular corpus with hooked gonial region; and an unfused mandibular symphysis.

The cheirogaleid + lorisid common ancestor possessed an ascending pharyngeal artery entering the cranial base through an anterior carotid foramen in front of the bulla. The bulla itself was penetrated by the internal carotid posteromedially, rather than posterolaterally. The postglenoid process was fused medially to the bulla, and in the nasal fossa a large first endoturbinal descended to overlap the maxilloturbinal. Dentally, this ancestor displayed a preprotocrista anterior to the paracone on the upper molars; in the lower molars the paraconid was lost, a hypocone was present on M^{1-2}, and a distinct paracristid ran down the anterior face of the protoconid to kink lingually up the metaconid.

The remainder of the Malagasy primate fauna, united in the cladogram, represents a highly diversified assemblage which at first glance seems to be a "not-A" group held together by the common retention of the primitive condition in the temporal arterial circulation. It is possible, however, to identify certain synapomorphies. Derived characters present in the common ancestor of the group include: lateral compression of the lower premolars with diminution of the talonid basins of these teeth, and an anterior lower premolar which was relatively low and mesiodistally elongate, and which showed a small heel. In the locomotor apparatus, this ancestor possessed a tibialis anterior muscle which was doubled in its central and distal thirds, and from which an abductor hallucis longus was derived.

The grouping *Lemur + Varecia* appears, surprisingly, to form the sister of a single "*Lepilemur*-group" clade, to which also belong *Indri* and *Hapalemur*, together with their relatives. Dental synapomorphies uniting the ruffed and true lemurs are numerous, including: loss of the paraconid; reduction of M_3; compressed, "filled-in" trigonids; cusps of the trigonid and talonid joined, creating two distinct basins; lack of distinct entoconid; anterolingual distension of upper molars; buccally displaced cristae obliquae; protocone lowered and lengthened, at least on M^2; and de-emphasis of the postprotocrista.

The assemblage formed by the *Lepilemur* group is much larger and adaptively more heterogenous, making it more difficult to identify in terms of synapomorphy. The common ancestor did, however, possess a derived state of the anterior upper molars wherein the preprotocrista swung anteriorly around the paracone.

Cheirogaleidae

The cheirogaleids are generally more plesiomorphic than the lorisids, but the following dental apomorphies distinguished the cheirogaleid common ancestor (fig. 6.2): buccal cingulum present on the trigonid of all lower molars; broad, parabolic protocristae on the upper molars, which also exhibited a cingulum around the protocone. Molar cusp relief was somewhat lowered, the occlusal surface of the upper incisors was enlarged, particularly on I^1, and the toothcomb was more gracile and procumbent.

Previously, we considered *Phaner* to be the sister of the other cheirogaleids. Certainly, this genus is strikingly different in its cranial gestalt. Reappraisal of dental characteristics, however, has led us to the conclusion that it is *Cheirogaleus* which is set apart from the other cheirogaleids by derived dental features such as the lowering of molar cusp relief, and that the *Mirza-Microcebus-Phaner* clade is united by the following apomorphies: posterior lower premolar lacks a talonid basin, and shows a strong buccal cingulum which arcs

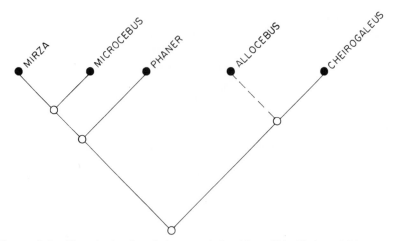

Figure 6.2. Hypothesis of evolutionary relationships within Cheirogaleidae.

from the posteriorly placed metaconid to the anterior face of a large protoconid. *Phaner* is otherwise highly autapormorphic, showing such features as the loss of the stapedial artery, lateral compression of the mandibular condyles, enlargement of the anterior upper premolar, reduction in size of cheek teeth relative to body size, extremely slender and procumbent dental comb, highly keeled and pointed nails, and so forth. In many of these features *Phaner* parallels *Euoticus*, which it also resembles behaviorally in its feeding specializations. *Mirza* and *Microcebus* are quite similar, and are generally plesiomorphic within the clade; in the dentition they are united synapomorphically by showing a small pericone on M^{1-2}, the reduction or loss of the paraconid on the posterior lower premolar, and a complete buccal cingulum on all lower molars.

Lepilemur-group

This diversified clade is the only one among the Malagasy primates which appears to have fossil (as opposed to subfossil) representation. The group is currently under study by Schwartz (in prep.); most fossil "lemuroids" seem to fall within a sister clade to that shown in figure 6.3. The common ancestor of this clade possessed upper molars with the cusps compressed laterally and the shearing crests emphasized; some stylar development, and a strong protocone fold, at least on M^1. The posterior aspect of the mandibular condyle was extended downward. These features may have been inherited from the common ancestor of these primates with such forms as *Pelycodus* and *Smilodectes*. Characters definitely synapomorphic for this group alone include the possession of a marked and obliquely oriented paracristid on the lower molars; yet stronger lateral compression of the cusps, and emphasis of the shearing surfaces on the lower molars; elongation and buccolingual narrowing of M^{1-2}; protocristae more broadly arced on the upper molars, with de-emphasis of the postprotocristae; and premolar development in the sequence P2→P4→P3 changing to eruption in the order P4–P2–P3.

As noted above, both *Hapalemur* and *Lepilemur* have been cited as bearing remarkable dental resemblances to adapids. We fully agree with Gingerich's observations already quoted, and conclude that on our evidence *Hapalemur* and *Adapis* must be placed within the same clade, one bearing a sister relationship to that containing *Lepilemur* and *Notharctus* (fig. 6.3). The dental similarities within these two pairs can be seen in figure 6.4. The *Notharctus-Lepilemur* group is united by the relatively deep posterolingual trough between the postcristid and entoconid on the first two lower molars, the *Hapalemur-Adapis-Megaladapis* clade by the presence of a small metastylid on M_1; a deep notch between the entoconid and metaconid regions on M_{1-3}, a protocristid parallel to the postcristid at least on the anterior lower molar, and by the presence of a submolariform posterior lower premolar. Within the latter clade, *Hapale-*

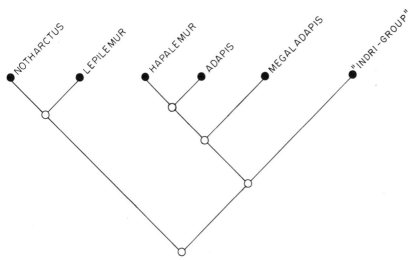

Figure 6.3. Hypothesis of evolutionary relationships within the "*Lepilemur*-group."

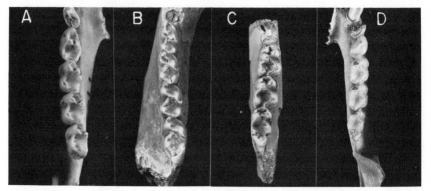

Figure 6.4. Lower cheek dentitions of: A: *Hapalemur griseus* (AMNH 170693); B: *Adapis parisiensis* (AMNH 13134); C: *Notharctus robustior* (AMNH 13134); D: *Lepilemur mustelinus* (AMNH 170575). Not to scale.

mur and *Adapis* are united by possession of a greatly molariform posterior lower premolar, and by the presence of a large metastylid on this tooth and on the anterior lower molars. Synapomorphies of the "*Indri*-group" and the *Hapalemur-Adapis-Megaladapis* clade include the loss of the paraconid from the lower molars, the presence of a small metastylid on the anterior lower molar, and a deep notch between the entoconid and protoconid regions on the lower molars.

Indri-group

This assemblage is extremely distinctive; almost entirely primitive in those characters it shares with the other Malagasy primates, it is otherwise highly autapomorphic. Derived character states present in the common ancestor of the entire group shown in figure 6.5, including *Daubentonia*, include: cranium subglobular; face short anteroposteriorly and deep superoinferiorly; horizontal ramus of mandible deep; gonial region expanded; small mesostyle present on M^1; expanded hypocone region on upper molars; and loss of one tooth from the toothcomb. The ancestral morphotype for the rest of the clade, excluding *Daubentonia*, is characterized by large, robust medial upper incisors; a large, sharp and distinct mesostyle on M^1, and a small mesostyle on M^2.

The relationship of the two major groups of extinct indriids vis-à-vis each other and the indriines has for long been a vexing problem. We now conclude that *Palaeopropithecus* and *Archaeoindris* form a sister clade to all other indriids, the common ancestor of which was

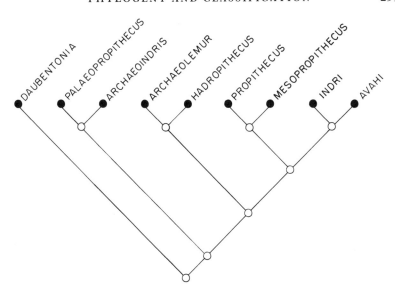

Figure 6.5. Hypothesis of evolutionary relationships within the "*Indri*-group."

characterized by a weak form of bilophodonty of the anterior molars. This arrangement of course demands that the loss of a premolar in each quadrant of the dentition occurred independently in the palaeopropithecines and in the indriines subsequent to the divergence of the archaeolemurines, but this is not hard to accept in view of the extreme autapomorphy of the palaeopropithecines in other characteristics. Notably, those characters shared uniquely by *Palaeopropithecus* and *Archaeoindris* include: regression of the auditory bulla, with the formation of a long, tubular, bony external auditory meatus; elongation and narrowing buccolingually of the upper molars; wrinkling of the molar enamel; and diminution of M³. The *Archaeolemur-Hadropithecus* clade is united by synapomorphies in the dentition; their cranial structure remains remarkably plesiomorphic for lemurs of such large size. Since *Hadropithecus* is so strongly autapomorphic in its molar and incisor teeth, these dental synapomorphies are found in the premolars, which in both, to a greater *(Archaeolemur)* or lesser *(Hadropithecus)* extent, are specialized into a shearing blade, the last tooth in the series also being molarized in its posterior moiety.

The common ancestor of the living indriines and *Mesopropithecus* possessed higher, more pointed molar cusps than did its ancestor with Archaeolemurinae. Postcranially, this form possessed a high greater femoral trochanter, a broad ilium and low humeral tuber-osities. Whether these features are truly apomorphic for this clade or whether they represent a holdover from an earlier stage in the evolution of the *Indri*-group is, however, difficult to assess at present. *Propithecus* and *Mesopropithecus* closely resemble each other in their cranial proportions, and are synapomorphically united in the diminution of their last upper molars, in which they parallel the palaeopropithecines. The common ancestor of *Indri* and *Avahi* retained the primitively somewhat long M^3, but was apomorphic in possessing a relatively long and slender toothcomb, elongate paroccipital processes, and deep mandibular diagastric fossae, and in having lost the protocone and talon basin of the last upper premolar.

Discussion

The bare recitation of linking apomorphies above quite obviously fails to reveal the difficulties involved in forming the sets defined. Most such difficulties have their roots in problems posed by parallelism. Invariably, whatever arrangement of taxa is accepted, parallelism raises its inconvenient if not downright ugly head. In accepting these cladograms we must accept, for example, parallelism in M^3 diminution in Palaeopropithecinae and in the *Propithecus-Mesopropithecus* clade, or in the anteriorly curving preprotocrista in both indriids and the cheirogaleid-loris group. But while it is relatively easy to accept parallel development of, say, nasal extension in the galagines and in *Loris/Arctocebus* (possibly because the occurrence of *Loris* in a separate continent), or between *Phaner* and *Euoticus* in cheek tooth reduction, elongation of the toothcomb, caniniformity of the upper anterior premolar, or keeling and pointing of the nails, one cavils instinctively at what appears to be the most startling claim for parallelism in the cladograms presented here: that the toothcomb, present in their common ancestor, was lost independently in *Adapis* and *Notharctus*. Not to accept this, however, demands the acceptance of what seems to us an even more striking parallelism in molar morphology between the former and *Hapalemur*, and the latter and *Lepilemur*. If the cladistic approach to phylogenetic reconstruction has shown anything, it is how rampant a phenomenon parallelism is in nature. It is impossible to form a cladogram using more than a few species or characters without running into problems created by parallel evolution; quite plainly this is a phenomenon we must accept more frequently than we might like, and in the light of which we must treat our conclusions with caution.

Doubt thus remains, and not without some justification. The cladograms presented above are those that seem to us to explain best the distributions among the Malagasy primates of the various states of the characters we have examined. I do not claim that they are definitive; indeed, they are couched in a format which invites falsification. Clearly, our perception of the evolutionary relationships of the Malagasy lemurs is incomplete, both in terms of our knowledge of the comparative anatomy of the living forms and of the abundance of the fossil record. Falsification of some, possibly all, of the hypotheses of relationship presented here is inevitable; but one trusts that attempts to do so will be made in terms of preferable cladograms based on more characters, and not on the basis of cavalier dismissals or the proposal of untestable scenarios, however elegant.

CLASSIFICATION

Phylogeny is a historical fact. For any group of organisms there is only one phylogeny, however imperfect our perception of it may be in practice. Classification, on the other hand, is a product of the human mind, and the range of possible classifications is hence limited only by the imagination of those producing the rules by which they are made. In this sense there are no classificatory systems which are inherently "right" or "wrong."

However, as Eldredge and Cracraft (1980) have cogently pointed out, the only type of zoological information that is logically equivalent to the Linnaean classificatory hierarchy is that contained in nested sets of the type represented by a cladogram. "Evolutionary" systematists have been at pains to claim that their classifications of organisms are devices to permit the storage and retrieval of information (e.g., Mayr 1969); but in fact, as Eldredge and Cracraft have stressed, "the information content of a classification is only the nested sets; nothing more can be retrieved" (1980:170). Classifications based on a hierarchical arrangement of monophyletic groups are thus the only ones from which any specific information can be derived; and from a properly formulated "phylogenetic" classification it is possible to reconstruct the cladogram on which it is based. This is in stark contrast to "evolutionary" classification in which structural and "genetic" similarities are thrown in with evolutionary relationships, and from which it is thus possible to derive no more than a vague idea of "overall similarity." Indeed, such classifications often effectively serve as devices for disguising the fact that no formal and exact phylogenetic hypothesis actually exists.

Does all this mean that use of the Linnaean classificatory system compels us to produce phylogenetic classifications? No, for the reason given at the start of this discussion. A moment's reflection upon the uses to which zoological classifications are normally put must surely lead to the conclusion that they are used almost all of the time simply to refer to organisms in the collective. This is their paramount function. And when one refers, say, to a family or to a suborder, it is vital that the precise species content of the taxon should be understood by all parties concerned. Obviously, a prime requirement for a classification permitting accurate communication of this sort is stability. The fatal practical problem with phylogenetic classifications is that they require a potential reordering whenever a new taxon (these days, mostly fossil) is included in the group, or when new analyses or the addition of new characters necessitate revision of the cladogram on which the classification is based. After all, cladograms are no more than "best" hypotheses which have so far resisted refutation. In an ideal world, everyone concerned would be aware of (if not necessarily in agreement with) changes in phylogenetic views and of the consequent impact on phylogenetic classifications. But, especially in the case of the primates, which interest many outside the immediate field, this simply cannot be the case; and a classification changing with every twist in phylogenetic thought, superior though it may be intrinsically, is unfortunately impractical. In any event, there exist far better and simpler ways of expressing phylogeny than through classification; after all, an explicit cladogram has to precede the construction of a phylogenetic classification. The practical advantage of nonphylogenetic classifications is precisely that which robs them of utility in the storage and retrieval of data: being imprecise, they are robust enough to survive changes, at least of low to moderate magnitude, in our understanding of the evolutionary history of the organisms classified. It may of course be objected that, in the absence of a clear-cut set of rules such as those governing the transformation of a cladogram into a classification, competing nonphylogenetic classifications may multiply indefinitely. Historically, however, this has not ordinarily been the case. As we have seen, Gregory's 1915 classification of the lemurs still stands almost unscathed as the accepted ordering of lemur groups.

Perhaps the best idea, then, would be to retain the conventional classification of the lemurs; certainly the cause of stability would hardly seem to be served by the addition of yet another classification to those already on offer. Unfortunately, however, the standard classification is by now so much at odds in certain respects with even the emerging general view of lemur relationships, let alone the more specific one espoused here, as to have outlived its usefulness. I thus propose below an alternative classification of the Recent Malagasy primate fauna, one based on the cladograms given earlier, but modified according to certain conventions to retain as many familiarly constituted taxa as possible, at the ranks at which they are commonly known. First, I have refrained from recognizing each successive node on the cladograms with a separate rank; second, I have avoided where possible the recognition of generically monotypic subfamilies; and third, I have used where appropriate the device of raising a taxon to the rank of the next larger clade which contains it. On occasion I have done this twice in succession within the same major clade, even, in one case, at the expense of recognizing a paraphyletic taxon. To the purists, I can only apologize. The classification is given to the level of the subfamily; the generic, specific and subspecific contents of the various subfamilies are as listed in chapter 3.

ORDER PRIMATES LINNAEUS, 1758

SUBORDER STREPSIRHINI POCOCK, 1918

Infraorder Lemuriformes Gregory, 1915
  Superfamily Lemuroidea Gill, 1872
    Family Lemuridae Gray, 1821
      Subfamily Lemurinae Gray, 1821
    Family Lepilemuridae Rumpler and Rakosamimanana, 1972
      Subfamily Lepilemurinae Rumpler and Rakosamimanana, 1972
      Subfamily Hapalemurinae Remane, 1960
      †Subfamily Megaladapinae Major, 1894
    Family Indriidae Burnett, 1828
      Subfamily Indriinae Burnett, 1828
      †Subfamily Archaeolemurinae Standing, 1908
      †Subfamily Palaeopropithecinae Tattersall, 1973
    Family Daubentoniidae Gray, 1870
Infraorder Lorisiformes Gregory, 1915
  Superfamily Lorisoidea Gray, 1821
    Family Cheirogaleidae Gregory, 1915

CHAPTER SEVEN

Behavior and Ecology

Brief observations have been made on the behavior and the use of the environment by a fairly large number of lemur species in a variety of different areas of Madagascar. Detailed field studies are much fewer in number, however, and have been carried out in a yet smaller number of localities. As a prelude to the following short review of the results of these studies, it may be useful to provide a brief introduction to the most important of these sites of intensive study. They are also listed in table 7.1, and their locations shown in figure 7.1.

Ampijoroa lies in the reserve of the Ankaranfantsika, about 80 km to the south and east of Majunga, the major port of northwest Madagascar. *Lemur mongoz* (e.g., Tattersall and Sussman 1975) and *Lemur fulvus fulvus* (Harrington 1975) have been studied in a plantation of mostly exotic plant species immediately surrounding the forestry station; *Propithecus verreauxi coquereli* has been studied nearby by Richard (e.g., 1978a) in indigenous mixed evergreen and deciduous forest of relatively low stature.

Analabé, also sometimes referred to as "near Morondava," or Beroboka, lies to the north of Morondava, the principal town of the west coast region. The forest at the site is seasonal, adapted to periods of prolonged dryness, and of rather low stature. Studies in this area have been principally of *Phaner furcifer* (e.g., Petter et al. 1971), of *Mirza coquereli* (e.g., Pagès 1978), and of *Microcebus rufus* (e.g., Martin 1972a).

Table 7.1. Sites at which major studies of lemurs have been carried out; locations are shown in figure 7.1. Survey and short-term sites not included.

Locality	Species studied
Ampijoroa	*Lemur fulvus fulvus, Lemur mongoz, Propithecus verreauxi coquereli*
Analabé	*Phaner furcifer, Mirza coquereli*
Analamazoatra	*Indri indri*
Antserananomby	*Lemur catta, Lemur fulvus rufus*
Berenty	*Lemur catta, Propithecus verreauxi verreauxi, Lepilemur mustelinus leucopus*
Hazafotsy	*Propithecus verreauxi verreauxi*
Mahambo	*Hapalemur griseus, Daubentonia madagascariensis*
Mandena	*Microcebus murinus*
Mavingoni	*Lemur fulvus mayottensis*
Tongobato	*Lemur fulvus rufus*

Analamazoatra, near Andasibé (Périnet), is situated several kilometers within the western boundary of the eastern humid forest strip, at an altitude of about 900 meters. Pollock (e.g., 1975a,b) has studied *Indri* in primary rainforest at this site; subsidiary observations were made at the higher elevation sites of Vohidrazana and Fierenana.

Antserananomby is the location of an intensive study of *Lemur catta* and *Lemur fulvus rufus* by Sussman (e.g., 1974). This forest lies to the south of Manja, in southwestern Madagascar. The main study area is covered by a decidous forest dominated by *Tamarindus indica,* and forming a closed canopy at a height of 7 to 15 m; this grades in areas of poorer soil into a drier "brush-and-scrub" formation of lower stature and with dense underbrush. Tongobato, at which Sussman studied *L. f. rufus* alone, is similar in physiognomy to the closed-canopy portion of Antserananomby.

Berenty is a classic site at which many authors have carried out detailed observations on *Lemur catta, Propithecus verreauxi verreauxi,* and *Lepilemur mustelinus leucopus.* Budnitz and Dainis (1975) have distinguished four distinct vegetational zones at this site, near the southern tip of Madagascar. Most of the vegetation is riparian gallery forest not unlike the closed-canopy forest at Antserananomby. Away from the river, however, the trees decrease in height and open spaces tangled with thorny vines become common. Another area of the reserve is roughly equivalent to Sussman's "brush-and-scrub" formation, while a final section contains the subdesertic spiny Didieraceae vegetation characteristic of the driest areas of the south of Madagascar. The forest reserve is now surrounded by sisal fields.

Hazafotsy, at which Richard (e.g., 1978a) studied *Propithecus ver-*

Figure 7.1. Principal sites of lemur field studies in Madagascar.

reauxi verreauxi, is a little to the northeast of Berenty and lies in an area of Didieraceae forest.

Mavingoni, a site where I have studied *Lemur fulvus mayottensis* (e.g., Tattersall 1977c), lies approximately in the center of the island of Mayotte, and is characterized by three distinct types of vegetation, none of which is indigenous. Along the river is found a mixed consociation dominated by *Mangifera indica;* contiguous with this is a formation composed almost entirely of very tall trees of the same species. Both of these formations are at an advanced stage of succession, in contrast to the remaining forest type, a young formation consisting largely of *Litsea glutinosa* saplings.

ACTIVITY PATTERNS

As Martin (1972b) has noted, the preponderance of nocturnal species in the surviving Malagasy primate fauna (and indeed, in the Malagasy fauna as a whole) reflects an imbalance resulting from the recent extinctions; the large extinct lemur species appear plausibly to have been diurnal (see chapter 5). However, the Malagasy primates nonetheless include the only diurnal extant strepsirhines, and indeed they embrace species which run the gamut of those major activity patterns so far documented within Primates as a whole.

Cheirogaleidae

Effectively, the cheirogaleids are exclusively nocturnal, although almost all have on some occasion been seen active a little before sunset. Martin (1972a) records that *Microcebus* species, which sleep in natural or artificial nests during the hours of daylight, normally emerge soon after sunset and return to their nests by dawn. Activity is not, however, evenly distributed throughout the night; sightings of mouse lemurs decrease around midnight, and peaks of activity are apparent during the first and final thirds of the night. Individuals may, however, be active at any time during the night. *Cheirogaleus* species probably show a similar pattern of activity, but detailed observations are lacking. Petter et al. (1977) note that dwarf lemurs spend all night away from their daytime sleeping places, but that from time to time during the night they may remain immobile, though not asleep, for extended periods. Pagès (1978) notes that *Mirza* tends to devote the first part of the night to maintenance activities, whereas the frequency of social behavior increases during the second.

Phaner, thanks to its noisier habits and its relative abundance at some sites in western Madagascar, has been somewhat better studied than the other large cheirogaleids (Petter et al. 1971; Pariente 1974). The time when *Phaner* leaves its daytime sleeping place is closely controlled by light level (Pariente 1974); activity commences when light intensity falls below two lux. Cessation of activity appears to be somewhat less regular, but remains quite closely light-dependent. *Phaner* erupts from its nest with great rapidity, and the first hour following its emergence seems to be the most active (Petter et al. 1971). Activity may be observed, however, and vocalizations heard, throughout the night.

Lepilemuridae

Lepilemur is rarely seen away from its daytime resting site before nightfall; but individuals may often be seen awake in daylight, notably during the late afternoon, staring out from the entrances to the tree hollows which, at least in some areas, are their preferred sleeping places. The only subspecies which has been studied in any detail is *L. m. leucopus* in southern Madagascar (Charles-Dominique and Hladik 1971; Hladik and Charles-Dominique 1974; Russell 1977); the results of these studied are unfortunately at variance with each other in certain respects. Charles-Dominique and Hladik, studying *L. m. leucopus* in Didieraceae bush, found that the animals left their sleeping sites regularly as night fell, after grooming themselves for periods of up to 20 minutes. They report that the subsequent activity consisted of short episodes of feeding interspersed with long periods of rest and brief bursts of locomotion. Individuals returned to their daytime sleeping place at the first hint of dawn. Pariente (1974) found that the onset of activity (defined as departure from the daytime shelter) was extremely closely keyed to light intensity, but that somewhat less regularity was displayed in the cessation of activity; this invariably occurred, however, at very low light levels, much lower than those triggering onset of activity.

A generally comparable pattern of activity onset/cessation was recorded by Russell (1977), who, when working in gallery forest found that activity began earlier relative to sunset as the dry season advanced; he suggested that in consequence warming conditions might thus have some effect as zeitgeber, a possibility denied by Charles-Dominique and Hladik and by Pariente. Activity periods, however, remained in proportion to night length; each of these declined by 14% over the four months (July–October, 1974) of Russell's study. A considerably more striking difference between Russell's results and those of Charles-Dominique and Hladik resides in the activity:rest ratio. Russell found his animals to be active during more than 40% of the total activity period, whereas Charles-Dominique and Hladik, while not providing a comparable figure, clearly imply in their discussion that the figure would be very much lower.

Activity in *Hapalemur griseus* has been reported by Petter and Peyrieras (1970b). These authors found *H. g. griseus* at Maroantsetra, in a rather degraded habitat, to be active at dawn, feeding. During the heat of the day the animals retreated to the shade of dense vegetation, only reemerging to feed again around 1600 hrs. This activity lasted at least until dusk and often continued after nightfall. A peak of activity was noted between 1700 and 1900 hrs. Petter and Peyrieras suggest a similar rhythm for the Alaotran subspecies also. In the case of *Hapalemur simus*, Petter et al. (1977) report activity regularly at daybreak and irregularly during the course of the morning. Activity appears to cease by noon and to recommence in the late afternoon, reaching a peak between 1600 and 1700 hrs. These authors believe it most likely that *simus* remains inactive during the night.

Lemuridae

Activity patterns are documented in greater detail in several species of *Lemur,* notably *L. catta,* certain subspecies of *L. fulvus,* and *L. mongoz.* Of these only one, the mongoose lemur, appears to be conventionally nocturnal (Tattersall and Sussman 1975; Sussman and Tattersall 1976). At Ampijoroa, in northwestern Madagascar, during July and August 1973, and in July and September 1974, these authors found mongoose lemurs to be active exclusively at night, leaving their sleeping places at nightfall and entering them at dawn. Activity commenced approximately 25 minutes after the official time of sunset, at light intensities of 2.8–22.0 lux, and ceased when the light level had increased to 1.4–22.0 lux: a very narrow range. Peaks of feeding activity followed sunset and preceded sunrise, and a prolonged resting period of 2–3 hours was typically recorded between 2200 and 0300 hrs. Limited observations at Ampijoroa a little earlier in the year by Harrington (1978) apparently suggest some diurnal activity, but further investigation is needed.

A nocturnal rhythm also applies to the mongoose lemurs inhabiting the warm seasonal lowland areas of the Mohéli and Anjouan islands of the Comoro group. It is not, however, displayed by the mongoose lemurs of the cool, moist highlands of Anjouan (Tatter-sall 1976c, 1978). Nocturnal observation was not possible in the latter habitat, but surveying during the day regularly disclosed groups of mongoose lemurs fully active in broad daylight. There appears to be a clear environmental correlation with activity pattern in this case, but the full physiological implications remain to be worked out.

Species active diurnally have understandably been studied in substantially more detail. A. Jolly (1966) reported that the *L. catta* she studied at Berenty in southern Madagascar woke before dawn and started moving at any time between 0530 and 0830 hrs, depending on the conditions. Travel at nightfall brought them to sleeping trees, from which they did not move during the hours of darkness. More recently, Sussman (1972, 1974) has conducted a quantitative study of activity in *L. catta*. Scan-sampling at 5-minute intervals yielded the number of animals engaged in feeding, grooming, moving, traveling, resting, and "other" activities. The overall activity:rest ratios between 0600 and 1830 hrs were 1.56 at Berenty (Jolly's study site), and 1.44 at Antserananomby in southwestern Madagascar. *Lemur catta* was active relatively consistently throughout the day although feeding peaked in the morning, and more markedly in the afternoon; and resting preponderated during the middle of the day. Sussman remarks that the activity of *L. catta* seemed to be relatively highly susceptible to seasonal differences in times of sunrise and sunset and in temperature; at Berenty, for instance, where sunrise was earlier and the temperature higher than at Antserananomby, the ringtails began to feed earlier, reached their resting period earlier, and recommenced feeding earlier in the afternoon. Similarly, the relatively high proportion of time spent "sunning" by *L. catta* early in the morning at the cooler Antserananomby suggested to Sussman that sunning is an important mechanism of thermoregulation in this species.

Lemur fulvus rufus, also studied by Sussman, showed a different pattern in its use of time, both at Antserananomby and at the nearby forest of Tongobato (Sussman 1972, 1974, 1975). This lemur was much less active than the ringtail, showing a daytime activity:rest ratio at Antserananomby of 0.79, and at Tongobato of 1.0. The relatively high ratio at the latter site may reflect the fact that the lemurs were hunted by locals and were thus impossible to habituate fully to observation. The relatively low activity level of the rufous lemur was reflected in its long afternoon resting period (60% or more of the rufous lemurs rested between 1000 and 1600 hours at Antserananomby, a figure equalled by *L. catta* only between 1030 and 1230 hrs). *Lemur catta* began feeding later in the morning than *L. f. rufus,* and ceased feeding earlier in the afternoon; but it fed and foraged far more consistently throughout the day.

Whereas neither *L. catta* nor, perhaps to a lesser extent, *L. f. rufus,* appears to exhibit a significant amount of activity at night, I found a remarkably different pattern to be characteristic of *Lemur fulvus mayottensis* on Mayotte, the most southerly of the Comoro islands. Daytime (0600–1800 hrs) activity records, collected in the same manner as Sussman's, revealed that the activity:rest ratio, already low in the wet season (February–May, 1975) at 0.52 (Tattersall 1977c), dropped during the dry season (July–August, 1977) to 0.40 (Tattersall 1979). Casual observations in 1975 had revealed that considerable nocturnal activity was characteristic of the Mayotte lemur, so groups were followed throughout the night on two occasions in 1977, and once in 1980. Since during the hours of darkness observation was affected by considerable visibility bias, the 5-minute scan-sample data were grouped into 30 minute periods from 1800–0600 hrs., and each half-hour period was scored according to whether observations of activity or resting predominated (fig. 7.2). The mean number per night of half hours during which activity exceeded rest was 6.8, with a range of 6–8; at the same season of the year, 16 full days of observation yielded a mean of 6.5, with a range of 4–10. Despite the small sample of nights, it is clear that the Mayotte lemur is active on a 24-hr rhythm, and that its low daytime activity:rest ratios reflect high activity at night compared to *L. f. rufus.* Whether the higher wet-season daytime activity:rest ratio implies a reduced level of activity at night, or merely an increased overall level of activity at that time of year, has yet to be determined; the latter is more plausibly the case, however.

Coordination of the activities of individuals within the group was imperfect in the Mayotte lemur, and routines varied considerably from day to day. It was thus impossible to predict with any certainty what the group, much less the individual, would be doing at a particular time of day. A general pattern emerges, however, of major feeding peaks in the early morning and evening, separated by a long resting period during the middle of the day. Feeding bouts also occurred at other times of day and night, but their timing was irregular. As a rule, the early morning feeding bout tended to be concentrated in the predawn period, while most of the evening feeding was accomplished before nightfall. Compared to the wet season, the dry showed an accentuation of both the peaks and troughs of activity. The tendency of the Mayotte lemurs to indulge

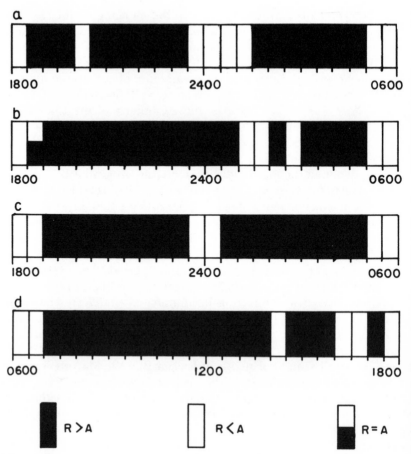

Figure 7.2. *Lemur fulvus mayottensis*: half-hour periods of observation characterized according to whether observations of resting exceeded those of activity or vice
versa. a: night of July 26–27, 1977; b: night of July 30–31, 1977; c: night of June 21–
22, 1980; d: a randomly chosen daytime (July 18, 1977). R: resting; A: activity.

in an extended period of rest during the middle of the day was yet
more pronounced, while the early morning and in particular the
late afternoon activity periods, were both longer and more regular.

The ecological basis of 24-hr "diel" activity remains unclear;
neither the type nor the intensity of activity in the Mayotte lemur
appeared to be affected by light levels or by temperature, and the
resources exploited by the lemurs at night were substantially the
same as those used during the day. The only exception to this was
noted in June 1980, when kapok flowers were frequently fed upon
by Mayotte lemurs at night, but were not observed to be eaten by

day. This is simply explained, however, by the fact that these flowers open only at night, remaining closed until shortly after dusk.

Conley (1975) has reported an activity pattern for captive *Lemur fulvus albifrons* which is in some respects similar to that found for the Mayotte lemur, while in the preamble to a study of the social behavior of *L. f. fulvus* in northwestern Madagascar Harrington (1975) remarked that this subspecies might exhibit a considerable amount of nocturnal activity. My own brief observations suggest that this also may be true for *L. f. fulvus* in the eastern rain forest of Madagascar, and for *L. f. sanfordi* on the Mt. d'Ambre. I have observed *L. coronatus* active both during the day and after dusk, but I do not know how active these animals tend to be later than about 2000 hrs. *Lemur macaco* may well exhibit a diel activity rhythm; again, I have seen them active after dark, but Petter (1962, Petter et al. 1977) states that the groups he studied on Nosy Komba were active in the predawn period rather than intermittently throughout the night. *Lemur rubriventer* has been observed active during the day; whether it displays any nocturnal activity is unknown. *Varecia*, unhappily, remains virtually unstudied in the wild, but again, it seems not impossible that its rhythm is diel; its daytime activity is well documented, and it may readily be heard vocalizing at night. Kress et al. (1978), however, state that they saw no evidence whatever of nocturnal activity in the captive ruffed lemurs they studied.

Indriidae

Indri may be the only Malagasy primate that is exclusively diurnal, having a short activity period ranging from 5 to 11 hours, according to the season and ambient conditions (Pollock 1975a,b). Activity, initially consisting of auto- or allogrooming, commences from two to four hours after dawn, and continues consistently until midafternoon, when it slackens and ceases well before dusk. Pollock (1975a) records that half of all activity among the animals he studied consisted of feeding, with almost all the rest of the activity period devoted to movement between food sources and to brief resting periods. Exceptionally, an *Indri* group would cease activity during the day for up to two hours, but this was rare, and at least on occasion seems to have been due to outside disturbance. Activity periods were longest during the austral summer and shortest during the winter, and in all cases were substantially shorter than day length. The difference was greater in winter, however, which implies that factors other than daylight may also influence this pattern.

Richard (1974, 1978) has noted a substantial seasonal variation in the activity pattern of *Propithecus verreauxi*. In the dry season the sifakas she studied often would not move from their sleeping places until one or two hours after sunrise, and then might sun in the treetops for another hour before departing to feed. Feeding then continued relatively consistently until the early afternoon (although her northern study animals reduced their activity somewhat around midday), when the sifakas would move to their sleeping stations. In contrast, during the wet season the feeding pattern was more reminiscent of that of *Lemur* than that of *Indri:* feeding commenced before sunrise, peaked between 7:00 and 9:00 a.m., then tapered off toward midday. A prolonged period of rest then intervened, after which feeding resumed and peaked once more in the later afternoon. Activity was very variable from day to day, however.

The diurnal *Propithecus diadema* has never been systematically studied; I suspect that its activity pattern may more closely approach that of *Indri* than that of its congener. The little-known *Avahi* is strictly nocturnal, appearing only after dark; but it may not be continuously active during the night.

Daubentoniidae

Daubentonia, is of course, nocturnal. Petter (e.g., 1977; Petter and Peyrieras 1970) states that it leaves its nest regularly at nightfall, and returns with equal regularity just before sunrise begins to lighten the sky. In December, this provides an active period of nine hours; at other times of year, rather more. Exactly what occurs between these two events remains obscure, although Petter believes that aye-ayes are consistently active throughout the night except during periods of heavy rain.

Discussion

Most recent considerations of the activity pattern putatively primitive for lemurs (e.g., Martin 1972b) have leaned heavily on the issue of tapetum (see chapter 4). While it does seem plausible for a variety of reasons that the ancestral strepsirhine was nocturnal, it is likely that nocturnality is secondary in certain lemurs, and the distribution of tapeta provides a confusing picture. The ancestral indriid, for instance, was probably diurnal; yet both *Propithecus* and *Indri,* as well as *Avahi,* possess highly reflective tapeta. On the other hand, the most nocturnal of the species of *Lemur, L. mongoz,* has only a very weakly reflecting eye, while the most diurnal of those species,

L. catta, displays the only well-developed tapetum to be found in the entire genus. Whether the ancestral *Lemur* was nocturnal or diurnal is thus problematical on this evidence, as is the significance of the diel activity of the Mayotte lemur and quite possibly others. If this rhythm represents a transitional stage between nocturnality and a strict diurnality newly made possible by the vacating of numerous diurnal niches by the extinct lemurs, it is difficult to explain the degenerative modification of the tapetum in species other than *catta,* if indeed they possess a true tapetum at all. More probably the diel rhythm is a stable pattern of long standing, in which case the possibility of its wide distribution among species of *Lemur* might even suggest that it represents the activity pattern ancestral for the genus as a whole.

SOCIAL ORGANIZATION

The social groupings displayed by the Malagasy primates vary widely, not simply between but in some cases also within genera and even species. Neither phylogenetic relationship nor ecology seems by itself to be an efficient predictor of the size or the organization of social units, although as a rule rain forest lemurs tend to live in smaller groups than do related forms inhabiting drier, more seasonal areas.

Cheirogaleidae

Although mouse lemurs are almost invariably sighted alone during their period of nocturnal activity, they are commonly found in sleeping groups during the hours of daylight. Such groups normally occupy nests which take the form either of tree hollows, sometimes lined with soft vegetable material, or of "leaf-nests," substantial spherical constructions of leaves hidden by dense foliage in the fine branches of trees. Martin (1972a) provides detailed descriptions of *Microcebus* nests, and notes that in general they are found low in the forest, at heights between two and four meters. Sampling outside the breeding season of nests of both species revealed that while males normally nested singly or in pairs, females formed larger groupings of up to 15, with a mean of 4.3 per nest (Martin 1972a). In only 3 of 39 nests examined were males found nesting with females, and in each case only a single male was involved. However, a survey conducted at the peak of mating activity produced a rather different result (Martin 1973): the size of female aggregations had decreased to a mean of 3.1 (and to a maximum of 7), while only 2 percent of nests yielded males alone. On the other hand, half of the nests examined contained males and females together, with a mean ratio of 2.1 females per male.

Both nest samples and night sightings and trapping suggested to Martin that mouse lemurs formed relatively stable "population nuclei," i.e., "localized concentrations . . . with sizeable gaps between them" (1972a:69). Within these concentrations, females outnumbered males by three or four to one. The discrepancy between this ratio and the approximately 1:1 ratio among juveniles found with their mothers is plausibly due to the peripheralization of most males relative to the population nuclei; Martin (1972a) notes that individuals sighted or trapped at the edges of the nuclei were usually male. Such peripheral males weighed less than central males, which may indicate that they were subadult or at a feeding disadvantage or both. Centers of the nuclei, on the other hand, were composed of a high proportion of females to males, the females ranging in relatively exclusive areas several of which might be overlapped by that of a single male. Martin (1973) believes that daytime sleeping associations among females may be relatively stable.

Almost nothing is known of social organization among *Cheirogaleus* spp.; individuals are invariably sighted singly during their nocturnal activity period, and pass the day in tree hollows or, less frequently and apprently only in the case of *C. major*, in constructed nests. Few samplings of nests have been made; Petter et al. (1977) report the presence of two adult *C. major* (sexes unspecified) in one nest examined, and that in captivity individuals of both species avidly seek the company of the maximum number of nest mates.

Quite elaborate daytime sleeping nests are constructed by *Mirza coquereli*, which appears not to exploit tree hollows (Petter et al. 1971, 1977); on one occasion two adult males were found in the same nest. Contacts between individuals during the night's activity are said by Pagès (1978) to be rare, although she suggests that relatively close associations between certain males and females may indicate a "loose pair bonding." Male and female ranges do not coincide, however, although they may overlap. Immature offspring share the mother's range and spend the day in the same nest, although they may forage separately.

Phaner furcifer do not appear to make nests of their own, although in areas of sympatry they may occupy those of *Mirza* during the day (Petter et al. 1971). They are known to occupy tree hollows, however, often in male-female pairs (Petter et al. 1977). According to Charles-Dominique (1978) the female is dominant to the male fork-marked lemur, who follows her for much of the night. Members of a pair retain contact when separated by means of loud vocalization. Charles-Dominique reports that up to seven individuals may converge during the night in zones of range overlap and forage close together, coordinating their activities by intermittent vocalization.

Lepilemuridae

Charles-Dominique and Hladik (1971; Hladik and Charles-Dominique 1974), studying *Lepilemur mustelinus leucopus* in *Didierea* bush at Berenty, found that males lived in ranges overlapping those of females but excluding other males, while female ranges excluded other females except for juvenile offspring, and possibly mature female offspring also. Female ranges were overlapped by the range of only one male. When males and females encountered one another during the night's activity, physical contact was not noted; one individual would vocalize, and both would continue on their way. The implication of Charles-Dominique and Hladik's discussion is that the sportive lemurs in their study area invariably slept alone during the day except in the case of mothers and young juveniles; ten-month old females would generally sleep independently of their mothers, while joining them for the night's activity.

Russell (1977) records a rather different pattern for *L. m. leucopus* in the same region, however. He found that heterosexual or adult female pairs might share all or most of each others' ranges, and associate quite closely from time to time during the period of activity. Feeding duos or trios of "range-mates" would form one to three times per night, for periods of 5 minutes to an hour, and episodes of mutual grooming and resting in contact were not uncommon. Some individuals were always seen sleeping alone during the day, but others would quite frequently form female-female or male-female daytime sleeping groups. In the mating season (May–June) three animals might be seen sleeping in contact during the day.

Both Charles-Dominique and Hladik's and Russell's observations do, however, agree in suggesting that in *Lepilemur* long-term bonds are formed between mothers and at least their female offspring. In neither study could a mother-daughter relationship be firmly demonstrated between females which tended to associate; but the frequency with which older and younger females were found to associate is at the very least suggestive.

Hapalemur is considerably more poorly known. *Hapalemur griseus* is most commonly seen in small groupings of three to five individuals (Rand 1935; Petter and Peyrieras 1970; Petter et al. 1977; Pollock 1979); these generally consist of one adult male, one or two adult females, and one or two juveniles (Petter et al. 1977). Petter believes that these units represent familial groups, although if so the frequency of groups of five suggests a higher twinning rate than that observed in captivity. Pollock (1979), who censused thirty-eight groups of *H. g. griseus*, arrived at a mean size of 2.8 individuals per group, with a maximum of 6. Georges Randrianasolo (quoted in Petter et al. 1977) has observed about 30 individuals of *H. g. occidentalis* in a group near Lac Bemamba; Petter *et al* (1977) note that Alaotran *H. griseus* is seen in small groups of 2 to 4 individuals in July, but in much larger groupings of 30 to 40 during the period of high water a little later in the year.

Petter et al. (1977) report seeing *Hapalemur simus* in groups of 5 or 6 individuals in February, but in a group of about 10 in April. In the former case the animals formed a polyspecific association with *H. g. griseus*, and in the latter with *Lemur fulvus fulvus*.

Lemuridae

The genus *Lemur* embraces a remarkable variety of group sizes and compositions, from pair-bonded units to large multi-male troops. Of all species of the genus, *Lemur catta* has been the most frequently studied. In 1963 Alison Jolly censused three groups of ringtailed lemurs in the gallery forest at Berenty, obtaining counts of 20, 16, and 12 individuals, and adult sex ratios of 0.7, 2.7, and 0.6 respectively. Subsequent censuses at the same site have yielded groups as small as 5 and as large as 23 (Jolly 1966, 1972; Budnitz and Dainis 1975; Mertl-Milhollen et al. 1979; Jolly et al. in press). Over the years the adult sex ratio at this forest has remained at approximately 1:1, but individual troop counts may yield a preponderance of either sex. Jolly (1972) found that recensusing of the reserve in 1970 yielded a similar density of ringtails to that recorded in 1963, but double the number of troops. One group of ringtails censused by Sussman (1972, 1974) at Antserananomby contained 19 animals, of which 7 were adult males and 8 adult females. This is very close to the mean size for all groups censused. At Antserananomby the larger group regularly split into smaller parties for foraging.

Although there is under most circumstances no obvious leader of a *L. catta* troop, both Jolly (1966) and Budnitz and Dainis (1975) concur that linear dominance hierarchies exist for both sexes, and that females are dominant over males. Females form the stable nucleus of the troop, forming permanent associations with the other female members; males, on the other hand, transfer quite readily from one troop to another (Budnitz and Dainis 1975). As an extension of this, it is the females who appear to be preponderantly concerned with the defense of the group range, although normal group progression may be led by almost any individual. Various authors have noted the tendency in *L. catta* toward the peripheralization of subordinate males (e.g., Jolly 1966; Sussman 1972, 1974; Budnitz and Dainis 1975); this may involve the exclusion from, or limited access to, favored resources. R. W. Sussman (pers. comm.) suspects, however, that at least part of the apparent dominance of females over all males may be an artefact of temperament. In any event, the linear hierarchy among males seems to collapse during the breeding season, when all adult males appear to have equal access to estrous females (Budnitz and Dainis 1975).

No dominance hierarchy at all is discernible in *Lemur fulvus rufus* (Sussman 1972, 1974, 1975), which has been studied at the southwestern sites of Antserananomby and Tongobato. Of 12 groups censused at the former forest and 5 at the latter the mean size was 9.5, with a range of 4–17 individuals. Adult sex ratios in both forests were approximately 1 male to 1.25 females, and all groups contained multiple adults of both sexes. Sussman (1972, 1975) speculates that a group of 13 adults and 4 juveniles, seen at the beginning of his study at Tongobato but not later, had split once the infants began to move independently. Another large group (15) seemed actually to be in the process of dividing; subgroups formed for feeding or daytime resting. These observations led Sussman to infer that rufous lemur groups divide when a critical size of 15–17 is reached.

Two groups of *Lemur fulvus fulvus* studied by Harrington (1975) in deciduous forest at Ampijoroa, in the Ankarafantsika, consisted of 12 individuals each. The proportion of males to females, 5:7 in one case, was reversed in the other. In the same region, Petter (1962) reported that larger groupings of up to 30 individuals occurred in the evenings, but I was unable to corroborate this in two seasons' fieldwork in the area. In the eastern rain forest, Pollock (1979) has censused groups of *L. f. fulvus* numbering between 3 and 10 individuals.

Lemur fulvus mayottensis, found only on the island of Mayotte and probably descended from imported *L. f. fulvus*, has a rather different form of social organization in which clearly defined groups are lacking (Tattersall 1977c). These animals are highly gregarious, virtually never being seen alone, and censusing of 100 groupings all over Mayotte yielded a mean of 9.1 individuals, with a range between 2 and 29. Taking together all cases where individuals could be identified by sex, the adult sex ratio was almost exactly 1:1. Intensive observation at the forest of Mavingoni revealed, however, that the groupings observed were not stable; groups under observation might divide at any time or be augmented by the arrival of new animals. Unhappily, it has not yet been possible to determine whether (as seems more likely) these units were simply subgroups of a larger, more exclusive, local population (perhaps equivalent to the "social networks" seen in spider monkeys [Klein and Klein 1975], or to the "regional populations" of champanzees [Nishida 1968]), or whether Mayotte lemurs live in groups that are truly "open." Independently moving groups often passed close to each other without incident; only very occasionally were distinctive intergroup encounters observed. The modal number of animals under observation at any one time during the intensive study was 9 or 10, but in the course of a day's fieldwork the group under observation might consist of fewer than 6 or as many as 16 individuals. No dominance relations could be discerned; agonistic interactions between individuals were rare, and individuals of any age/sex class might initiate or lead group progression.

The mongoose lemur, *L. mongoz,* has been studied in northwest Madagascar, at Ampijoroa (Tattersall and Sussman 1975; Sussman and Tattersall 1976), and on Mohéli and Anjouan islands of the Comoro Archipelago (Tattersall 1976c, 1978). In Madagascar, this species was found to live in small family groups composed of an adult male and female together with their immature offspring. This pair-bonding may be at least quasi-permanent; we have observed adult males and females to associate exclusively over two successive years. A 21-month-old juvenile female in one group did not associate consistently with her parents during the first year of study, and was gone from the group in the next, having been replaced by a new infant. She appeared, however, to be under no pressure to depart, associating peacefully with and leaving and rejoining her parents during the night's activity apparently at will. No member of the group appeared to exercise dominance over any of the others, and aggressive interactions between group members were virtually nonexistent.

Some variation in mongoose lemur group structure was, however, found in the Comoros. The size and composition of the groups censused in Anjouan were concordant with the hypothesis of pair-bonding within this population, although the mean group size was slightly larger than on the mainland (3.1 as compared to 2.6). But on the island of Mohéli only 45% of the groups censused could be interpreted as normally constituted family units; and even on the supposition of an abnormally high rate of twinning, this figure could be raised only to 73%. The remaining groups counted were totally at variance with a pair-bonded structure; one contained 6 adult-sized individuals (4 males, 2 females) and one juvenile. It seems possible that on Mohéli there is some seasonal change in group structure, smaller dry-season groupings coalescing in the wet season; but the evidence is inconclusive.

Lemur macaco was studied by Petter (1962), who found that 10 groups censused on Nosy Bé and Nosy Komba yielded a mean of 9.6 individuals with a range of 4 to 15. The sex ratio overall was found to be 1.4 males per female, the excess of males being characteristic of all groups counted except one. Petter indicates that certain groups foraged as distinct units during the day but came together into larger aggregations at night.

The remaining species of *Lemur* have not received adequate study. *Lemur coronatus* lives in relatively large groups containing multiple adult males and females, while *Lemur rubriventer* is found in small groups which probably represent family units. No aggregation larger than 5 has ever been reported. *Varecia* likewise appears to live in pair-bonded units. Groups of more than three individuals have only exceedingly rarely been reported, and Petter et al. (1977) suggest that the relatively rapid development of young ruffed lemurs allows them to leave their parents at the end of their first year. Sexual maturation of males appears to be rather late in this lemur, however. The uniformly small size of counts of *Varecia* groups is somewhat surprising in view of the marked tendency towards multiple births in this lemur; one possibility is that a relatively high level of infant mortality is associated with the "parking" of infants noted by Petter (1962).

Indriidae

Jolly's (1966) original census of *Propithecus verreauxi verreauxi* at Berenty in 1963 produced a mean count of 5 individuals per group, with a range from 2 to 8. Subsequent censuses at that locality have yielded similar results (Jolly 1966, 1972; Jolly et al. in press; Richard

1973, 1974a, 1978a). Surveys conducted by Richard (1973, 1974a, 1978a) at four sites (including Berenty) in southern and southwestern Madagascar, and at Ampijoroa in the northwest, produced a range in group size of from 3 to 12. Mean group sizes at the various forests ranged from 5.0 (Ampijoroa: *P. v. coquereli*) to 7.8 (Antserananomby: *P. v. verreauxi*); the latter was the only site not to yield a mean between 5 and 6. Adult males were initially found by Jolly (1966, 1972) to outnumber females slightly but consistently from troop to troop at Berenty, but later censuses have shown this to have been a temporary phenomenon at the site. Richard (1973, 1978a), on the other hand, found a very slight overall excess of females at all forests except Antserananomby, but suggested that all deviations so far recorded from the 1:1 ratio represent normal fluctuations about that point. This judgment is borne out by later censuses at Berenty in which the excess of males had disappeared (Jolly et al. in press). Richard notes, moreover, that the ratio of males to females varied so widely from group to group at all the sites surveyed that she could establish no norm of group composition. Of the seven groups accurately censused at Ampijoroa, for instance, one contained 3 adult males and one adult female, while another had only 2 adult males but 5 females.

Comparison of census results from 1963/4 and 1970 led Jolly (1972) to conclude that sifaka groups remained relatively stable in composition, although she noted the likelihood of the reassortment of troops (Jolly 1966). More recently, specific observations have been made of the interchange of individuals between troops. Thus Richard (1973, 1978a) has noted that while she saw no adult females or immature individuals of either sex entering or leaving any of her study groups over a period of eighteen months, males did indeed do so, both temporarily and permanently, outside as well as within the breeding season. Mertl and Gustafson (Jolly et al. in press) have further noted "chain migration" among sifakas at Berenty. The death of an old male, for instance, led not only to the reconstitution of his group but also to changes in the composition of 5 neighboring groups. These changes entailed the migration of 6 out of the total of 12 adult males, and of at least 3 of the 15 adult females belonging to the troops involved.

Jolly (1966) found no dominance order to exist in *P. verreauxi* groups when females were not either in estrus or carrying infants. She remarked on the extremely low frequency of agonistic interactions between group members, but suggested that aggression might increase during the breeding season, an inference later confirmed by Richard (1973, 1974a,b, 1978a), whose study overlapped a mating season. However, Richard also found that using priority of access to feeding resources as a criterion of social dominance did yield a clearly defined hierarchy within the group. In the three groups she studied intensively, the dominant individual in the feeding hierarchy (the same rank order did not necessarily stand up when other criteria of dominance were used) was female, but she concluded that feeding dominance was not necessarily a function of sex since some males were dominant over some females. In other ways, however, the structure of *P. verreauxi* groups was rather diffuse, their total organization depending on a host of factors such as interindividual relationships and the raised level of aggression displayed by infant-carrying females. Moreover, group structure tended to change seasonally. During the mating season and the period just preceding it intragroup aggression increased; in particular, normally subordinate males might display aggression toward previously dominant ones, disrupting the established feeding hierarchy. Further, females would permit sexual access only to males dominant at the moment of mating; these might have been subordinate earlier in the year. Changes in the dominance status during the mating season often exercised a lasting effect on group organization; thus a previously subordinate male who established himself as mating-season dominant to other males, either in his own group or in another encountered during the "roaming" behavior also displayed by males at this period, might retain that position of dominance after the mating season ended.

Propithecus diadema is normally seen in groups of 2 to 5 (Pollock 1979a). I have seen a group of 6 in dry forest to the north of the eastern rain forest strip, and Albignac (quoted by Pollock 1975b) has reported seeing a group of 6, including 2 infants, at an unspecified locality.

Pollock (1975a,b) states that censusing of 18 groups of *Indri indri* produced a mean group size of 3.1 individuals, with a range of 2 to 5. All groups in which sexes could be accurately determined (except for one consisting of two males) contained one adult male and female, and three size-classes of immature animals could be recognized. Pollock concurs with Petter (1962) in interpreting these units as pair-bonded family groups. Pollock (1975a, 1979b) notes that the adult females he studied were groomed more frequently than were the adult males, and often displaced the latter at feeding sites while the reverse never occurred. Further, adult males tended to feed much lower in the canopy than adult females, hence in areas of lower food availability. From these and other observations, Pollock concluded that the adult female was the dominant individual of the group, feeding and moving at will while to some extent controlling the movements and feeding of the male. However, he did note the possibility of changes during the mating season (Pollock 1979b).

Of *Avahi* extremely little is known, although group sizes have invariably been reported at between 2 and 4 individuals. Petter (1962) believes that woolly lemurs live in family groups, and my own limited observations support this.

Daubentoniidae

Almost all aspects of the life of the aye-aye are poorly known, and social organization is no exception. Petter (e.g., Petter and Peyrieras 1970; Petter et al. 1977) reports having observed only isolated individuals, although at Mahambo 3 adults were sighted within 50 meters of each other. Presumptive mother-offspring pairs were not uncommonly found close together. Petter (1977; Petter et al. 1977) suggests that the social organization of *Daubentonia* resembles that of *Microcebus*, but this must for the moment remain relatively uninformed speculation. Olfactory marking is highly important in the aye-aye's behavioral repertoire, and it is clear that this lemur is social, if not gregarious. Petter reports that an individual seen trying to enter an occupied daytime nest was rudely repulsed; but captive aye-ayes associate tranquilly.

Discussion

Although in most cases the observer in the field rarely encounters difficulty in recognizing "groups" of animals which tend to associate over longer or shorter periods of time, the concept of the "group" eludes ready definition. In a sense, it is easy to point to the variety of group structures exhibited by the gregarious, mostly diurnal, lemurs in which social organization is at all well known: at a superficial level one can without much difficulty distinguish the "family" groups of *Indri* or *Lemur mongoz*: the "small multi-male" groups of *Lemur fulvus rufus* or, perhaps, of *Propithecus verreauxi*; or the "large multi-male" groups of *Lemur catta*. One may even rank groupings thus categorized in order of supposed social complexity, and seek within such groups for structures which express some underlying continuity of social organization. But, as Richard (1978a) has eloquently argued, to do so implies not only a view of the "group" as a much more closed and static entity than observations warrant, but also the assumption of an excessively simplistic set of relations between individual group members.

In the gregarious lemurs, as among higher primates, the movement of individual animals, particularly males, between groups is becoming increasingly well documented. Brief though it is, the mating season seems to have a disproportionate importance in this regard: in both *Lemur catta* and *Propithecus verreauxi,* the best-known members of the Malagasy primate fauna, it is this period which witnesses the highest level of individual exchange between units which remain at least relatively stable in composition at other times of year (although the "chain migration" of *P. v. verreauxi* noted earlier took place in May, outside the breeding season). We should thus be wary of overstressing the importance of the "group" as a reproductive unit: it is highly suggestive that the only copulations witnessed in *Propithecus verreauxi* by Richard (1974, 1978a) and in *Lemur mongoz* by Tattersall and Sussman (1975) involved individuals belonging to different groups. Similarly, there is as yet no evidence that in those lemurs which seem to live in family groups, for instance *Lemur mongoz* and *Indri,* any compulsion to leave is exercised on the maturing offspring by the adult pair; yet the departure of adolescents would seem to be mandated by the maintenance of monogamy. Should one then say that the "local population" or "neighborhood" (Jolly 1966a), not the "group" as seen by the observer, is the primary reproductive unit among the lemurs? Perhaps; but in reality such populations are extremely hard to recognize, let alone to define, except in cases where habitat destruction has artifically created local isolates in patches of surviving forest. Even in the case of the Mayotte lemur, where it is quite clear that the groupings under the observer's eye at any one moment are no more than ephemeral, if perhaps repetitive, the local population has yet to be recognized as a consistent entity.

In this context it is as important to understand the nature of intergroup relations as it is not to oversimplify the relationships exhibited between individuals. Indeed, the two may well be interlinked, a possibility broached by Richard (1978a), who found that a consistent pattern of intergroup dominance existed between the two groups of sifakas she observed in her southern study area. These two groups interacted equally with all contiguous groups outside the mating season, but during that period they interacted disproportionately with each other. At all times, however, one group consistently displaced the other following confrontations. Richard believes that the cause of this unequal relationship lay in the cross-group recognition of individuals and of their status. Thus the dominant male in the subordinate group permitted sexual access to a female by a high-ranking member of the other group, but not by one of low rank. On the other hand, within-group relations were not necessarily consistently structured when different criteria were examined, and situations were noted where, for instance, frequency of aggression did not correlate positively with an individual's position in the feeding hierarchy. Indeed, Jolly (1966a) noted that the rank order of male *Lemur catta* when measured in terms of the frequency of aggressive behavior was the reverse of that derived from considering the frequency of sexual behavior. In any event, as Richard proposed, it is important to understand the relationship between individual animals, rather than merely between age-sex classes, when considering social structure. Unhappily, the level of individual recognition achieved in most studies of gregarious lemurs has been inadequate to allow this.

Charles-Dominique (1978) visualized a continuum of complexity in the social organization of the nocturnal lemurs. Noting that among many nocturnal mammals males and females move in individual but overlapping territories, he proposed that the social organization of *Lepilemur* represents a modification of a relatively primitive condition represented by *Microcebus*, and perceived elements of "pregregariousness" in the social organization of *Phaner*. It seems equally probable, however, that this spectrum exhibits adaptive divergence, rather than linear progression, from an ancestral pattern, and that the various social groupings seen among the diurnal lemurs are separately derived without necessarily having progressed through a "pre-gregarious" phase, at least of the kind exhibited by *Phaner*. Quite evidently, social organization among lemurs is highly responsive to external conditions; although within some polytypic species such as *Propithecus verreauxi* such organization appears to remain at least uniformly variable over large geographic areas, the same cannot be said for others. *Lemur fulvus* provides one example of this; the monotypic *Lemur mongoz* presents us with an even more dramatic illustration of the dangers inherent in speaking of "species-specific" behaviors. Phylogenetic inheritance presumably specifies the limits within which a species may respond behaviorally to environmental pressures; but to discover in the case of any one lemur the nature of these pressures and how they act to influence the myriad aspects of behavioral response would require an intimacy of knowledge far exceeding anything of which we can boast at present.

RANGES AND POPULATION DENSITIES

Cheirogaleidae

Mouse lemurs apparently occupy very limited ranges; Martin (1972a) observed that known individual *Microcebus* were never found more than 50 meters distant from the point at which they had been identified originally. The greater density of females in the centers of population nuclei (see "Social Organization") may suggest that they range over smaller areas than do central males, as may also the regularity of their nest-sharing; similarly, the generally more rudimentary nature of nests occupied by single males may imply that their ranging is less limited. It is not known, however, to what extent ranging behavior is constrained by nesting patterns. Peripheral males may, over the medium term if not on a nightly basis, range further than central males; if Martin's inferences as to population structure are correct, and a certain level of exogamy is maintained by migration of peripheral males between nuclei, these animals are likely to migrate over relatively large distances in accomplishing this end. The degree of stability in the ranges of individual mouse lemurs is uncertain; comparison of the results of his 1968 and 1970 studies at Mandena, near the southwestern tip of Madagascar, led Martin (1973) to conclude that a degree of continuity probably existed, although he did note certain changes, for instance in the apparent splitting of one range of associating females into two.

Mouse lemurs appear to be adapted to forest-edge habitats. Martin (1972a, 1973) notes that *Microcebus* are overwhelmingly sighted within a few meters of paths and tracks and rarely deep in the forest; he gives a host of reasons why this distribution is not simply an artefact of visibility bias. It seems unlikely that the mouse lemurs Martin studied were active primarily on the forest fringes uniquely because their major plant food resources happened to be located in such areas in his study locality; there does appear to be a clear behavioral preference for such environments.

This habitat preference tends to throw into some doubt the population density estimates made for the rufous mouse lemur by Petter and Petter-Rousseaux (1964) at Mahambo, on the east coast of Madagascar. Using a strip-censusing technique along a road, these authors found 20–21 individuals in a 4 km strip totalling about 80,000 m², giving an average population density of approximately 2.5 individuals/ha. If extrapolations from Mandena are appropriate, this must be an overestimate if all *Microcebus* living along the road were counted; on the other hand, the figure is substantially lower than the 13–36 gray mouse lemurs/ha reached by Martin for his study area at Mandena. In any event, the apparent discontinuity of mouse lemur population nuclei makes it difficult to arrive at reliable averages. In arid *Didierea* forest near Bemarivo, not far from Mandena, Charles-Dominique and Hladik (1971) found gray mouse lemurs occurring at a density of 3.6 individuals/ha.

Mirza coquereli has most recently been studied by Pagès (1978). Working in dry deciduous forest near Morondava in western Madagascar, she found that the home ranges of adults of opposite sex overlapped partially, that of the female (averaging 10 ha) exceeding that of the male (8 ha). Most time was, however, spent in much smaller core areas, averaging 3 ha for females and 2 ha for males. On the basis of rather limited observations, Pagès concluded that the core area was defended, at least by males. Outer areas were patrolled primarily during the latter half of the night, and travel distances within them were relatively long; males might move direct distances of up to 400 m, while females tended to circle the range periphery. Travel during a single night could total as much as 1,500 m. Encounters between conspecifics in the outer range areas were infrequent, but generally friendly. Overall population density in the study area was approximately 0.3 individuals/ha (Petter 1978).

Virtually nothing is known of ranging behavior in either species of *Cheirogaleus*. Petter et al. (1977) have reported "considerable" densities of *C. major* in certain east coast areas of high feeding resource concentrations, while in deciduous forest near Morondava *C. medius* is estimated to occur at a density of 3 to 4 individuals/ha (Petter et al. 1977; Petter 1978). Charles-Dominique (1978) regards *Phaner furcifer* as being "territorial," but notes that up to 7 individuals may gather and vocalize in zones of range overlap. Heterosexual pairs are said to occupy the same range, the female being dominant to the male and having priority at favored feeding sites. Members of a pair remain in constant vocal contact, and the male regularly marks the female with his throat gland; olfactory marking seems, however, to be unimportant in range delimitation. No actual ranges have been published, but a decade ago Petter et al. (1971) recorded a density near Morondava of 6.8 individuals/ha. More recently, this has been scaled down to a maximum density of 1 to 2 identified individuals per hectare in the same area (Petter 1978).

Lepilemuridae

Studying *Lepilemur mustelinus leucopus* in *Didierea* bush near Berenty, Charles-Dominique and Hladik (1971) found the animals living in very small territories, those of males (0.20 to 0.36 ha; mean, 0.30 ha) being rather larger than those of females (0.15 to 0.32 ha; mean, 0.18 ha). Ranges were generally exclusive of other adults of the same sex, but male ranges overlapped those of a variable number of females. Females shared their ranges with their immature offspring, and Charles-Dominique and Hladik felt it possible that relationships between females and their female offspring persisted over long periods. With range diameters typically in the order of 50–50 m, the entire range could be surveyed by its occupant from a single high point; such "surveillance" behavior was noted particularly commonly among males, who might survey each other mutually for hours. Removal of a male from his territory, which overlapped that of 5 females, resulted in the immediate invasion of his range by neighboring males, each of whom began associating with one of the females. Range boundaries did not appear to be delimited by olfactory marking, but by vocalizations, displays, and surveillance.

One night's following of an adult male *Lepilemur* by Charles-Dominique and Hladik yielded a total distance covered of 270m, achieved in about 180 separate leaps. A high level of inactivity was found during the night, but these authors suggested that energetic limitations were nonetheless operating on activity (see "Diet and Feeding Behavior" later in this chapter).

Russell (1977), who later studied the sportive lemur at the same site, arrived at rather different conclusions, however. He found that heterosexual or female-female pairs often shared most or all of their range, which was more or less exclusive if rarely overtly defended. Vocalizations peaked during the mating season, and were heard relatively infrequently at other times of the year; some individuals hardly ever vocalized at all. Surveillance behavior was rarely noted. Three monthly ranges calculated by Russell for female sportive lemurs varied from 0.1 to 0.2 ha, and one male monthly range was 0.3 ha; these ranges are closely comparable to those published by Charles-Dominique and Hladik. Nightly travel distances were found by Russell to vary from about 320–500 m, but no consistent correlation obtained between monthly range area and average nightly distances travelled. Russell concluded that relative to body size *Lepilemur* travels significantly farther than other lemur species. This is in direct opposition to Charles-Dominique and Hladik's finding of exceptionally low activity, but is based on more extensive observations.

Lepilemur occurs at high densities in several areas of Madagascar. Charles-Dominique and Hladik (1971) recorded a density of 2.0–3.5 individuals/ha in the *Didierea* bush near Berenty, and of 2.7–8.1 individuals/ha in the nearby riparian forest. Sussman (1972) noted a density of 2.0 individuals/ha at Antserananomby, while Pollock counted 5 *Lepilemur* in a 2 hectare area at Analamazoatra, in eastern Madagascar. A census at this latter site by Charles-Dominique and Hladik (1971) had yielded a similar density of 2.0 individuals/ha.

No quantified information on the ranging or population density of either species of *Hapalemur* is available, apart from the observation of Pollock (1979a) that two 18 ha areas at Analamazoatra each contained 17–22 *H. griseus,* giving densities of 1.1 to 1.2 individuals/ha. *Hapalemur* groups appear to maintain a relatively high degree of spatial cohesion while foraging and travelling.

Lemuridae

The ranges of *Lemur catta* troops at Berenty have been measured on several occasions, and vary from a minimum of 5.7 ha (Jolly 1966) to a maximum of 23 ha (Budnitz and Dainis 1975). These latter authors concluded that range size was inversely proportional to the richness of the environment, larger ranges sizes occurring in the poorer areas of the reserve. Periodic restudies between 1963 and 1975 (Jolly 1966; Klopfer and Jolly 1970; Budnitz and Dainis 1975; Mertl-Milhollen et al. 1979; Jolly et al. in press) indicate that at Berenty home ranges in general have stayed relatively stable over 12 years, although between births, mortality and male migration, there clearly has been much change in group compositions. Seasonal differences in range have been noted, however; for example, during January and February groups living adjacent to the river at Berenty were observed to stay much closer to the water than at other times. Similarly, in 1970 Jolly (1972) found that approximately the same number of ringtailed lemurs was present in her study area at Berenty as in 1963–4, but that they were grouped into twice as many smaller units which ranged without respect to earlier boundaries. She also found that range overlap had increased dramatically, and that major feeding resources were shared on a "time-plan" basis rather than being defended. It seems plausible to ascribe these changes at least in part to the season of study: in 1970 the ringtails were studied in late September, close to the birth season which Budnitz and Dainis (1975) and later authors have shown to be the moment of maximal change in group composition and resource partitioning. Short-term environmental influences in the wake of an unusually hard dry season cannot be ruled out as causes, however.

Home range boundaries were defended at Berenty, at least at most times a year, although the general pattern seemed to be one of mutual avoidance when groups encountered one another at the periphery of their ranges. Jolly (1972) records that in her 1970 study intertroop "battles" occurred on several occasions, and Mertl-Milhollen et al. (1979) found that such battles typically involved physical contact between females, the primary antagonists. The hierarchical structure of the troop was most clearly displayed in its spatial arrangement when it was moving to repel invaders or making forays outside the home range; in such cases the dominant female invariably led (Budnitz and Dainis 1975).

Sussman (1972, 1974) studied *Lemur catta* at both Berenty and Antserananomby. At the former, day ranges averaged 965 m, and at the latter, 920 m. Home ranges were 6.0 and 6.8 ha respectively. Travel was found to account for 11.3% of time at Antserananomby, and 6.9% at Berenty. Jolly (1966) observed the ranging behavior of *Lemur catta* at Berenty to consist typically of three or four days' exploitation of one segment of the range, followed by a similar concentration on another part. Within a week or ten days a troop would in this manner have visited all parts of its range. Sussman's study group at Antserananomby did not display this pattern, however, moving more consistently thoughout its home range. Ringtailed lemurs move and forage relatively consistently throughout the day, and most progressions are more or less linear, headed by adult females, juveniles and central males, with the "drones' club" of subordinate males bringing up the rear. The latter may thus be excluded from access to food or water at certain places of limited supply. Around the birth season, these males become increasingly dissociated from the group.

Early estimates of the population density of *Lemur catta* at Berenty were arrived at by extrapolation and seem to have been rather high. Budnitz and Dainis (1975) were the first to census the entire 95 ha reserve. In 1972 they arrived at a figure of 153 individuals, and in 1973 at 152. The same process was repeated in 1975 (Mertl-Milhollen et al. 1979), yielding a figure of 155 individuals. Budnitz and Dainis point out that population density is not uniform within the reserve, the relatively rich riverine forest supporting a greater density of animals than the poorer areas of brush and scrub into which it grades. If bias due to this factor explains the high estimates of population made by Jolly (1966:350 individuals) and Sussman (1972, 1974:250 individuals), then the population of *Lemur catta* at Berenty would seem to have remained remarkably constant over more than a decade. Certainly it has done so over a period of four years. Sussman (1972, 1974) arrived at a population density for Antserananomby of about 2.15 individuals/ha, somewhat in excess of the 1.6 individual/ha of Berenty; the difference is likely to be due to the different carrying capacities of the two forests, since Antserananomby was quite exhaustively censused.

It is noteworthy that a birth season intervened between the 1972 and 1973 censuses of Berenty by Budnitz and Dainis. In this period, 47 infants were born. The net loss of one ringtailed lemur over the year thus implies a very high level of mortality. Budnitz and Dainis showed that this loss was largely due to infant, rather than adult, mortality.

Lemur fulvus rufus, studied by Sussman (1972, 1974, 1975) at Antserananomby and Tongobato had very much smaller home and day ranges than *Lemur catta,* and much higher population densities. The mean ranging area size at the former site was 0.75 ha, and at the latter, 1.0 ha. These group ranges were not defended, and over-lapped considerably. Encounters between groups might, however, elicit some inter-group aggression. Day ranges averaged around 125–150 m. Travel occupied 2.4% of time at Antserananomby; the substantially higher figure for Tongobato of 12.1% is probably greatly inflated because of the low level of habituation achieved at that forest. Sussman estimated population densities for the rufous lemur of around 12.0 and 9.0 individuals/ha at Antserananomby and Tongobato, respectively.

Harrington (1975) has recorded a home range of about seven hectares for one group of *Lemur fulvus fulvus* in dry deciduous forest at Ampijoroa. Neighboring groups showed some overlap of home range, but displayed "territorial defense" when both groups were in the area of overlap at the same time. Mutual avoidance was more common, however. Pollock (1979a) reports a population density of 0.4–0.6 individuals/ha for this subspecies in the eastern rainforest.

Lemur fulvus mayottensis was found in 1975 to exist at Mavingoni in a high density of around 10.0 individuals/ha (Tattersall 1977c). A similar result was obtained at this forest in 1977, but in 1980 the density of lemurs at the site was somewhat diminished although this may have been due to habitat disturbance (Tattersall in prep.). Home range size was impossible to quantify because of the fluid social organization shown by this lemur, but day ranges varied from 450 to 1150 m (mean: 800 m) in the wet season, (Tattersall 1977c) and from 350 to 840 m (mean: 560 m) in the dry (Tattersall in prep.). Travel on average occupied some 4.7% of the hours of daylight during the wet season, and perhaps surprisingly because of the generally lower distances covered, 5.3% in the dry.

Petter (1962) found 29 *Lemur macaco* inhabiting about 50 ha on Nosy Komba, giving a density of ca. 0.58 individuals/ha. Range size, daily or annual, is unknown, but Petter noted that certain groups which maintained separate daytime ranges tended to congregate at night. Petter also recorded that group movements were generally led by an adult female, with a male bringing up the rear.

Lemur mongoz at Ampijoroa was found to live in small, overlapping home ranges (Tattersall and Sussman 1975). During a two-month study the group most intensively observed was not seen outside an area of 1.15 ha, within which other groups were also sometimes observed. It seems highly likely that this is a seasonally limited range and that the annual range of the group was somewhat larger. Night ranges varied from 460 to 750 m; progression could be initiated or led by any individual. Different groups might feed in the same tree during the course of the night, but at different times.

As in the case of other *Lemur* species and *fulvus* subspecies, the ranging behavior of *Varecia* remains completely obscure.

Indriidae

At Berenty, Jolly (1966) found *Propithecus verreauxi verreauxi* living in small home ranges of about 2 to 3 ha, with much of this area (0.6–1.8 ha) being defended against neighboring troops. Jolly described territorial "battles" as "formal affairs" resembling an arboreal chess game, members of the groups involved occupying sections of tree rather than attacking the opposition. Territorial disputes were quite frequent although travel within the range was somewhat limited; groups would rarely move more than 200–300 m in a day. In the course of a week or so, all parts of the range would be visited. Mertl-Milhollen (1979) has presented evidence from Berenty that olfactory marking is significantly more common at the periphery than in the center of ranges, and suggests that this is important in territorial demarcation. The restricted size of the home ranges of the sifakas at Berenty is probably related to the relatively high density (1.1–1.5 individuals/ha) of the animals in this forest: a density which seems to have remained more or less constant over more than a decade of periodic recensusing (Jolly et al. in press). This stability seems to be the product of a relatively high rate of infant mortality, especially since some 40 percent of adult females give birth annually.

The ranging behavior of *P. verreauxi* has been documented in some detail by Richard (1973, 1974b, 1978), who studied *P. v. verreauxi* in arid forest at Hazafotsy in southern Madagascar, and *P. v. coquereli* in seasonal forest at Ampijoroa. Two groups at each site ranged over areas from 6.75 to 8.50 ha, and in each case there was some overlap between the ranges of neighboring groups. However, those in the south had exclusive use of 87% and 91% of their total home ranges, whereas the corresponding figures for the northern groups were only 46% and 43%. Areas of exclusive use did not correspond to the "core areas" in which the groups spent most of their time; core areas might be found in zones of overlap, and did not necessarily consist of a single block of territory. All groups spent at least 60% of their time in areas of 2.5 ha or less. As might be expected, intergroup encounters occurred more commonly in the north than in the south; but intergroup "battles" were proportionately more frequent in the south. Richard concluded that while the southern groups of *Propithecus verreauxi* could properly be considered territorial, the northern ones could not.

In both areas sifaka groups visited most parts of their home range within the space of 10 to 20 days. Daily ranges in the dry season were 750 m in the north and 550 m in the south, and in the wet season 1,100 and 1,000 m, respectively; the wet-season increase in travel appeared to be due to faster rates of movement over the same parts of the range rather than to ranging over larger areas.

During the mating season, but rarely outside it, "roaming" behavior was observed by Richard in adult males who, singly or in pairs, might leave their own groups and make long excursions into the home ranges of neighboring groups.

Pollock (1979a) has reported that part of the range of one of three groups of *Propithecus diadema diadema* which he observed measured at least 20 ha. The rarity with which this animal is seen argues for an extremely low population density, but to what extent this is due to human interference is uncertain.

Ranging behavior in *Indri indri* has been described by Pollock

(1975a,b, 1979a). Groups defended large central portions of their ranging areas, which totalled 17.7 and 18.0 ha in two cases at Analamazoatra. Overlap between the ranges of neighboring groups was confined to narrow bands of 30–50 m in width; intergroup encounters were observed only rarely, and it may be that group spacing is achieved largely through vocalization. Daily ranges varied from 300 to 700 m, and Pollock observed two distinct patterns of range use. One of these applied at times when concentrated food sources were available, and involved relatively regular movement; wider, more unpredictable ranging was seen at other times. Certain areas of the range of both major study groups were preferred over others; respectively, they spent 48% and 45% of their time in 20% of their home ranges, and 78% and 75% of their time in 50% of their ranges. In general, a group would occupy a particular part of its range for some time, then abruptly switch to another area, not necessarily adjacent. Pollock suggests that economy of movement within the range may reflect the familiarity of its occupants with it, and thus their length of tenure.

Pollock's estimates of *Indri* density at his study areas of Analamazoatra, Vohidrazana, and Fierenana, whether calculated from direct or indirect evidence, all fell between 0.09 and 0.16 individuals/ha. He notes, however, that at Betampona, near Tamatave, densities were certainly much lower than this.

Avahi laniger probably occupies relatively restricted home ranges, but no detailed information is available. Pollock (1979a) encountered five groups within 150 m at Analamazoatra, and it is probable that densities are locally quite high.

Daubentoniidae

Effectively nothing is known about the ranging behavior of aye-ayes. Petter et al. (1977) report finding 2 individuals in a 5 ha patch of forest near Mahambo.

DIET AND FEEDING BEHAVIOR

"Diet" is a term which has proven extremely difficult to define in the context of primate field studies, and although a large number of observations has been made on lemur feeding, the published record does not constitute a homogeneous body of literature. Much of this literature by now consists of quantitative reports on feeding by lemurs, but few of these contain directly comparable figures because of the differences in the techniques of data collection employed. Some studies have quantified the time spent feeding on various resources, either through direct time recording or indirectly through time sampling; some have tried to record feeding rates, and others have attempted to establish actual quantities of food ingested. Each of these procedures is subject to unknown and probably unsystematic error beyond that of the visibility bias inherent in observing animals in dense vegetation; they yield totally noncomparable figures; and in at least the first two cases there is no obvious relationship between the figures obtained and nutritional intake. For all these reasons I have preferred not to tabulate available quantitative data on lemur feeding; such tabulation would inevitably imply a degree of comparability which the data simply do not possess.

Cheirogaleidae

Martin (1972a) has provided the most detailed account of mouse lemur feeding. He noted that although fruit constituted the major part of the plant diet of *Microcebus* spp., flowers were fairly often consumed and leaves occasionally. Mouse lemurs were seen to trap and eat insects on a variety of occasions, but examination of the stomach contents of two individuals from each of four study areas (including both *M. murinus* and *M. rufus*) revealed that animal food had accounted for under half of the food intake of these animals. Beetles provided the bulk of the animal diet, and a variety of other insects the rest; in captivity, however, mouse lemurs will eat small vertebrates such as tiny tree-frogs and chameleons, and it is probable that these, and small mollusks, are occasionally taken in the wild. Martin concluded that earlier reports that it is highly insectivorous are misleading, and that the mouse lemur is best regarded as an omnivore. Available evidence indicates that mouse lemur plant diets are quite specialized by locality, at least seasonally; thus in the eastern rain forest *M. rufus* primarily exploited the berries of *Clidemia hirta* at a number of sites, while at Mandena *M. murinus* was most frequently seen to eat the fruit and flowers of *Vaccinium emirnense*. Similarly, stomach contents of gray mouse lemurs taken at Ampijoroa were predominantly composed of an unidentified small fruit. In contrast, the exploitation of animal food seemed to be more opportunistic.

According to Pagès (1978), *Mirza coquereli* at Analabé showed great seasonal variation in diet. During the wet season these lemurs fed omnivorously, on fruit, flowers, and insects, and probably also on eggs and small vertebrates. In contrast, during the dry season, when forest productivity diminished markedly, they fed mainly on the secretions produced by colonies of homopteran larvae.

The dietary habits of *Cheirogaleus* spp. are poorly known, but Petter et al. (1977) state that dwarf lemurs feed primarily on ripe fruit, nectar, and pollen; insects are only occasionally taken. No *Cheirogaleus* has ever been recorded to eat leaves, either in the wild or in captivity. Petter (1978) claims that *C. medius* at Analabé feeds mainly on flowers from November to January, and on fruit in February and March. Occasional observations have been made of *C. medius* licking homopteran larval secretions.

Phaner furcifer appears to be highly specialized for the exploitation of gums exuded from the trunks and larger branches of certain trees. Its keeled and pointed nails, which permit it to scale broad tree trunks, and its long, robust dental scraper, are plausibly adaptations to this end. Petter (1978) states that the gums exuded during the day from parasite scars of other wounds on trees are "avidly" licked at night by fork-marked lemurs, and that these secretions appear to provide the bulk of the diet throughout the year. During the wet season, however, the diet of gum is supplemented by flowers and fruits, while at the end of the dry season homopteran larval secretions are also sought. Animal prey, generally insects, are also taken, normally during the second half of the night after gum-feeding has been largely completed.

Lepilemuridae

In a brief but famous study, Charles-Dominique and Hladik (1971; Hladik and Charles-Dominique 1974) conducted short-term but intensive observations on *Lepilemur mustelinus leucopus* living in Didieraceae bush near Berenty. The dominant floral components at the site were the tall, spiny *Alluaudia procera* and *A. ascendens*. The leaves and flowers of these two species made up the bulk of the diet, with the leaves of four other species and the green fruit of another composing the remainder. Study of feeding rates, combined with direct recording of feeding time, permitted an estimate of about 60 g to be made of the total weight of food ingested per night. Analysis of food composition then allowed Charles-Dominique and Hladik to calculate a very low energetic value for the readily assimilable components of this nightly diet, about 13.5 kcal. An adult male *Lepilemur* followed for one night covered 270 m in 180 leaps, otherwise remaining motionless for most of the time. The minimum energetic expenditure involved in leaping was calculated to be 2.16 kcal, while even a low estimate of basal metabolism required a daily energy consumption of 20–30 kcal. These sundry calculations revealed a considerable gap in the energy budget of the sportive lemur, which Charles-Dominique and Hladik filled by invoking caecotrophy, the reingestion of a certain proportion of the feces. They suggested that this took place during the daytime resting period, when sportive lemurs were sometimes seen licking their anogenital regions. The feces involved were said to be a fraction of the food partly broken down in the caecum and rapidly excreted; this, higher in protein and hemicellulose than the food in its original form, was estimated to produce the additional 10–15 kcal needed to balance the animal's apparent energy deficit.

Russell (1977) has, however, contested the existence of caecotrophy, a behavior he never saw in four months of study at the same site, although he did observe anogenital grooming in the posture described as typical of it. In the course of his study, Russell observed sportive lemurs to feed on 24 different plant species, although two species, notably *A. procera*, accounted for almost 65% of feeding time. Overall, he found 91% of feeding time to be devoted to leaves, and 6% to flowers and fruit together. Russell's estimates of weight of food ingested are considerably higher than Charles-Dominique and Hladik's, monthly averages ranging from about 90–120 g per night, but it was nonetheless found that the quantities of food available to the lemurs were always substantially in excess of consumption. Russell estimated daily expenditures for *Lepilemur* of 40–53 kcal, and found no reason to conclude, as Charles-Dominique and Hladik had done, that any behavioral limitations were imposed by the poverty of the environment; the energetic cost of locomotion came to little more than 2–3% of estimated total daily energy expenditure, even though per unit body weight he found the sportive lemur to travel long distances compared to other lemurs.

Petter (1978) reports that *Lepilemur mustelinus ruficaudatus* at Analabé ate mostly leaves, supplemented by buds and fruit; he claims that home ranges at this site are very small "due to the utilization of ubiquitous food resources" (p. 217).

Petter and Peyrieras (1970b) found that *Hapalemur griseus griseus* fed primarily in the early morning and the evening, and that its diet was essentially confined to young bamboo shoots and leaves. Milton (1978) has drawn attention to a dental specialization apparently related to this dietary preference. The upper canine tooth is short and robust, the anterior premolar behind it is relatively large, and the two are not separated by a diastema. Close observation of captive individuals feeding showed that bamboo shoots were clamped between these teeth and then pulled sideways, the tender young shoot thus being liberated from its tougher sheath. The shoot was then reinserted from the side of the mouth, and crushed. Milton notes that young shoots, in addition to containing a lower portion of relatively undigestible structural materials, probably contain more protein than older ones, and lower concentrations of unpalatable secondary compounds. Typically, a gentle lemur will scissor off a bamboo shoot with its premolars, then sit in an erect posture to eat it as already described; it seems possible that the characteristic truncal uprightness of this form is at least as much a feeding as a locomotor adaptation.

Alaotran *Hapalemur* is said by Petter et al. (1977) to subsist on the leaves and young shoots of *Phragmites* reeds, and on the buds and pith of *Papyrus*. *Hapalemur griseus occidentalis* is found in forests characterized by the presence of bamboo vines ("viky") and presumably feeds off these; Petter et al. report having seen these lemurs eating the fruit of *Flacourtia ramontchi* and wild dates. They also say that *Hapalemur simus* may feed on bamboo stems of up to 20 cm in diameter at periods when young shoots are rare. The feces of one individual also contained the seeds of a palm, *Dypsis* species.

Lemuridae

At Berenty, Jolly (1966) noted *Lemur catta* feeding on 24 different plant species, of which 11–15 were exploited at any one time. She never saw them eating insects, and recorded them feeding on fruit in 70% of her observations and on leaves in 25%. More detailed but shorter-term observations on feeding by ringtails at this site have been made by Sussman (1972, 1974, 1977c), who scan-sampled at five-minute intervals, noting the number of individuals feeding at each interval and what they were feeding on. He found that Berenty *L. catta* spent 31.1% of their time feeding on a total of 24 plant species (in only a single month), and that fruit accounted for 59.3% of feeding time, leaves for 24.4%, flowers for 6.1%, and herbs for 5.5%. At Antserananomby, on the other hand, where 24 plant species were also exploited (although only three food species were common to both forests), ringtailed lemurs spent 24.9% of their activity on feeding, of which 42.6% was devoted to leaves, 36.6% to fruits, 14.7% on herbs, and 8.1% on flowers. The figures for the two sites are thus substantially different; but so also were the phenological states of the forests at the times of study: far more trees were in fruit at Berenty than at Antserananomby.

Better evidence of seasonal variation in the diet in a species of *Lemur* is available for the Mayotte lemur, studied at Mavingoni for four months (February–May) of the wet season (Tattersall 1977c), and for two months (July and August) of the dry (Tattersall in prep.) Data were collected in a manner comparable to Sussman's. In the wet season, the Mayotte lemurs devoted 11.7% of the daytime (0600–1800 hrs) study period to feeding. Overall, 67.4% of this time was spent feeding on fruit, with monthly figures ranging from 48.2% to 78.7%; 27.3% on leaves (20.1% to 36.1%); and 5.0% on flowers (1.2% to 14.2%). The two dry-season months saw an increase in feeding activity (14.6% of the time) as well as substantially different patterns of feeding: in July, 51.8% of daytime feeding was devoted to flowers and floral buds, 38.8% to leaves, and only 9.0% to fruit. Corresponding figures for August were 24.0%, 65.3%, and 10.0%, respectively. Thus, depending on the time of year, *Lemur fulvus mayottensis* might be found subsisting primarily on fruit, on leaves, or on flowers. Given the 24-hour activity of this subspecies, it is possible that daytime feeding patterns are not fully representative of feeding over the full activity period. Nocturnal observations during July and August did not suggest that this was the case (Tattersall 1979), but brief observations conducted in June 1980 do indicate a difference between nocturnal and diurnal feeding patterns in that month. During one complete night of observation the lemurs fed virtually exclusively on the flowers of the kapok tree (*Ceiba pentandra*), whereas during two complete days they fed on 65.5% leaves, 6.6% fruit, and only 27.9% flowers and floral buds (none of them kapok). The flowers of *Ceiba pentandra* open only after dusk; it seems reasonable to assume that they are eaten by the lemurs only at night because of this, and that outside the flowering period of this tree nocturnal and diurnal diets are generally similar. Flower-feeding by the Mayotte lemur was almost invariably destructive; flowers were bitten, not licked.

No detailed record of forest phenology was kept at Mavingoni, but it was clear that the lemurs preferred to eat fruit when it was available, and that they fed at least on certain leaves only when relatively few alternatives existed. For instance, during August the leaves of *Litsea glutinosa* occupied 60% of feeding time whereas they were virtually never eaten during the wet season (0.1% of feeding time), despite being abundantly available. Unfortunately, no information exists on any possible seasonal variation in secondary compound content of these leaves.

During the wet-season study, *Lemur fulvus mayottensis* was seen to feed on over 48 different food resources, involving the fruit, flowers and leaves of at least 32 plant species. In July, only 13 plant species provided 19 different foods, and in August the corresponding figures were 21 and 29. In all months except May, however, a mere three or four resources each accounted for more than 5% of time spent feeding, and in each month a single resource was predominant: in February and March, *Saba comorensis* fruit (37.9% and 32.7% respectively); in April, *Adenanthera pavonina* fruit (39.1%); in May, *Litsea glutinosa* fruit (29.1%); in July *Erythrina fusca* flowers (47.2%); and in August *Litsea glutinosa* leaves (60.0%). Thus in both seasons Mayotte lemurs fed on a wide array of plant species and plant parts, but gained the bulk of their sustenance from very few.

This feeding pattern is somewhat at variance with what Sussman (1972, 1974, 1977c) observed in *Lemur fulvus rufus* at both Tongobato (wet season) and Antserananomby (dry season). In each forest (hence both seasons) leaves accounted for the bulk of feeding time: 52.1% and 89.2%, respectively; and in each many fewer plant species were eaten: eight at Tongobato and 11 at Antserananomby. As might be expected, a very small number of plant species provided most of the diet: at Tongobato, three species accounted for over 80% of feeding time, and at Antserananomby the rufous lemurs ate the mature leaves of the single species *Tamarindus indica* for about 75% of feeding time. Fruit was more commonly eaten at Tongobato than at Antserananomby (42.4% vs. 6.8%), possibly reflecting a seasonal difference more than one of forest composition; and at neither did flowers account for more than about 4 or 5% of feeding time.

The diet of *Lemur mongoz* at Ampijoroa is highly specialized, at least seasonally (Tattersall and Sussman 1975; Sussman and Tattersall 1976). In July and early August 1973, we observed these mongoose lemurs to feed on only five plant species. Flowers composed much the greatest part of the diet, kapok flowers alone accounting for 64% of feeding time, and the flowers of two other species for a further 39% of feeding time. The flowers were not eaten whole; instead they were licked, presumably to obtain nectar. An additional 14% of feeding time was spent in eating the extrafloral nectaries of *Hura crepitans*. The lemurs would simply snip off the base of the leaf containing these structures, and discard the rest. A total of 81% of feeding time was thus devoted to nectar-containing plant parts, and even though visibility bias may have favored observation of this sort of feeding, it is nonetheless clear that nectar provided the bulk of the diet at the time of the study. Of the remaining feeding time, fruit (of *C. pentandra*) accounted for 17.5%, and leaves 1.5%. It is important to note that the study forest consisted largely of introduced tree species, and that this diet is highly unlikely to be typical for this primate; but its characteristic nondestructive flower-visiting behavior, which contrasts strongly, for instance, with the destructive flower feeding of *Lemur fulvus*, suggests that the mongoose lemur may play a major role as pollinator in its area of occurrence.

Feeding techniques are fairly uniform among the various species of *Lemur,* and normally do not involve much manipulation of food items. In most cases, a branch bearing food will be pulled towards the mouth with one hand, and the food eaten directly from it (fig. 7.3). When eating large fruits in this fashion these animals often employ both hands, and in the case of very large ones such as jackfruit they may simply cling with all extremities to the object being consumed. Small fruits, on the other hand, notably those which commonly occur among fine branches, may on occasion be scissored from the stem with the premolars and be carried to large branches to be eaten; during this latter process the fruit is normally held in both hands, its possessor sitting on its haunches (fig. 7.2). *Lemur* show great agility in reaching virtually all areas of the forest canopy in search of food. Manual suspension is relatively rare, but pedal suspension is a common component of the very wide postural repertoire which characterizes these animals while feeding.

No systematic observations have been made on the feeding behavior of *Varecia*: Petter et al. (1977) claim them to be frugivorous, which agrees with inferences from their dental morphology.

Indriidae

Thanks largely to the efforts of Richard (1973, 1977, 1978a,b), we know more about seasonal and geographical dietary variation in *Propithecus verreauxi* than in any other Malagasy primate. Two groups of *P. v. coquereli* were studied in the relatively rich mixed forest of Ampijoroa, and two of *P. v. verreauxi* in the semi-arid Didiereaceae forest of Hazafotsy, some 600 miles to the south. Richard followed focal individuals which she regarded as feeding if they were chewing or ingesting, and sampled their behavior at one-minute intervals. Data collected in this way revealed seasonal differences in both study areas in time devoted to feeding. In the south, feeding occupied 32.8% of total time during the wet season, but only 24.2% during the dry; corresponding figures for the north showed a less marked difference, at 37% vs. 30%.

The northern study forest was richer in plant species than the southern one, and in the north both study groups fed on more different species (85 and 98: 34.5% and 41% of all plant species potentially available) than did either southern study group (77 and 65: 64% and 52.5% of plant species potentially available). No more than four plant species were identified as common to both forests; of these only two were fed on in both areas, and then in greatly disparate proportions. Richard also found considerable differences in the species composition of the diets of neighboring groups in each area, and further noted that this composition changed almost totally between seasons; very few species were fed on throughout the year. Those which were consistently exploited generally produced leaves at all seasons, while species of short-term dietary importance tended to be deciduous seasonal producers of flowers and fruit.

Quite notable variation was displayed by the northern and southern study animals in the plant parts preferred. In the north, immature leaves were preferred overall to mature ones (ca. 26% vs. ca. 20%); in the south, vice versa (ca. 6.5% vs. ca. 33%). In the north, mature leaves marginally exceeded fruit (ca. 32%) as the largest single dietary component; in the south, fruit was by far the largest item in the diet (ca. 46%). Seasonal dietary variation was equally marked. Consumption of immature leaves peaked in the north early in the wet season, around October; this was followed by an increase in the exploitation of flowers in November and fruit in December. Consumption of mature leaves predominated during the dry season in the north, as it did in the south where fruit and flowers were also very commonly eaten during the wet season. Peak consumption of mature leaves by one of the southern study groups hit almost 70% in one month of the dry season; peak consumption of fruit by the same group reached almost 80% of time in one month of the wet. A similar pattern was noted in the north. Much time was spent feeding on bark and/or dead wood during certain periods of the year. In one month of the dry season one southern study group spent over 15% of total feeding time eating the bark and soft cambium of one thin-barked tree species. The sifakas used their tooth-scrapers to gouge out strips of the moist wood, leaving scars up to one centimeter deep and four long. Little bark was eaten in the north, but long periods were spent gnawing at dead tree stumps. Seasonal dietary changes quite closely reflected local phenological changes in

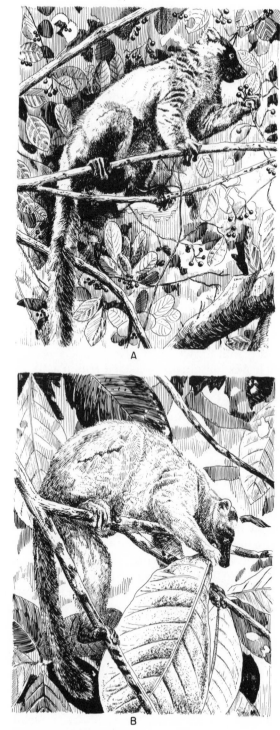

Figure 7.3. Typical postures of *Lemur fulvus mayottensis* when feeding on: A: *Litsea glutinosa* fruit; B: *Terminalia catappa* leaf petioles; C: *Adenanthera pavonina* fruit; D: *Saba comorensis* fruit. From Tattersall (1977c).

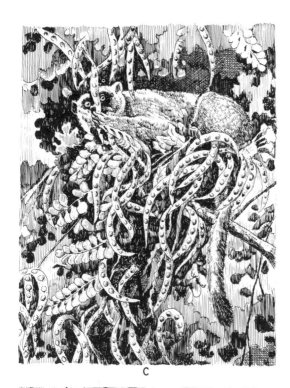

C

D

both study forests, and although both commonly and rarely occurring foods were eaten, almost all abundantly occurring plant species were exploited, whereas more than half of the rare species present were never fed on.

Food is normally consumed by sifakas where it is found; Richard records that she rarely observed food-carrying. Use of the hands during feeding was thus generally limited to pulling the food-bearing branch to the mouth, and most feeding was done in areas of small supports (where most food occurs) rather than on trunks or large branches. In 75% of feeding observations Richard recorded sifakas in sitting or vertically clinging postures; but in the remaining 25% the animals were observed in an enormous variety of feeding postures, often suspensory. This versatility allowed exploitation by these relatively large animals even of the most peripheral areas of the forest canopy.

Pollock (1975a,b, 1977) studied the feeding ecology of two groups of *Indri indri* in rainforest at Analamazoatra. On the basis of 15-minute focal-animal samples, he found that on average feeding occupied 39.2% and 36.6% respectively of the total activity periods of the two groups. However, over the year the proportion of the activity period devoted to feeding varied from 30% to 60%, and at least in one case time spent feeding increased and decreased in proportion with seasonal change in the length of the activity period.

At least 62 plant species were fed on by *Indri* during Pollock's study, but a fair degree of selectivity was noted. One group devoted 50% of feeding time to the top five food specie and 13.9% to the top one; the other spent the same proportion of time on the top five, but 20% on the top one. Young leaves and leaf shoots accounted for 36.1% of the feeding time of one group and of 32.2% of that of the other; the plant part next most commonly consumed was fruit (26.4% and 23.8%), followed by flowers (2.3% and 0.0%) and mature leaves (0.9% and 0.2%). The plant part involved was unidentified in the remainder of Pollock's feeding observations. Pollock states that seasonal peaks in the consumption of the various foods were normally due to frequent visits by the babakotos to a few localized food resources, and that each group's diet normally changed completely every two or three months in response to effectively random changes in the phenological state of major resource trees. On most days of observation babakotos would descend to the ground to eat small quantities of the earth exposed by the roots of fallen trees. Between five and ten such sites were regularly visited by each study group.

Indri generally ate leaves, flowers, leaf buds, and some fruit directly from branches pulled to the mouth. Small fruits were, however, sometimes severed dentally from the stem and gnawed from the fist, while large fruits were bitten from their stalks and carried to a stable support for eating. The tooth-scraper was regularly used to scoop out the green pulp of unripe fruit. Thin saplings, unable to bear the weight of this heaviest of the surviving lemurs, were often pulled in toward vertical trunks which then supported the individual during feeding.

Daubentoniidae

No systematic study exists of the natural feeding behavior of *Daubentonia*, but several remarkable observations have been made by Petter and others (e.g., Petter and Peyrieras 1970; Petter 1977), who claim that all nocturnal activity is related to food collection. These authors state that fruit, in particular coconuts, provide a substantial proportion of the diet of the aye-aye. Immature coconuts are chosen while still on the tree, and the specialized incisors are used to gnaw through the outer husk. Through the small opening thus produced, the milk and the still-soft pulp are extracted using the remarkable third digit. Mangoes are said to be another favored fruit, and Petter notes that a mango may be carried in the mouth for several meters before the aye-aye, hanging by its feet and holding the fruit in both hands, eats it.

Extreme morphological specialization such as the enormous external ear and the long, thin third finger, are often cited as adaptations related to the aye-aye's propensity for seeking insect larvae in dead branches. Petter reports that aye-ayes will carefully advance along a dead branch, sometimes suspended from above, listening or scenting for larvae in its interior. On detecting the presence of a larva, the animal attacks the bark with its incisors and when a hole has been made probes the interior of the branch with its third digit. Often the larva is pulped by this finger before being transferred to the mouth. According to Petter, another source of larvae is provided by the insects which infest the nuts found within *Terminalia catappa* fruit. Once again, the fruit is opened with the incisor teeth, and the larvae are extracted by the third finger. Adult insects are apparently never taken. Sap and, as already noted, the pulp of fibrous fruits are also consumed; aye-ayes appear habitually to manipulate fruits far more than other lemurs do, a propensity which is probably a product of the importance of the third digit in feeding. Cartmill (1974a) has drawn attention to the way in which certain cranial and other adaptations of the aye-aye parallel those of the diprotodont marsupial *Dactylopsila*. He suggests that in their respective regions these two mammals, occupying as they do the only extensive areas of the world lacking woodpeckers or woodpeckerlike birds, at least partly fill the niche occupied elsewhere by the latter.

Discussion

One of the most striking results of those studies of lemurs which have produced data on diet at different times of year lies in the seasonal variations revealed. Thus, while Richard (1978a) has produced cogent arguments for regarding *Propithecus verreauxi* as essentially a folivore, calling attention to its activity and ranging patterns as well as to its diet, it is nonetheless true that the proportions of leaves and fruit in the diet of the groups she studied were almost exactly reversed between the "most folivorous" and "most frugivorous" months of the year. Similarly, the Mayotte lemur may in one month be found to be eating a preponderance of fruit, in another, leaves, and in yet another, flowers. It seems highly likely that increasing knowledge of feeding patterns would point up a basic opportunism in lemur diets which would tend yet further to obscure the distribution of major dietary "adaptive types" within the Malagasy primate fauna. Moreover, as Richard emphasizes, while folivorous primates do tend to display rather depressed activity levels which are reflected in rather low frequencies of social interaction as well as in the time devoted to maintenance activities, folivory is not associated with any particular "type" of social organization. She concludes that it is unproductive to seek "a close correlation between specific social systems and broad niche categories" (1978a: 191), and none of the field studies of Malagasy primates published since she wrote has done anything to weaken this generalization.

However, the correlation just noted between general activity levels and dietary and foraging behaviors does appear to have some validity. Although on the one hand home range sizes and daily ranges within lemur species clearly vary locally according to the productivity of the vegetation and in response to the distribution of food resources, while on the other lemurs as a whole tend to exist at high densities in small ranges when compared to other primates, it is nonetheless possible to identify specific adaptive strategies in various lemur populations. Sussman (e.g., 1974) in particular, has been struck by the apparent behavioral adaptations to distinctly different

niches shown by the two species of *Lemur* which he studied in southern and southwestern Madagascar. *Lemur fulvus rufus* is confined to
closed-canopy forests, where it inhabits the upper levels, feeding
primarily upon leaves during the early morning and late afternoon;
day ranges are small, home ranges limited and overlapping, and
population densities are high. *Lemur catta*, in contrast, is found in a
wider variety of forest formations, and at all levels of the forest; it
feeds on a greater diversity of plant species and parts; it forages
quite consistently throughout the day, over much longer distances
and within larger ranges which are generally defended, and population densities are considerably lower. Sussman proposes that the
behavioral pattern of *Lemur fulvus rufus* is adapted to the exploitation of restricted areas of continuous-canopy forest where food is
abundant and evenly distributed, whereas *Lemur catta* is adapted to
hotter, more arid conditions where the forest is less continuous and
more variable in character, and where food resources are relatively
sparse and discontinuously distributed. The fact that the characteristic behavioral traits of these two species were equally pronounced
whether or not the animals occurred in sympatry adds considerable
weight to Sussman's scenario, and would seem to reinforce the notion that these behavioral characters are in some way "fixed." But
the level at which the scenario should apply is less obvious. Thus the
Mayotte lemur, a close relative of *Lemur fulvus rufus*, shows a substantially different pattern of resource exploitation: one that in certain respects more closely resembles that of the more distantly related ringtail. The pattern exhibited by the rufous lemur was quite
consistent between Sussman's two study forests, but so was the environment. We still do not know to what extent the rufous lemur
might modify its behavior in response to different environmental
conditions, or whether limitations in behavioral response might in-
hibit this lemur from invading certain other environments. And if
we did, we still would not know whether we could then generalize to
the probable responses of other lemurs, even conspecifics.

USE OF THE VERTICAL HABITAT

Cheirogaleidae

Martin (1972a) concluded as a result of his field studies that it
was the nature of the substrate which dictated the vertical distribution of *Microcebus* in the forest, rather than a preference for any particular level. In bushy vegetation mouse lemurs were found from

the ground up to the tops of the bushes, while in dense primary forest they were seen in the canopy at heights over 30 m. In secondary forest lacking low vegetation and offering many thin, vertical supports, the mouse lemurs were commonly seen at intermediate levels, between 2 and 10 m in height, in the leafy branches of young trees or among lianas. Particularly where fine branches and thick foliage provided cover to ground level, mouse lemurs showed no reluctance to descend to the ground. Mouse lemur nests of all kinds are usually sited low in the forest: Martin found most to occur between 2 and 4 m above the ground.

Martin also reported that *Cheirogaleus* species were rarely found at heights below 3 m; he related this to the behavioral preference of dwarf lemurs for large branches as opposed to the fine-branch niche favored by mouse lemurs.

Pagès (1978) found that during the dry season *Mirza coquereli* at Analabé were active primarily in the remaining zones of foliage, from 1 to 6 m above the ground.

According to Petter et al. (1977) *Phaner furcifer* most commonly moves in the levels of the forest where the maximum number of horizontal branches is found; they say that depending on forest type, these levels are found between 3 and 4 and 10 to 12 m above the ground. However, fork-marked lemurs have been observed both on the ground and at levels exceeding 15 m.

Lepilemuridae

Lepilemur clearly prefers vertical supports, and its movements within the forest are conditioned by this preference. No quantitative data exist on the levels favored by this lemur in any forest, but visits to the ground seem to be rare. *Hapalemur* is similarly unstudied in this respect. The smaller gentle lemurs, in particular, are adept jumpers between vertical bamboo stalks, and the Alaotran subspecies moves between reed stems in this way. Gentle lemurs do not seem reluctant to descend to the ground, and are often seen in low levels of the forest; Petter et al. (1977) claim that *H. g. alaotrensis* is an accomplished swimmer.

Lemuridae

Interspecific variety in the utilization of the various levels of the forest is better known in the genus *Lemur* than in any other, thanks largely to the work of Sussman (1972, 1974). This author recognized five forest levels: the ground (level 1); the shrub layer, from 1 to 3 m above the ground (level 2); the intermediate layer, from 3 to 7 m (level 3); the continuous canopy, from 5 to 15 m (level 4); and the emergent crowns of trees above 15 m high (level 5). All of Sussman's study forests (Antserananomby, Tongobato, and Berenty) were primary formations in which these levels were quite distinct. On the basis of 5-minute scan-samples, Sussman found that *Lemur fulvus rufus* spent over 90% of its time in levels 3, 4, and 5, with the continuous canopy predominating. When level 4 was not the most visited layer, level 5 was, while rufous lemurs were seen on the ground in under two percent of observations. Eighty-eight percent of travel at Antserananomby took place in level 4, and 85% at this level at Tongobato; otherwise, use of the various levels remained roughly constant regardless of activity or time of day. *Lemur catta,* on the other hand, spent more time on the ground (36% at Berenty and 30% at Antserananomby) than at any other level, and fully 71% (Berenty) and 65% (Antserananomby) of travel was on the ground. Ringtailed lemurs fed regularly at all levels of the forest, doing up to 30% of their feeding on the ground, whereas the rufous lemurs spent around 85% of feeding time in levels 4 and 5, and almost never fed on the ground. Each species displayed its characteristic pattern of use of the forest levels regardless of the presence or absence of the other species in the study forest.

I studied the Mayotte lemur at Mavingoni in a forest in which several vegetal formations were represented, only one of which was comparable in structure to Sussman's study forests (Tattersall 1977c). Nonetheless, like the rufous lemur, *Lemur fulvus mayottensis* displayed a clear preference for level 4, spending over 56% of its total time, and 63% of its feeding time, in the canopy or in levels equivalent to it. However, at those times when level 4 was not the most frequently visited it was level 3, not level 5, which predominated. This may have been largely a product of the structure of the Mavingoni forest, where an immature *Litsea glutinosa* formation in which the intermediate level predominated physically provided many of the feeding sources most commonly used by the lemurs. The Mayotte lemurs spent under one percent of their time on the ground, normally descending only to feed on fallen ripe fruit, or, rarely, to drink briefly at a pool. They were uncharacteristically nervous at such times, and would flee on the ground only as far as the nearest tree.

Sussman and I (Tattersall and Sussman 1975) observed *Lemur mongoz* almost exclusively in the upper strata of the forest at Ampijoroa, at heights from about 10–15 m. Mongoose lemurs would descend from these levels only when feeding in the lower branches of large trees or when their path of arboreal travel necessitated it. We never saw *Lemur mongoz* on the ground at this site; this is in accord with observations at other sites in Madagascar and the Comoro islands. Comorian mongoose lemurs are not uncommonly found, however, in low secondary vegetal formations.

Lemur coronatus appears in certain ways to occupy the niche in the arid areas of northern Madagascar which *Lemur catta* occupies in the dry south. No quantitative data are available on forest level utilization by this species, but Petter et al. (1977) state that crowned lemurs may travel substantial distances over the ground, and this is borne out by my own observations. Terrestrial flight is not unusual among these lemurs, and they may often be observed at lower levels of the forest. Conversely, largely anecdotal reports suggest that *Lemur rubriventer* confines its activities primarily to the higher forest strata, although Petter et al. record having seen red-bellied lemurs on the ground. These authors also report that *Lemur macaco* rarely move terrestrially although they may do so. My impression is that black lemurs prefer to travel in the continuous canopy where it is available.

Ruffed lemurs appear to prefer the upper forest strata, although captive *Varecia* on an island with trees in the Tananarive zoological park showed no hesitation in descending to the ground and spending considerable amounts of time there.

Indriidae

In her study of *Propithecus verreauxi*, Richard (1973, 1978) established eight absolute height categories in each of her study forests, without identifying structural levels. One-minute, focal-animal sampling showed that within each season at Ampijoroa there was little difference in the heights at which different activities were performed, but that there was a considerable between-season difference in feeding heights. During the dry season, the sifakas spent much time feeding high in the forest, often in emergents, whereas during the wet season most feeding time was spent in the dense continuous levels between three and ten meters. On the other hand, the southern study groups showed little seasonal variation in feeding heights, spending most feeding time between 2 and 5 m in both seasons. The relatively low feeding heights of the southern groups reflect the low stature of the dry forest formation they inhabited. In the cooler dry season, the southern sifakas exhibited an increased incidence of sunning behavior in the tree tops. Jolly (1966) stated that sifakas at Berenty rarely travelled any distance on the ground, descending there only in areas where rapid access to trees was possible. She also noted that bushy vegetation was avoided. Effectively nothing is known of forest level preferences in *Propithecus diadema,* although the low stature of the forest inhabited by *P. d. perrieri* and by the Daraina population obliges these animals to live closer to the ground than appears to be typical of the subspecies inhabiting the eastern rainforest.

Thirty-minute scan-sampling of individual heights in the forest at Analamazoatra produced a mean height above the ground of 14 m for *Indri indri* (Pollock 1975a, 1977). Throughout the year, babakotos were observed feeding at all levels of the forest, from 2 to 40 meters, and occasional descents to the ground to eat earth were noted. Pollock reported that about two-thirds of stationary time was spent in the forest canopy; time spent in the middle canopy was approximately equal to that spent in the canopy top and bottom combined, but individual animals showed quite large differences in the forest levels at which they preferred to feed. In both of Pollock's study groups the adult females regularly displaced the adult males at feeding sites, obliging them to feed at lower, and less favorable, levels of the forest.

No systematic study exists of the use of the forest strata by *Avahi*; woolly lemurs are most commonly observed at the upper forest levels.

Daubentoniidae

Little is known about the preferences of aye-ayes for the various forest levels. A wide range of arboreal substrates is used by these animals, and Petter (1977) has noted that they frequently descend to the forest floor, even traveling substantial distances on the ground in areas of degraded vegetation.

SOCIAL BEHAVIOR

As Pierre Charles-Dominique pointed out for lorisids several years ago, all lemurs are social, even those which are not particularly gregarious. The structures and sizes of lemur populations and groups have already been reviewed briefly under "Social Organization" in this chapter; the behaviors considered here as social are those which act to maintain the integrity of those populations and groupings. The most obvious of these are communication behaviors and interindividual interactions such as contact, grooming, play, and aggression. As in other areas, our knowledge of such behaviors is very unevenly distributed, some species being relatively well known while others remain totally unstudied.

Cheirogaleidae

Social behavior in this family of lemurs is more poorly known than in any other. Olfactory communication is clearly of great importance to mouse lemurs, in which punctuated urination, urine washing, and fecal marking have been observed among other similar behaviors (Petter et al. 1977; Schilling 1979). Data on context are generally lacking, which impedes interpretation of such marking behaviors; some, however, appear to play a role in "familiarization" of the environment, while others seem to function in range definition for the benefit of conspecifics. Vocalizations of *Microcebus* are extremely high pitched, thus sometimes difficult to identify; most have been interpreted (e.g., by Petter et al. 1977; Petter and Charles-Dominique 1979) as related, either directly or indirectly, to sexual activity. Certain "whistling" calls have, however, been characterized by these authors as alarm vocalizations, and a further set of calls, made by infants, as demands for mother contact.

Many of the vocalizations of *Mirza coquereli* have been interpreted as paralleling those of *Microcebus,* but further investigation seems to be warranted. Pagès (quoted in Petter et al. 1977) has noted scrotal marking by male *Mirza* towards the birth season, and anal marking in both sexes; but the most common form of marking in this lemur appears to be salivary. These animals also periodically emit a strong odor, of uncertain source; in the wild, this occurs in both the periphery and in the most commonly frequented parts of their range. Pagès (1978) has drawn particular attention to the importance of tactile communication in *Mirza.* Tactile contact between captive males and females increases as the brief period of sexual activity nears, and wanes following it, but nonetheless continues at a low level all year round; Pagès believes that successful reproduction depends on the maintenance of a pair bond which is reinforced by such tactile signals. Social activities, which include side-by-side resting, mutual grooming, play, and encounters, generally take place in the wild in the second half of the night, after most feeding is completed, and in the outer areas of the individual ranges. During the months of Pagès' field study such contacts were relatively infrequent, occurring only once every two nights or so.

Neither species of *Cheirogaleus* is notably vocal, but olfactory communication is clearly important to both. In particular, the unusually positioned and protuberant anus underlines the importance to dwarf lemurs of fecal marking, this structure being ideally adapted to "dragging" defecation (Schilling 1979). Such behavior is very commonly observed among captive dwarf lemurs, and Petter et al. (1977) have characterized it as territorial signaling.

Uniquely among the cheirogaleids studied, *Phaner furcifer* does not appear to practice urine or fecal marking. However, males, in particular, possess a well-developed neck-gland, with which they frequently mark their females as well as branches. Male-female pairs maintain contact with loud vocalizations, as do neighbors (see "Social Organization" earlier in this chapter). Petter et al have proposed that the relatively complex pelage markings characteristic of this lemur indicate that visual signals are of greater importance in both communication and in individual recognition among *Phaner* than among other cheirogaleids.

Lepilemuridae

The most detailed account of *Lepilemur* social behavior has been provided by Russell (1977), who found that whichever focal *L. m. leucopus* he followed would be joined between one and three times per night by its "range-mate." Such meetings took place in specific food trees, in which one animal appeared to wait for the other's arrival. Greeting took the form of nose-touching, usually initiated by the arriving individual. This was normally followed by silent autogrooming, then by allogrooming, then by mutual grooming. Following this, the individuals would rest in contact for up to an hour, and in about a third of Russell's observations this resting bout was followed by one to six minutes of silent rough-and-tumble play. In the remaining cases, the individual first reawakening would discreetly depart, sometimes after briefly autogrooming. Only once did Russell observe a violent encounter, between a male and a female who occupied neighboring ranges, and on rare occasions what appeared to be avoidance behavior was exhibited by two females when a neighboring male entered their range. Vocalization was infrequent, but was most common during the breeding season; individual *Lepilemur* varied greatly in their propensity to vocalize, however. Petter and

Charles-Dominique (1979) provide sonagrams of a variety of calls. Urine marking was rarely seen, and only in "rendezvous" trees well inside the range boundary; Russell concluded that such behavior functioned in the maintenance of the social bond between range-mates rather than in range delimitation.

Virtually nothing is known of the social behavior of *Hapalemur* species and subspecies in the wild state. Captive *H. griseus* are highly sociable; pairs sleep in close contact and spent large amounts of time in allo- and mutual grooming (Petter et al. 1977). Gentle lemurs are equipped with specialized marking glands, and readily mark their enclosures anogenitally, with urine, and with the brachial gland. Petter et al. state that marking by captive females is provoked even by relatively minor disturbances, and that in general males mark only after sniffing the marks left by the female. The vocalizations of *Hapalemur griseus* are generally rather discreet, although diverse; several sonagrams are provided by Petter et al. (1977) and Petter and Charles-Dominique (1979). The calls of *Hapalemur simus* are said to be quite distinctive.

Lemuridae

Jolly (1966a) described a variety of agonistic intratroop behaviors in *Lemur catta*. The least complex of these behaviors was simple staring. This was sometimes supplemented by movement of the starer towards the recipient; dominant males often did this with a characteristic "swaggering" gait. The agonistic gesture most frequently observed was the cuff, in which one animal swiped a hand toward the other's face. During the breeding season, in which aggression was at its commonest and most intense, a cuff often ended with the tearing out of a tuft of the recipient's fur. A common accompaniment to agonistic interactions was the "spat call," in which several loud squeaks were uttered in succession. This was given in almost all kinds of such encounters except for the most intense, such as chasing and jump-fights: behaviors seen by Jolly only during the breeding season. Aggression reached its most intense level in jump-fights, where ringtails hopped bipedally around each other endeavoring to gain a height advantage from which the opponent could be slashed with a downward sweep of the long, sharp canines. Jolly suggested that this was the kind of fighting that accounted for the worst injuries she observed during the breeding season. Spat calls were also generally not given during supplantation, a low-intensity encounter where one individual displaced another simply by nosing it away.

Scent-marking characterized some of the types of agonistic encounter described by Jolly. Of these the most spectacular was the tail-marking and tail-flicking sequence displayed by males. An adult male, facing another animal, would stand on his hindlimbs, touch the antebrachial to the brachial gland ("anointing"), protrude his tail forward between his legs, then pull it backwards several times against the antebrachial gland. Resuming a quadrupedal posture, he would then flick the tip of his tail forward over his head, thus propelling towards his opponent the odor of both glands. Male recipients would normally spat and run, females spat and cuff. Jolly found that dominant males indulged in this behavior much more frequently than subordinate ones. Tail-flicking was often accompanied by palmar-marking, in which males grabbed slender vertical stems and rocked their bodies from side to side, leaving palmar secretions spread on the twig. Jolly coined the term "stink fights" for sequences of palmar-marking and tail-flicking by two males towards each other. Such sequences sometimes continued for up to an hour, punctuated by episodes of yawning and autogrooming. Schilling (1974) regards tail-marking behavior as analogous to brachial branch marking, in which the lemur applies the spurred antebrachial gland to a branch (usually thin and vertical), then strongly pulls the arm back. The spur cuts a clearly visible comma-shaped scar on the stem, impregnated with the scent of the antebrachial gland. Schilling found that such marking was heavily concentrated in two limited areas of the range of his study group at Berenty, one of them peripheral and frequently the site of territorial disputes. Mertl (1977) suggested more specifically that repeated marking of boundaries by ringtailed lemurs serves to reduce aggression between neighboring troops at range boundaries.

Jolly noted that palmar-marking by males also occurred independently of agonistic sequences, as did genital-marking of branches by both sexes. In the latter case, individuals would stand facing away from more or less vertical small stems, which were then grasped by the feet as high up as possible. Following this the genitalia would be rubbed several times against the stem while the hands remained on the ground or grasped the horizontal support branch. Marks thus left are clearly visible on the bark, and according to Evans and Goy (1968), deposited scent remains attractive to conspecifics for several days.

In an experimental captive study, Mertl (1976) found that both visual and olfactory cues were important in intermale interactions. In the absence of visual contact, male olfactory signals emitted either directly by the body or indirectly via scent marks failed to elicit a response in other males. When one male could see another, his response differed according to whether he could also smell him or not. Clearly, there is a subtle relationship between olfaction and vision in transmitting and receiving social signals. Mertl (1975) also reported that ringtails were able to distinguish between antebrachial and brachial odors emitted by different conspecifics.

Like all members of their genus, ringtailed lemurs are highly gregarious. They spend much time in bodily contact with conspecifics, and during resting periods several individuals will often clump together. Initial contact between individuals is frequently accompanied by nose-touching, which seems to be a generalized greeting gesture (Jolly 1966a). Interactions between individuals most commonly take the form of social grooming with the dental comb. Sussman (1972, 1974) found that grooming of all kinds occupied almost 5% of total time at Antserananomby, and almost 11% at Berenty; it was quite evenly distributed throughout the hours of daylight (Sussman 1972, 1974) as well as through the year (Jolly 1966a). All parts of the body are groomed during mutual grooming, not merely those parts inaccessible to autogroomers.

Play is another important type of behavior in establishing social bonds. Infrequently observed between adult *Lemur catta,* it is common among juveniles. After birth, the infant is carried longitudinally beneath the ventrum of the mother, in contrast to other *Lemur,* in which the infant clings across the mother's belly. The relatively precocious ringtail infants begin to move actively about the mother's body within a few days of birth (Sussman 1977b), and by the age of four to six weeks are tasting solid foods and actively playing with juveniles and other infants. At this stage females often exchange or "kidnap" infants, and reciprocal nursing is often seen (Klopfer 1974). Apparent "aunting" is shown by females, who may carry the infants of others for several minutes at a time (Sussman 1977b). Regular play groups are established by the time the infants are three months old, and shortly thereafter the young animals seem to be more a part of juvenile-infant subgroups than of mother-infant pairs, even though weaning is not yet complete.

Several of the calls of *L. catta* seem to be directed toward group cohesion. Jolly (1966a) reported that before a troop began to move, "clicks" were emitted by group members, often followed by "moans," "wails," or "meows." When a moving or foraging troop is surprised on the ground, its members normally scatter silently and melt into the undergrowth. When the danger has passed, contact between group members is reestablished by means of soft mewing calls. The "howl" call, characteristic of males and audible at distances of up to 1,000 m, is believed by Jolly to function as a group spacing mechanism. Petter and Charles-Dominique (1979) present sonagrams of certain *L. catta* vocalizations.

Harrington (1975) studied social interactions among *Lemur fulvus fulvus* at Ampijoroa. He found body contact to be frequent among these animals, and observed grooming in 57% of their contacts. The frequency of contacts between the various sex-age classes did not vary significantly over the period of the study, which included a mating season. Female-young and young-young contacts were more frequent than expected, other contacts less so; female-female contacts were, however, substantially more common than male-male contacts. Agonistic interactions were infrequent and rarely intense, being largely confined to cuffing, with or without contact, and to the "head-threat," where an individual would lunge with its head towards another, sometimes grunting or baring its teeth. Threats of this kind were rarely returned, and only once did Harrington see one animal chase another. Most agonistic gestures were directed by adults towards immature animals, and under 20% were associated with sexual behavior. No order of social dominance was discernible. Play was observed only between adult females and young and between immature animals. Groups of up to five individuals would grapple, play bite, lunge at each other, jump, or chase.

Olfactory marking was most noticeable in the context of sexual behavior. Most such marking was anogenital, and was more noticeable among males than among females; it was often stimulated in males by sniffing female marks or genitalia. However, experiments upon sexual discrimination by scent in captive brown lemurs (e.g., Harrington 1977) produced rather equivocal results: males sniffed the scents of other males more than those of females, but did not mark more as a result. Apparently the stimulus to specific behaviors of this kind is not simply olfactory. Males frequently anogenitally marked the backs of females, or spots where a female had sat or marked. Males also head-marked or palmar-marked. Sexual behaviors, in the form of clasping, following, anogenital sniffing, and scent-marking were almost entirely confined to adult male–female combinations; females were generally passive in such sequences. However, females tended to be more vocal than males, responding more readily to stimuli such as the approach of humans.

Sussman's study (1972, 1974) of *Lemur fulvus rufus* at Antseran- anomby and Tongobato revealed that over 11% of total time at the former site, and over 5% at the latter was devoted to grooming, most of it social. Play was most frequent at the early stages of infant independence from the mother, after about three months of age, and became less frequent as the young animals matured. Since the rufous lemur groups were small, they contained few young, and most play observed was between young animals and adults. New- born infants formed a focus of group attention, and were regularly groomed by adults other than their mothers although in contrast to *Lemur catta,* infant sharing was unusual. Agonistic encounters be- tween group members were rare, occurring at a rate of about 0.1/hr (Sussman 1975); most occurred when one individual attempted to sit in contact with or groom another. Such incidents were mild affairs, usually involving no more than a lunge or a cuff, although short chases might ensue. Scent-marking increased in frequency during the mating season, but aggressive encounters did not (Sussman and Richard 1974).

Lemur fulvus mayottensis at Mavingoni similarly exhibited mini- mal aggression (Tattersall 1977c). The commonest social interaction was social grooming (which occupied by far the largest part of the 4.3% and 3.7% of daylight time devoted to grooming in the wet and dry seasons, respectively), and individuals usually rested in bodily contact with one or more others. Play behavior, never a strikingly prevalent activity, decreased in frequency as young animals matured.

Social grooming was not commonly seen in *Lemur mongoz* at Ampijoroa (Tattersall and Sussman 1975), but the group was a coh- esive unit, within which agonistic behavior was rarely exhibited. Ol- factory marking was also apparently uncommon, and in about one- third of cases was associated with intergroup encounters. Males were seen to mark such more often than females. In contrast, mongoose lemurs are highly vocal; group cohesion seemed to be enhanced during leisurely progression by soft grunting, and during rapid travel by the loud, sharp "creaking door" call, which may also have functioned in intergroup spacing.

Virtually nothing is known of the behavior of *Varecia* in the wild. Kress et al. (1978), who studied two caged groups, reported virtually no agonistic interactions between adult males or males and females, but found that adult females would not easily tolerate other females during estrous periods or in the presence of young infants. Staring, advancing, cuffing, and occasional biting were all noted. Central adult males marked with the throat and the angle of the jaw regularly and continuously throughout the year; adult females marked much less frequently, and almost always anogenitally, as did immature animals of both sexes. Social grooming was common, but was rare or absent between adult males and central females. The most common grooming pairs were formed between adult females and young.

Indriidae

Jolly (1966a) reported an extremely low level of intratroop aggression in *Propithecus verreauxi* outside the mating season. Richard (1973, 1974, 1978a; Richard and Heimbuch 1975) collected more detailed quantitative data at both her northern and southern study sites, using the technique of continuous recording, and found that in her various study groups agonistic encounters occurred at average rages of from 0.25 to 0.44 per animal hour. These figures reflect the considerable difference found between individual groups; no such variation was found between the two study areas. The vast majority of such interactions occurred over access to feeding sites. Generally the aggressor supplanted the other at a feeding place, or was already feeding when the recipient approached too close. Most other aggressive encounters involved disputes over a resting place, or when a mother rebuffed an attempt to handle her newborn infant. Aggressive behavior included staring, lunging, cuffing, biting, and simple displacement by approach. Gestures of submission occurring in response to such aggressive acts included baring the teeth, rolling up the tail between the legs, and back-hunching; they were normally accompanied by the "spat" vocalization, similar to that of *Lemur catta* but fainter.

Aggression, in parallel with general activity, was more commonly seen during the wet season than the dry. But it was during the brief mating season, including the immediate precopulatory period, that agonistic behavior was at its most frequent and intense. At this time it involved aggression by males who ranked low in the feeding hierarchy against those of higher rank; aggression of this kind was not stereotyped, but involved savage fighting in which serious wounds were often received. Such behavior often led to lasting changes in the feeding hierarchy, even after agonistic behavior had returned to its normal low level.

The preponderant nonagonistic social behavior seen by Richard was social grooming. Most allogrooming was directed towards the head, face, and back of the recipient: areas inaccessible to autogrooming. Seventy percent of grooming episodes observed were unidirectional: one individual groomed the other throughout the bout. Reciprocal (alternate) grooming characterized four percent of bouts; reciprocal and mutual grooming together, 26%. Most grooming took place during rest periods, when individuals commonly rested in physical contact with others, often forming chains of animals facing in the same direction: what Jolly (1966a) termed "locomotives." Richard never saw play during the dry season, but noted it frequently during the wet. It was usually initiated by young individuals; adults, however, quite readily responded to such approaches, although most play bouts were between young animals. Individual sifakas may "greet" each other by nose touching, which was occasionally seen throughout the year and accounted for 5.7% of the nonagonistic social interactions recorded by Richard.

Scent marking by male and female sifakas was seen in all seasons by both Jolly and Richard. Mertl (1979) found that most marking was done in peripheral areas of the range, and suggested that it thus had a territorial significance; but olfactory marking is clearly also important in maintaining social bonds, as the difference between male and female marking behaviors makes clear. Richard reported that while females marked either with urine or by rubbing the anogenital region against the support, males usually showed a more complex sequence. A spot would be rubbed with the throat gland, then with the tip of the penis which usually emitted some urine, and finally with the perineal region. The entire sequence was displayed in most episodes of marking, although any individual component might be performed alone. Spots just marked by females were often "endorsed" by males in this manner. The "sniff-approach and mark" was another behavior displayed only by males, who would carry out the marking sequence on a trunk just below a clinging female after nose-touching her anus. These specifically male marking behaviors increased in frequency during the mating season in one of the two groups Richard studied at that time, but not in the other.

Pollock (1975a,b) has described social behavior in *Indri* as constituting a "minute proportion" of the day's activities, although at 0.8 per hour social events took place at the same rate as that estimated by Jolly (1966a) for *Propithecus verreauxi* at Berenty. Agonistic interactions were almost always seen in the context of feeding, and most commonly occurred when a number of individuals were feeding in the same tree. Displacement of one animal by another through simple approach occurred more often than positive acts of aggression consisting of wrestling, kicking, or biting, usually accompanied by "grunting," "kissing," or "wheezing" vocalizations. In one of Pollock's study groups, 28 displacements of the adult male were seen, 22 of them by the adult female; but of these, only 37% involved aggressive behavior. In the other group, 34% of 107 displacements involved such behavior. Ritualized submissive gestures appeared to be lacking in the behavioral repertoire of *Indri*. The adult female was the dominant individual in both groups, but when one was badly injured and unable to travel with the others, the rest of the group appeared to range normally, returning toward the female only at day's end, attracted by her loud calling.

Grooming in *Indri* was confined to those body parts which could not be self-groomed. Allogrooming was usually solicited by the recipient and was never mutual, although Pollock occasionally observed rapid role-switching between partners in reciprocal grooming bouts. One group devoted 4.0% of its activity to autogrooming and 0.7% to allogrooming; in the other, corresponding figures were 1.5% and 0.4%. In one group the adult male participated in 37% of observed allogrooming bouts, and groomed his partner in 60.6% of these. The adult female, in contrast, groomed the other animal in only 27% of the 39% of all allogrooming episodes in which she was observed to take part. Pollock only saw play during the long activity periods of the austral summer; all play bouts, which lasted from a few seconds up to 15 minutes, took the form of silent wrestling. Young individuals were usually involved, and play was commonly solicited by a formalized reaching gesture which normally met with a positive response. Play, which accounted for 1.2% of the activity of one group, sometimes led to bouts of allogrooming and occasionally terminated in aggressive displacements.

Pollock observed scent marking (cheek and anogenital) in *Indri* at a rate of about 0.3–0.4 episodes per hour. Adult males were responsible for about 70% of the marking seen, and marked both spontaneously and in response to disturbance. All parts of the home range were marked at approximately equal frequencies. Pollock remarked on the contrast between the sympatric *Propithecus diadema*, a discreet lemur which marked frequently and intensely, and the noisy *Indri*, in which marking was rare. Pollock suggests that in some cases the "song" of *Indri* may have functioned to promote group reaggregation when individuals had become widely dispersed; in any event, audition appeared to function more importantly in the maintenance of group cohesion than did olfaction.

Daubentoniidae

Aye-ayes of both sexes urine mark frequently (Petter and Peyreieras 1970a; Petter 1977). The penis or clitoris is brought into contact with the substrate during such marking, and a trail of urine may be "dragged" over a distance of several feet. Marking behavior may be elicited by disturbance or when an individual encounters a conspecific. A captive female was seen to "mark" a branch on several occasions by ripping its bark with her incisors. Six adult aye-ayes kept together in the same cage for several days apparently failed to exhibit any aggression, but Petter (1977) has observed agonistic behavior involving "an aggressive posture and a hissing noise". This author also noted abrupt transitions between play and aggressive behaviors. A "ron-tsit" call, given at various intensities, often accompanied emotional arousal.

Discussion

It is now close to two decades since Alison Jolly (1966b) made the most important single point about the social behavior and organization exhibited by the gregarious diurnal lemurs: that they are similar in all essential respects to those of modern nonstrepsirhine primates, even in the absence of the manipulative and problem-solving abilities characteristic of monkeys and apes (although subsequent work [Ehrlich et al. 1976] has suggested that the lemurs have quite possibly been maligned to some extent on the latter score). From this Jolly concluded that in primate evolutionary history, a complex social existence probably both preceded and directed the evolution of the type of object-directed intelligence associated with anthropoids. All that the basic observations actually demonstrate, however, is that the former can exist independently of the latter, and although few students of the primates have chosen to quarrel with Jolly's larger conclusion, it does present certain problems when considered in the specific context of primate evolutionary relationships. The implication of the argument would seem to be that social behavior and organization of this sort were characteristic of the common ancestor of strepsirhines and anthropoids. But if this is so, the types of sociality exhibited by the cheirogaleids, say, must have been secondarily derived: a conclusion in conflict with prevailing views on the evolution of strepsirhine social organization. These considerations open up a multitude of entertaining possibilities, probable or improbable: that the dogma on behavioral evolution is wrong; that at least some group of diurnal gregarious lemurs shares a common ancestry with the anthropoids, to the exclusion of the others; that "primate-type" social organization and behavior evolved independently in the two major groups of living primates; and so forth. It is evident that the essential sociality shown by all the lemurs is indeed a fundamental primate trait, and the most straightforward hypothesis is that in its simplest form it is preadaptive to the more complex types of sociality seen today throughout the primates. As discussed in chapter 8, the common suite of adaptations displayed by all "primates of modern aspect" implies that a vast range of ecological opportunities was opened to the primates at a very early date. If only a fraction of these were exploited, it seems likely that the variety seen today in social organization and behavior among the strepsirhines is of ancient origin.

CHAPTER EIGHT

===

Retrospect and Prospect

ORIGIN OF THE MALAGASY PRIMATE FAUNA

The question of the origin of the mammal fauna of Madagascar is closely tied up with the history of the island itself, but is unhappily little clearer for that. It has never been seriously in doubt, despite occasional claims to the contrary (e.g., Gingerich 1975), that Africa was the source of this fauna; but there is no compelling reason to believe that any mammals other than the precursors of the pygmy hippopotamus (now extinct), and the living Malagasy viverrid carnivores[1] have ever managed to cross the Mozambique Channel in either direction since that water barrier assumed its present dimensions. Moreover, the lack of an appropriate fossil record both in Madagascar and in pre–Oligocene Africa makes it difficult to construct scenarios of the populating of Madagascar by mammals with

[1]The Malagasy viverrids are fissiped carnivores of highly primitive aspect, whose ancestors were not present in the Old World until the early Oligocene, at which time they and many other groups entered Africa (Van Couvering and Van Couvering 1975). They must therefore have arrived in Madagascar via an accidental (waif) dispersal event, probably during the major glacio-eustatic regression of the mid-Oligocene (ca. 35 m.a.). The premise of this discussion is that waif dispersal events of this kind should never be invoked in biogeographical hypotheses except where the evidence is compelling, as in this case when many other taxa existed in Africa at this time which nevertheless failed to accompany the carnivores. The Malagasy rodents and the aardvark may also, however, eventually be proved to have been introduced in this way, and probably also at this time.

any confidence, especially since the geological history of the Mozambique Channel is also so poorly understood.

Nonetheless, certain of the phylogenetic relationships among the lemurs proposed in chapter 6, notably those within the *Lepilemur*-group, suggest that a significant part of the diversification of the lemur fauna must have taken place before its definitive isolation from the early African primate fauna, and thus apparently before some time in the earliest Tertiary. Even if these particular relationships prove eventually to be unsubstantiated, much the same is implied by the more firmly based cheirogaleid-lorisid affinity. This circumstance has one of two possible implications: either that multiple introductions from Africa occurred early on (in terms of primate evolution) via some kind of filter, or that at a relatively late stage (in terms of plate tectonics) Madagascar possessed more or less exactly the same mammal fauna as Africa. In either case, the question arises as to why there is now such a low diversity of major mammal taxa in Madagascar, since in the former case multiple primate introductions would imply a much less severe filtration effect than one would need to assume from a single introduction.

This question is quite plausibly answered, at least in part, by the high probability that the African mammal fauna in the early Tertiary both resembled that of Madagascar and was substantially different in composition and diversity from that represented in the Oligocene and subsequent fossil records. The Fayum mammal fauna, the only Oligocene land fauna known from Africa, has a high component of forms with Eurasian affinities, and indeed it is probable that prior to the middle Eocene (at which time creodonts and anthracotheres may have arrived via a Balearic route) African mammals were limited to highly endemic populations of tethytheres (proboscideans, hyracoids, sirenians, etc.), tenrecs, primates, and possibly condylarths (J. A. Van Couvering pers. comm.). A low diversity of major mammal taxa in Africa at that time may well have been associated with high diversity within those taxa represented, just as it is in Madagascar today.

At this point, it may be appropriate to look once more at the geological record to determine whether or not times of possible faunal interchange between Africa and Madagascar may be calibrated with any confidence. As we saw in chapter 2, this record is at present somewhat equivocal, but the evidence points to at least intermittent separation of Madagascar from Africa by a sea barrier from the middle Jurassic onwards, with definitive separation by some time quite early in the Tertiary. However, it is reasonable to assume that the floor of the Mozambique Channel has been marked since its formation by continuing thermal subsidence, and that at the beginning of the Tertiary its depth was considerably less than it is now. Matthews and Poore (1980) have convincingly proposed that available volumetric and oxygen isotopic data best fit a model of periodic major glacio-eustatic sea-level fluctuations extending back to the Cretaceous and probably beyond, while Kauffman (1975) has found evidence for repeated dramatic cooling together with sea level lows in the Interior Cretaceous basin of the United States at various times during the Cretaceous. Most interesting from our point of view is the evidence for major lows during the Cenomanian (ca. 92 m.a.) and the Maestrichtian (ca. 65 m.a.). Either of these events of drastically lowered sea level might have produced a continuous land bridge across a relatively shallow Mozambique Channel; failing that, it is not hard to imagine at either time a substantial narrowing of the channel, or possibly the creation of an island chain of no more than moderate effectiveness as a faunal filter. This effect may also, at a later time, have facilitated waif dispersal of the ancestral Malagasy viverrid(s) (see footnote 1 in this chapter).

However, if Madagascar in the late Cretaceous or earliest Tertiary by one means or another shared at least a large part of its mammal fauna with Africa, why is it so depauperate in major mammalian groups today, even in terms of the relatively low mammalian diversity of Africa at the beginning of the Tertiary? One obvious possibility exists. It is generally accepted that on an island the probabilities of the extinction of any particular taxon are increased; and although Madagascar is certainly an enormous island, from the late Cretaceous to the present world faunas as a whole have undergone several major episodes of severe stress, with large-scale extinctions world wide. It seems reasonable to hypothesize that on the island of Madagascar such episodes may have resulted in the demise of groups which survived on the mainland. From our point of view the two greatest of these events, those at the end of the Cretaceous and at the end of the Eocene (Cavelier 1979), may have been especially significant. It is well documented that dinosaurs, giant ammonites, and so forth, existed in Madagascar in the late Cretaceous. And unquestionably, like their relatives in other parts of the world, they failed to survive the Mesozoic. On the other hand, those forms that did survive, in Madagascar as elsewhere, included the ratites and small mammals such as the tenrecs and primates. Any other groups that may have survived this episode in Madagascar must then have succumbed to the climatic stresses which characterized the end of the Eocene.

This scenario, while admittedly highly speculative, has the merit of simultaneously accounting for the small number of major mammalian taxa in Madagascar today, and for the fact that certain modern Malagasy strepsirhines which are closely related to each other appear to be related even more closely in each case to non-Malagasy Eocene forms. At the same time, it requires a very early origin in Africa of at least two modern mammal groups, the primates and the tenrecs. Further, a considerable amount of evolutionary stasis is indicated in some lines of lemurs together with substantial diversification in others. None of these assumptions is in itself hard to accept, and that of an early diversification of mammals in Africa is supported by other evidence, albeit indirect. The critical evidence still lies buried, however; and pending the recovery of early Tertiary (or earlier) mammal fossils in Madagascar and Africa, it is possible that corroboration or refutation of the hypothesis may emerge from the earliest Tertiary record of India, and possibly also South America or Antarctica, all of which promise to become better known in the course of the next decade.

LEMURS AS EARLY PRIMATES

If there is one general attribute of the lemurs upon which all students of the primates agree, it is that they provide a living model of the early stages of primate evolution. This quality has rarely been put to explicit use, however, and it is not difficult to see why. Traditionally the lemurs (and strepsirhines in general) have been regarded as "low men" on the *scala naturae* of primate evolution. This view has received considerable reinforcement from the conventional view of the founding of the primate population of Madagascar, colorfully expressed recently by Alison Jolly when she envisaged "a raft of floating tree trunks or tangled branches [carried from Africa across the Mozambique Channel] where a pop-eyed ancestral lemur clung with all its hands onto the wave-washed twigs" (1980:15). But just what was this unhappy voyager like? As long as little was known about the spectrum of living lemurs it was not difficult to believe that such an ancestor existed; but in the absence of hard information it was regrettably impossible to be very specific about its characteristics. But now that we are beginning to realize how adaptively heterogeneous an assemblage the lemurs actually constitute, the problem of specifying the attributes of such an ancestor seems hardly to be ameliorated. The construction of any *scala naturae* depends among other things on our ability to recognize reasonably well-defined stages; and the more we learn about the lemurs the more difficult it becomes to resolve them down to a stereotype.

Tacitly recognizing this, some recent authors have turned toward recognizing a *scala naturae* within the lemur fauna itself. Thus, for instance, Martin (1972a) has envisioned a possible evolutionary sequence in forms of social organization among these animals. But while one would expect in any array of related forms that certain species would retain more primitive character states—behavioral or otherwise—than would others, it does not appear to be very profitable to consider the distribution of such states from this point of view outside the context of an explicit phylogeny. Moreover, approaching our knowledge of the lemurs from a "vertical" standpoint of this sort tends largely to obscure the "horizontal" variety which is becoming increasingly well documented; after all, each species existing today itself represents a sort of endpoint.

The evidence discussed elsewhere in this volume suggests that a) only part of the diversification of the lemurs took place in isolation on Madagascar, and that b) the remainder of that diversity is of very ancient origin. The adaptive spectrum represented in Madagascar today is enormous: in social organization, in activity patterns, in locomotion, in diet and the digestive system, and in other ways. And what all this points to is that adaptive variety on this scale, or something close to it, much have characterized the Order Primates very early on in its evolution. The common ancestor of the strepsirhines, like its common ancestor with the anthropoids, must have possessed grasping extremities with claws replaced by nails, convergent orbits defined by a postorbital bar, and a brain: body size ratio large in comparison to that of other mammals existing at the time; in other words, the major prerequisites for entering the adaptive zone characteristic of "primates of modern aspect." If entering that adaptive zone meant anything at all in terms of ecological opportunity, it meant that a vast spectrum of niches was opened up to exploitation, and it would be unreasonable to suppose that such new opportunities were not grasped with alacrity by early strepsirhines.

The replacement of an oversimplified stereotypic image of the lemurs with an appreciation of their adaptive and phylogenetic variety leads, then, to the conclusion that as far back as the earliest Tertiary (and possibly even earlier) the Order Primates probably existed in similar variety (a variety not fully reflected in the geographically selective fossil record), and that a monolithic view of "early primates" is as inappropriate to them as it is to the lemurs. This is so even if the diversification of Strepsirhini (at a very early date) took place subsequent to the origin (cladistically if not adaptively speaking) of Anthropoidea. Nonetheless, it remains true that some of the most interesting biological questions about the lemurs relate to those factors in which they have remained relatively uniform. For example, at any given body size lemurs tend to exist at high biomass in small home ranges when compared to higher primates, while for all the adaptive variety of the lemur fauna none of its members has totally abandoned olfactory communication or increased its brain:body size ratio to anything approaching that characteristic of higher primates. Their sum total does not add up to a very satisfying morphotype for the ancestral strepsirhine; but the ways in which these animals have remained primitive are without question at least as intriguing as their specializations.

DISTRIBUTION AND SPECIATION

Madagascar presents us today with a complex pattern of faunal distributions. This reflects a history of speciation in the relatively recent past about which we are at present able to do little more than speculate. Not only is precise information lacking on the present distributional and habitat limits of virtually all lemur species, but the modern Malagasy ecotope merely represents the fragmented remains of what existed prior to the advent of man. One major generalization that can be made about the lemurs is that while genera are in most cases island-wide in their distributions, species as well as subspecies show largely disjunct ranges. We know from subfossil evidence that only a thousand years ago representatives of *Lemur, Varecia, Lepilemur, Hapalemur, Propithecus, Avahi, Indri,* and *Cheirogaleus* inhabited parts of the central plateau (see chapter 5); but whether or not these subfossils belonged to species now existing in the eastern or western coastal regions is uncertain. The large extinct subfossil genera are generally represented on the plateau by species distinct from those found in the sites of the south and west, but in the absence of subfossil lemurs from the eastern region it is impossible to know whether the plateau subfossil fauna was distinct in species from that of the east. Under present conditions, however, by far the clearest faunal division in Madagascar at the species level is that between the eastern and western regions, although it is by no means absolute.

In a characteristically stimulating review, Martin (1972b) divided Madagascar into seven zones (roughly north, northeast, southeast, south/southwest, central west, northwest, and central plateau) on the basis of certain lemur distributions. Calling attention to Morat's (1969) use of Emberger's "pluviothermic quotient" to distinguish four Malagasy "bioclimatic stages" corresponding to zones of plant distribution, Martin suggested that his lemur distribution zones accord reasonably well with regions delimited on botanical and climatological grounds. Recognizing that speciation among the lemurs must have been allopatric, he then sought potential geographical barriers which may have served to isolate the distribution zones from one another, on the assumption that many species now restricted to coastal regions could not have survived the low winter temperatures of the elevated central plateau region when that area was forested. Thus hypothesizing the existence of coastal species "wreaths" around the central plateau of Madagascar, Martin proposed that a number of large rivers (e.g., the Betsiboka, Mangoro, and Tsiribihina) and the Anosy mountain chain, which between them approximately define the limits of his various lemur distribution zones, could have acted as such barriers. Dispersal across these, he suggested, would have been highly selective if they did indeed demarcate the boundaries of climatic/vegetational zones.

The biogeographical hypothesis depends, among other things, on two propositions: that the central uplands of Madagascar constituted a barrier to lemur migration at the species level, and that the posited peripheral barriers indeed functioned as such. The first proposition, as the foregoing discussion demonstrates, is untestable on the basis of current evidence. The second is somewhat bedeviled by the fact that any river in Madagascar which today seems to represent the boundary of distribution of any lemur species or subspecies is crossed by another; and broad as they might be toward the sea where most of the forest is now confined, it is highly unlikely that these rivers posed substantial obstacles to faunal migration farther inland when the forest, more continuous than its remnants are today, extended further into the interior. Moreover, rivers are notorious for changing their courses over relatively short periods of time, and passive "transfer" of local mammal populations from one side of a river to another is an event of quite high probability. Finally, few if any of the putative barriers appear to define such clearcut ecological limits that the probabilities of dispersal of a lemur species away from the barrier, once crossed, would have been significantly reduced.

It seems justifiable, then, to inquire whether an alternative model might be invoked for lemur speciation/subspeciation in the geologically recent past. This is especially the case since Martin's model implicitly assumes that relatively stable climatic conditions have prevailed in Madagascar for a substantial time. It is now well established not only that the earth as a whole has been subjected to considerable climatic fluctuation during the last several hundred thousand years, as indeed earlier, but also that the tropical environment, against most a priori expectation, has been far from stable. Temperate latitudes were not alone in being affected by the factors which produced the glacial advances and retreats of the Pleistocene, and in the tropical and subtropical zones the effects of these events were far from being confined to glacio-eustatic sea-level changes.

Livingstone (1975) has summarized the considerable geomorphological, paleohydrological, and palynological evidence for late Pleistocene and Holocene changes in African temperature and humidity, while Hamilton (1976) has discussed such climatic changes in the context of floral and faunal distributions within that continent. Expansive phases of tropical forest alternated with periods of contraction, with local faunas being confined to limited forested refugia during the latter, and presented with opportunities for range expansion during the former. Grubb (1978) has emphasized the role of such climatic and vegetational changes in facilitating the isolation and dispersal of populations necessary for mammal speciation to occur.

It would be unrealistic to suppose that Madagascar remained insulated from the climatic oscillations which wrought such profound changes in the floral and faunal distributions within its parent continent. The boundaries of the vegetational zones in Madagascar must have moved up and down, south and north, as temperatures and humidities waxed and waned. This would not have been a regular process; Livingstone (1975) has pointed out that plant communities do not react in a uniform manner to climatic changes but that species instead react in different ways. In an island of Madagascar's varied topography, such zones would not only have changed somewhat in character but would probably also have broken up and become discontinuous. The physiography of Madagascar is extremely different from that of Africa, as is its size and the species content of its vegetation, so direct extrapolation from the continent to the island would be inappropriate. Evidence exists, however, in the form of ancient *lavaka*, erosion gullies of the type which disfigure so much of the plateau today, that at least certain areas of central Madagascar were denuded of forest at some time in the past (Petit and Bourgeat 1965); and it is not hard to imagine that broad swaths of grassland characterized parts of Madagascar during periods of greater coolness and dryness than the present. It seems highly probable that most if not all of the vegetational zones occupied today by lemurs have been reduced at times to fragmented refugia. Haffer (e.g., 1974) has convincingly argued that the biota of the Amazon forest consists of a mosaic of differentiated populations, each representing expansion to a greater or lesser extent from a forested Pleistocene refugium. Investigation of Malagasy faunal distributions from this viewpoint would be well worthwhile, albeit a task of substantial dimensions.

AFTER MANY A SUMMER, DIES THE LEMUR?

It comes naturally to separate the living and the recently extinct lemurs into distinct compartments of the mind. In one sense, of course, the distinction is all too real, with the former being quite properly the concern of neozoologists, and the latter that of paleontologists. But on another level the distinction is artificial: as Standing (1908a) pointed out many years ago, all of these primates are biological contemporaries. Moreover, all are or have been unwilling participants in the same process: the confrontation between man and nature in Madagascar. Those lemurs which have so recently disappeared are merely those which were the most immediately vulnerable to the threats posed by man. Worse, the pressures that led to their extinction did not disappear with them but on the contrary exist today in exacerbated form. Simple extrapolation of current trends suggests that the lemurs will all but totally disappear well within the next century.

What are these pressures, and can anything be done to ameliorate them? The problems, at least, are clear, and Madagascar's hardly differ from those anywhere else in the tropic world today. Essentially, threats to the lemurs are of two kinds: direct human predation, and habitat destruction. While the former was presumably responsible for the extinction of the larger subfossil lemurs, the smaller and fleeter survivors are probably in the long run more severely menaced by the latter. Dramatic regional differences exist, however, in their relative importance.

Hunting of lemurs is carried out in many areas of the island, both using traditional methods (nest raiding, snaring, tree felling, stone throwing, and so forth), and more modern means, notably guns. In certain regions of the island particular lemur species (nowhere all) have traditionally been protected from predation of this kind by *fady* (taboos). *Lemur macaco* and *Propithecus verreauxi* are, for example, *fady* among the northern Sakalava, as is *Indri* among the Betsimisaraka. The human population is becoming increasingly mobile, however, and as Malagasy with no ritual scruples against killing lemurs move into areas where such *fady* exist, these taboos are increasingly broken down. One which seems to have survived unscathed, unfortunately, is the Betsimisaraka belief which demands the killing on sight of aye-ayes, since should one of these animals enter a village, death will shortly follow. Richard and Sussman (1975) have correctly pointed out that traditional forms of hunting, lethal though they may have been for *Palaeopropithecus* and *Megaladapis,* probably do not at present constitute a critical threat to more than a handful of lemur species or subspecies (such as Alaotran *Hapalemur,* slaughtered every year in their hundreds as they flee from fires set in the reeds in which they live). They are concerned, however, that shooting of lemurs by townsfolk is a growing menace; Alison Richard once met a hunter who had shot twelve sifaka in a single afternoon. There are places, especially in northeast Madagascar, where magnificent primary forest still covers large areas but where all is silent, for the forest is almost entirely empty of mammals and birds. Only hunting could explain this. I have never been asked to join a lemur hunt, but it was perhaps in a similar spirit that I was once invited by some residents of Antalaha to go fishing at a spot several miles from town. No rods were packed; no lines, no hooks. The equipment turned out to be dynamite.

In most areas of Madagascar hunting seems nonetheless to be subsidiary to habitat destruction as a threat to the survival of those lemurs which still exist. Traditional Malagasy agriculture depends in most areas on the practice of *tavy* (slash and burn). Every year new areas of forest are cleared, and old fields are abandoned, rarely if ever to return to their original state. Every year the savanna burns, to encourage new green growth for the cattle to graze on. And the fire licks remorselessly at the edges of the forest that remains. Every year trees are felled for firewood, for construction, or simply to expose more fertile ground for planting.

More highly technological exploitation is also taking its toll. Richard and Sussman (1975) have emphasized the negative effects

on the indigenous Malagasy fauna and flora of the activities of
large-scale logging operations and of the clearance of vast areas of
Madagascar to create plantations. I recall in 1974 being taken to see
a new project near Sambava, funded by the Common Market, which
involved the clearance of nine hundred hectares (three and a half
square miles!) of virgin littoral forest to allow the planting of coconut palms, of all uneconomic things. Heavy machinery was making
short work of the forest, but at the time of my visit virtually no
planting had been done. The exposed soil had returned to sterile
sand, identical to that underlying thousands of hectares of adjoining
degraded bush, which could have been much more inexpensively
cleared. But without cost, of course, where are the benefits?

In the light of this steady encroachment, what is the current situation of the lemurs? The truth of this is, in most cases we simply
don't know. Our knowledge of distributions, abundances, and ecotope specificities is so inexact as to leave us largely guessing. All that
is clear is that under present conditions virtually every lemur species
is on an inexorable road to extinction, and that some have progressed further down this road than have others. Some species, for
instance *Allocebus trichotis*, barely known to science, *Daubentonia madagascariensis*, possibly more widely distributed than once believed but
nonetheless at a dramatically low ebb, or *Hapalemur simus*, now
known only from a single locality, have almost certainly dropped below the critical population size necessary for long-term survival.
Others, among them *Lemur rubriventer* and the subspecies of *Propithecus diadema* and *Varecia variegata*, have dramatically declined in
abundance over the past several years and may be approaching the
same point (*P. d. perrieri* has almost certainly reached it already). Yet
others, such as most subspecies of *Lemur fulvus* and *Propithecus verreauxi*, still exist in adequate abundance to ensure their survival (under present conditions, not necessarily if these worsen) for a few
decades yet. The one potential exception to this grim catalog is provided by the mouse lemurs, tiny animals which seem to be more successful in secondary regrowth than they are in climax forests, and
which in some regions may actually have benefited from habitat disturbance. But even here we really do not have the background information necessary for reaching this conclusion, and recent observations have tended to cast some doubt on it (R. J. Russell, quoted
by Richard and Sussman 1975). Information pertinent to the conservation status of various species may be found in the summary table published by Richard and Sussman (1975), and in chapter 3 and
elsewhere in this volume. But at the present deplorable state of our

knowledge the only reasonably sure bet is that the populations of virtually all primate species and subspecies in Madagascar (with the exception of *Homo sapiens*, which is increasing at about three percent annually) are consistently declining every year. The trend is everywhere the same; only the rates differ.

In the face of this dismal situation, what, if anything, can be done to ensure the survival of at least some lemur species in the wild? Unfortunately, as Richard and Sussman have cogently pointed out, it is easy to make recommendations but very much the reverse to reconcile them with the pressing needs of an underdeveloped country whose population is expanding at an alarming rate. At present, legislation exists which forbids the killing of lemurs (for other than "scientific purposes"), but it is virtually unenforced. This is hardly surprising when one realizes that very few Malagasy, whether those charged with enforcing the law or those upon whom it is (or is not) imposed, are aware of the uniqueness of the fauna of their island; that they are the guardians of an irreplaceable and rapidly disappearing patrimony. Most Malagasy, particularly those from the capital and the surrounding plateau region, have never seen a lemur; and to those who live among them they are normally no more than a natural, and expendable, part of the scene, much as squirrels are to us. Such ignorance is neither astonishing nor reprehensible if one considers the fact that while those Malagasy who receive an education may learn about the flora and fauna of western Europe, they are taught nothing about those of their own island. Only very recently has a textbook which treats such matters been printed in the Malagasy language, and that is far from readily available. There is no evidence that eating lemurs contributes in any significant way to the sustenance of any group of Malagasy, so no great sacrifice would be entailed if the hunting and trapping of lemurs were to be abandoned; but the educational problem is one of immense proportions, and one is hard put to see how the appropriate awareness could be instilled on the necessary scale within the limited time available. Ultimately, however, as Richard and Sussman have stressed, "the success or failure of conservation efforts depends . . . on the willingness of people to implement measures taken . . . [and] . . . that willingness is dependent on education" (1975:349).

Alison Jolly (1980) has recently provided a moving account of the human needs which underly the practice of *tavy*, and it is clearly unrealistic to hope for the protection from this type of exploitation of very much of the primary forest which remains. Certain areas of Madagascar have, however, been designated as reserves. At present, eleven Natural Reserves exist (a twelfth, on the eastern side of the Masoala Peninsula, was decommissioned several years ago to permit timbering), as well as two National Parks and one Special Reserve (supported, but not policed, by the International Union for the Conservation of Nature). In addition there are two private reserves established by M. Henry de Heaulme, of which the status under present circumstances in unclear. The locations of these reserves are shown in figure 8.1, and the lemur species occurring in them are listed in table 8.1.

The existence of a system of reserves is a fortunate starting point for new conservation efforts, but unhappily most of these reserves are only nominally protected. Many are home to substantial populations of Malagasy who continue their traditional lifeways within them, cutting trees for firewood, practicing *tavy*, and grazing their cattle (Richard and Sussman 1975). Even when the population is relatively sparse, and where large tracts of forest are more or less undisturbed, hunting can be a major problem; for example, parts of the most recently created Natural Reserve, that of Marojejy, appear largely bereft of larger mammals and birds. This general situation may be at least in part a result of a failure to consider demographic factors when the reserves were designated, and it might be profitable at this point to investigate whether alternative sites might be found where the interests of the flora and fauna are less directly in conflict with the ways of life of local residents.

For the present, though, it must be admitted that we know far too little about the condition of the reserves, and of the states of the animal populations within them, to make rational recommendations as to how to proceed in the future. Andriamampianina and Peyrieras (1972) have summarized what is known. Richard and Sussman have pointed out that careful cost-benefit studies (in both financial and social terms) should precede the formulation of any strategy for the conservation of the Malagasy flora and fauna, in addition to the necessary surveys of the reserves and of other areas which might appropriately figure in developing such a strategy. Clearly, there is no simple solution; and equally clearly, any solution which is found must include some economic incentive for Madagascar itself. No third-world government can afford the expense of maintaining a system of reserves without any economic justification for doing so. This is especially true when enforcing the protected status of reserves might involve considerable social as well as financial cost. Most difficult of all, any economic incentive must be an ongoing one that pays in the long term; this eliminates grant-based arrange-

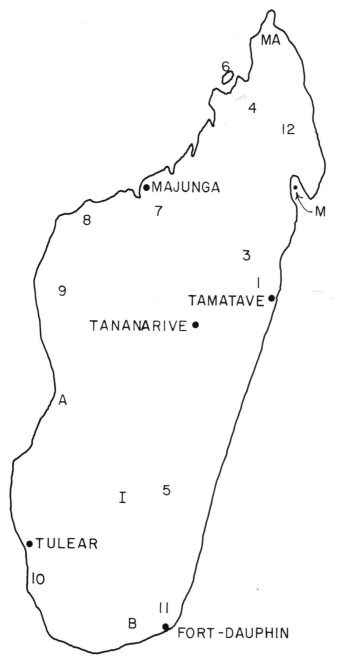

Figure 8.1. Locations of Natural and Special Reserves, National Parks and Private Reserves in Madagascar. Natural Reserves are numbered as in table 8.1; A: Analabé; B: Berenty; I: Massif d'Isalo; M: Nosy Mangabé; MA: Mt. d'Ambre.

Table 8.1. Natural and Special Reserves, National Parks, and Private Reserves of Madagascar, with provisional listings of the lemur genera occurring in each. (Natural Reserve no. 2 no longer exists).

Name	*Lemur Genera*
Natural Reserves	
Betampona (No. 1)	*Lemur, Varecia, Hapalemur, Lepilemur, Avahi, Indri, Propithecus, Microcebus, Cheirogaleus*
Zahamena (No. 3)	*Lemur, Varecia, Propithecus, Avahi, Indri, Microcebus, Cheirogaleus.*
Tsaratanana (No. 4)	*Lemur, Hapalemur, Lepilemur, Cheirogaleus, Phaner*
Andringitra (No. 5)	*Lemur, Varecia, Microcebus*
Lokobé (No. 6)	*Lemur, Lepilemur, Microcebus*
Ankarafantsika (No. 7)	*Lemur, Lepilemur, Avahi, Propithecus, Microcebus, Cheirogaleus*
Namoroka (No. 8)	*Lemur, Lepilemur, Propithecus, Microcebus*
Bemaraha (No. 9)	*Lemur, Propithecus, Microcebus, Mirza*
Tsimanampetsotsa (No. 10)	*Lemur, Lepilemur, Propithecus*
Andohahela (No. 11)	*Lemur, Lepilemur (?), Propithecus, Avahi (?), Microcebus*
Marojejy (No. 12)	*Lemur, Propithecus, Hapalemur, Microcebus, Cheirogaleus (?)*
National Parks	
Mt. d'Ambre	*Lemur, Lepilemur, Microcebus, Cheirogaleus, Phaner*
Isalo	*Lemur catta (?)*
Special Reserve	
Nosy Mangabé	*Lemur, Varecia, Daubentonia*
Private Reserves	
Berenty	*Lemur, Lepilemur, Propithecus, Microcebus*
Analabé	*Lemur, Propithecus, Lepilemur, Mirza, Microcebus, Cheirogaleus*

SOURCE: Data from Andriamampianina and Peyrieras (1972) and elsewhere.

ments. It is difficult to think of anything other than the development of tourism (which has apparently sustained Berenty over recent years) that might provide such an incentive. For even when, for example, abundant economic justification might exist for preserving forest cover to promote soil protection and water conservation (as in the case of the Ankarafantsika reserve), such justification does not self-evidently extend to the expense of ensuring the protection of any fauna the forest might contain.

The prospects, then, are far from rosy. But room for optimism remains, if only because, ultimately, it would be as tragic for the Malagasy to lose their environment as for the world to lose the unique and fascinating biota which it shelters.

References

Adams, W. E. 1957. The extracranial carotid rete and carotid fork in *Nycticebus coucang*. *Ann. Zool.* (Agra) 2:21–37.

Andriamampianina, J. and A. Peyrieras. 1972. Les réserves naturelles intégrales de Madagascar. *C. R. Conférence Int. Consérv. Nat. et Ressources à Madagascar.* U. I. C. N., Morges:103–123.

Andriamiandra, A. and Y. Rumpler. 1968. Rôle de la testostérone sur le déterminisme des glandes brachiales et antebrachiales chez le *Lemur catta. C. R. Soc. Biol.* 162:1651–1655.

Andriantsiferana, R. and R. Rahandraha. 1973a. Influence de quelques facteurs intrinsèques (espèce, âge, gestation, et lactation) sur la consommation alimentaire de *Microcebus murinus* et *Microcebus murinus rufus. C. R. Acad. Sci.* (Paris) 277:1893–1896.

Andriantsiferana, R. and T. Rahandraha. 1973b. Variations saisonnières de la température centrale du Microcèbe *(Microcebus murinus). C. R. Acad. Sci.* (Paris) 277:2215–2218.

Andriantsiferana, R. and T. Rahandraha. 1974. Effets du séjour au froid sur le Microcèbe *(Microcebus murinus* Miller, 1777). *C. R. Acad. Sci.* (Paris) 278:3099–3102.

Andriantsiferana, R., Y. Rarijaona, and A. Randrianaivo. 1974. Observations sur la réproduction du Microcèbe *(Microcebus murinus* Miller, 1777) en captivité à Tananarive. *Mammalia* 38: 234–243.

Appert, O. 1968. La distribution géographique des lémuriens

diurnes de la région du Mangoky au sud- ouest de Madagascar. *Bull. Acad. Malgache* 44 (for 1966):43–45.

Battistini, R. 1976. La biogéographic des Nymphalidés, et spécialement des Charaxes (Lépidoptères). *Rev. Geogr. Madagascar* 28: 123–131.

Battistini, R. and G. Richard-Vindard. 1972. *Biogeography and Ecology of Madagascar.* The Hague: W. Junk.

Bergeron, J. and J. Buettner-Janusch. 1970a. Hematology of prosimian primates: *Galago, Lemur* and *Propithecus. Folia Primat.* 13:155–165.

Bergeron, J. A. and J. Buettner-Janusch. 1970b. Hematology of prosimian primates. II. A comparative study of lemuriformes in captivity in Madagascar and North Carolina. *Folia Primat.* 13:306–313.

Berner, R. A. 1968. Calcium carbonate concretions formed by the decomposition of organic matter. *Science* 159:195–197.

Besairie, H. 1973. Précis de géologie malgache. *Ann. Geol. Madagascar* 36:1–141.

Bishop, A. 1964. Use of the hand in lower primates. In J. Buettner-Janusch, ed., *Evolutionary and Genetic Biology of Primates,* 2: 133–225. New York: Academic Press.

Bourlière, F. and A. Petter-Rousseaux. 1953. L'homéothermie imparfaite de certains Prosimiens. *C. R. Soc. Biol.* 147:1594–1595.

Bourlière, F., J.-J. Petter, and A. Petter-Rousseaux. 1956a. Le dimorphisme sexuel de la glande sous-angulo-maxillaire d'*Avahi laniger* (Gmelin). *Mém. Inst. Sci. Madagascar* (ser. A) 10:299–302.

Bourlière, F., J.-J. Petter, and A. Petter-Rousseaux. 1956b. Variabilité de la température centrale chez les lémuriens. *Mém. Inst. Sci. Madagascar* (ser. A) 10:303–304.

Brenon, P. 1972. The geology of Madagascar. In R. Battistini and G. Richard-Vindard, eds., *Biogeography and Ecology of Madagascar,* pp. 27–86. The Hague: W. Junk.

Budnitz, N. and K. Dainis. 1975. *Lemur catta:* ecology and behavior. In I. Tattersall and R. W. Sussman, eds., *Lemur Biology,* pp. 219–235. New York: Plenum Press.

Buettner-Janusch, J. and J. B. Twichell. 1961. Alkali-resistant haemoglobins in prosimian primates. *Nature* 192:669.

Buettner-Janusch, J. and R. L. Hill. 1965. Evolution of hemoglobin in primates. In V. Bryson and H. J. Vogel, eds., *Evolving Genes and Proteins,* pp. 167–181. New York: Academic Press.

Buettner-Janusch, J. and A. Hamilton. 1979. Chromosomes of Lemuriformes. IV. Karyotype evolution in *Lemur fulvus collaris* (E. Geoffroy, 1812). *Amer. J. Phys. Anthrop.* 50:363–365.

Buettner-Janusch, J., J. L. Washington, and V. Buettner-Janusch. 1971. Hemoglobins of lemuriformes. *Arch. Inst. Pasteur Madagascar* 40:127–136.

Buettner-Janusch, J., V. Buettner-Janusch, and D. Coppenhaver. 1972. Properties of the hemoglobins of newborn and adult prosimians (Prosimii: Lemuriformes and Lorisiformes). *Folia Primat.* 17:177–192.

Bugge, J. 1972. The cephalic arterial system in the insectivores and the primates, with special reference to the Macroscelidoidea and Tupaioidea and the Insectivore-Primate boundary. *Z. Anat. Entwickl.- Gesch.* 135:279–300.

Bugge, J. 1974. The cephalic arterial system in insectivores, primates, rodents, and lagomorphs, with special reference to systematic classification. *Acta Anat.* 87 (Suppl. 62):1–160.

Carleton, A. 1936. The limb bones and vertebrae of the extinct lemurs of Madagascar. *Proc. Zool. Soc. Lond.* 110:281–307.

Cartmill, M. 1972. Arboreal adaptations and the origin of the order Primates. In R. Tuttle ed., *Functional and Evolutionary Biology of Primates*, pp. 97–122. Chicago: Aldine Atherton.

Cartmill, M. 1974a. *Daubentonia, Dactylopsila*, woodpeckers and klinorhynchy. In R. D. Martin, G. A. Doyle, and A. C. Walker, eds., *Prosimian Biology*, pp. 655–670. London: Duckworth.

Cartmill, M. 1974b. Pads and claws in arboreal locomotion. In F. A. Jenkins, ed., *Primate Locomotion*, pp. 45–83. New York: Academic Press.

Cartmill, M. 1975. Strepsirhine basicranial structures and the affinities of the Cheirogaleidae. In W. P. Luckett and F. S. Szalay, eds., *Phylogeny of the Primates: A Multidisciplinary Approach*, pp. 313–354. New York: Plenum Press.

Cartmill, M. 1978. The orbital mosaic in prosimians and the use of variable traits in systematics. *Folia Primat.* 30:89–114.

Cartmill, M. 1979. The volar skin of primates: its frictional characteristics and their functional significance. *Amer. J. Phys. Anthrop.* 50:497–509.

Cartmill, M. and R. Kay. 1978. Cranio-dental morphology, tarsier affinities, and primate sub-orders. In D. Chivers and K. Joysey, eds., *Recent Advances in Primatology*, 3:205–214. London: Academic Press.

Cauche, F. 1651. Relation du voyage que francois Cauche de Rouen a fait à Madagascar, isles adiacentes, & coste d'Afrique. In A. Courbe, ed., *Relations veritables et curieuses de l'isle de Madagascar, et du Brésil.* Paris.

Cauche, F. 1711. A voyage to Madagascar, the adjacent islands, and coast of Afrik. No. 7 in John Stevens, ed. and tr., *A New Collection of Voyages and Travels.* London.

Cavelier, C. 1979. La limite Eocène-Oligocène en Europe occidentale. *Sci. Géol.* 54:1–272.

Charles-Dominique, P. 1978. Solitary and gregarious prosimians: evolution of social structures in primates. In D. Chivers and K. Joysey, eds., *Recent Advances in Primatology,* 3:139–149. London: Academic Press.

Charles-Dominique, P. and M. Hladik. 1971. Le *Lepilemur* du sud de Madagascar: écologie, alimentation et vie sociale. *Terre et Vie* 25:3–66.

Charles-Dominique, P. and R. D. Martin. 1970. Evolution of lorises and lemurs. *Nature* 227:257–260.

Chauvet, B. 1972. The forest of Madagascar. In R. Battistini and G. Richard-Vindard, eds., *Biogeography and Ecology of Madagascar,* pp. 191–200. The Hague: W. Junk.

Chu, E. H. Y. and M. A. Bender. 1962. Cytogenetics and evolution of primates. *Ann. N.Y. Acad. Sci.* 102:253–266.

Chu, E. H. Y. and B. A. Swomley. 1961. Chromosomes of lemurine lemurs. *Science* 133:1925–1926.

Clark, W. E. Le Gros. 1931. The brain of *Microcebus murinus. Proc. Zool. Soc. Lond.* (for 1931), pp. 463–486.

Clark, W. E. Le Gros. 1936. The problem of the claw in primates. *Proc. Zool. Soc. Lond.* (for 1936), pp. 1–24.

Clark, W. E. Le Gros. 1945. A note on the palaeontology of the lemuroid brain. *J. Anat.* 79:123–126.

Clark, W. E. Le Gros. 1947. Deformation patterns in the cerebral cortex. In W. E. le Gros Clark and P. B. Medawar, eds., *Essays on Growth and Form Presented to d'Arcy Wentworth Thompson,* pp. 1–22. Oxford: Clarendon Press.

Clark, W. E. Le Gros. 1959. *The Antecedents of Man.* Edinburgh: The University Press.

Conley, J. M. 1975. Notes on the activity pattern of *Lemur fulvus. J. Mammal.* 56:712–715.

Conroy, G. C. and J. R. Wible. 1978. Middle ear morphology of *Lemur variegatus:* Some implications for primate paleontology. *Folia Primat.* 21:81–85.

Cooper, H. M. 1974. Learning set in *Lemur macaco.* In R. D. Martin,

G. A. Doyle, and A. C. Walker, eds., *Prosimian Biology,* pp. 293–300. London: Duckworth.

Cronin, J. E. and V. M. Sarich. 1978. Primate higher taxa—the molecular view. In D. J. Chivers and K. A. Joysey, eds., *Recent Advances in Primatology* 3:287–289. London: Academic Press.

Crook, J. H. 1966. Gelada baboon herd structure and movement: a comparative report. *Symp. Zool. Soc. Lond.* 10:237–258.

Crook, J. H. 1967. Evolutionary change in primate societies. *Sci. Jour.* 3:66–72.

Darracott, B. 1974. On the crustal structure and evolution of southeastern Africa and the adjacent Indian Ocean. *Earth Planet. Sci. Lett.* 24:282–290.

Decary, R. 1950. *La Faune malgache.* Paris: Payot.

Dene, H. T., M. Goodman, and W. Prychodko. 1976. Immunodiffusion evidence on the phylogeny of the primates. In M. Goodman and R. Tashian, eds., *Molecular Anthropology,* pp. 171–195. New York: Plenum Press.

Dixey, F. 1960. The geology and geomorphology of Madagascar: A comparison with eastern Africa. *Quart. J. Geol. Soc. Lond.* 116:255–268.

Donque, G. 1972. The climatology of Madagascar. In R. Battistini and G. Richard-Vindard, eds., *Biogeography and Ecology of Madagascar,* pp. 87–144. The Hague: W. Junk.

Doyle, G. A. and R. D. Martin, eds. 1979. *The Study of Prosimian Behavior.* New York: Academic Press.

Drenhaus, U. 1975. Ein Beitrag zur Taxonomie der Lemuriformes Gregory 1915 unter besonderer Berücksichtigung der Gattung *Lepilemur* I. Geoffroy 1851. Ph.D. Thesis, Christian-Albrechts-Universität, Kiel.

Eaglen, R. H. and K. J. Boskoff. 1978. The birth and early development of a captive sifaka, *Propithecus verreauxi coquereli. Folia Primat.* 30:206–219.

Ehrlich, A., J. Fobes, and J. King. 1976. Prosimian learning capacities. *J. Hum. Evol.* 5:599–617.

Eldredge, N. and J. Cracraft. 1980. *Phylogenetic Patterns and the Evolutionary Process,* New York: Columbia University Press.

Eldredge, N. and I. Tattersall. 1975. Evolutionary models, phylogenetic reconstruction, and another look at hominid phylogeny. *Contrib. Primat.* 5:218–242.

Elliot, D. G. 1913. *A Review of the Primates,* vol. 1. New York: American Museum of Natural History.

Embleton, B. J. J. and M. W. McElhiny. 1975. The palaeoposition of

Madagascar: Palaeomagnetic results for the Isalo Group. *Earth Planet. Sci. Lett.* 25:329–341.

Evans, C. S. and R. W. Goy. 1968. Social behavior and reproductive cycles in captive ring-tailed lemurs *(Lemur catta)*. *J. Zool.* 156:181–197.

Fiedler, W. 1959. Über das Baculum des Cheirogaleinae. *Zool. Anz.* 163:57–63.

Filhol, H. 1895. Observations concernant les Mammifères contemporains des *Aepyornis* à Madagascar. *Bull. Mus. Natl. Hist. Nat.* (Paris) 1:2–14.

Flacourt, E. de. 1658. *Histoire de la grande Isle Madagascar composée par le Sieur de Flacourt.* 2 vols. Paris: Chez G. de Lvyne.

Flacourt, E. de. 1661. *Histoire de la grande Isle Madagascar. Avec une Relation de ce qui s'est passé és années 1655, 1656 & 1657, non encor veuë par la première impression.* Troyes, N. Oudot; Paris, Pierre L'Amy. Pp. 1–202 (Histoire); pp. 203–471 (Relation).

Flores, G. 1970. Suggested origin of the Mozambique Channel. *Trans. Geol. Soc. S. Afr.* 73:1–16.

Flower, M. J. F. and D. F. Strong. 1969. The significance of sandstone inclusions in lavas of the Comores Archipelago. *Earth Planet. Sci. Lett.* 7:47–50.

Foerg, R. 1978. Das Verhalten von *Lemur variegatus* in Gefangenschaft. M.A. Thesis, University of Tübingen.

Foerg, R. In preparation. Reproductive behavior in *Varecia variegata*.

Forbes, H. O. 1894. *A Hand-Book to the Primates.* 2 vols. London: W. H. Allen.

Francis, T. J. G., D. Davies, and M. N. Hill. 1966. Crustal structure between Kenya and the Seychelles. *Phil. Trans. Roy. Soc. Lond.,* (ser. B.), 259:240–261.

Gillette, R. G., R. Brown, P. Herman, S. Vernon, and J. Vernon. 1973. The auditory sensitivity of the lemur. *Amer. J. Phys. Anthrop.* 28:365–370.

Gingerich, P. 1975. Dentition of *Adapis parisiensis* and the evolution of lemuriform primates. In I. Tattersall and R. W. Sussman, eds., *Lemur Biology,* pp. 65–80. New York: Plenum Press.

Godfrey, L. R. 1977. Structure and function in *Archaeolemur* and *Hadropithecus* (subfossil Malagasy lemurs): The postcranial evidence. Ph.D. Thesis, Harvard University.

Grandidier, A. 1875–1921. *Histoire physique, naturelle, et politique de Madagascar.* 32 vols. Paris: Hachette.

Grandidier, G. 1899. Description des ossements de lémuriens disparus. *Bull. Mus. Natl. Hist. Nat.* (Paris), 5:272–276, 344–348.

Grandidier, G. 1900. Note sur des ossements d'animaux disparus. *Bull. Mus. Natl. Hist. Nat.* (Paris), 6:214–219.

Grandidier, G. 1902. Observations sur les lémuriens disparus de Madagascar: Collections Alluaud, Gaubert, Grandidier. *Bull. Mus. Natl. Hist. Nat.* (Paris), 8:497–502.

Grandidier, G. 1905. Recherches sur les lémuriens disparus et en particulier ceux qui vivaient à Madagascar. *Nouv. Arch. Mus. Natl. Hist. Nat.* (Paris) (4 sér), 7:1–142.

Grandidier, G. 1929. Une variété du *Cheiromys madagascariensis* actuel et un nouveau *Cheiromys* subfossile. *Bull. Acad. Malgache* 11 (for 1928):101–107.

Green, A. G. 1972. Sea-floor spreading in the Mozambique Channel. *Nature Phys. Sci.* 236:19–21.

Gregory, W. K. 1915. 1. On the relationship of the Eocene lemur *Notharctus* to the Adapidae and other primates. 2. On the classification and phylogeny of the Lemuroidea. *Bull. Geol. Soc. Amer.* 26:419–446.

Gregory, W. K. 1920. On the structure and relations of *Notharctus,* an American Eocene primate. *Mem. Amer. Mus. Nat. Hist.* (n.s.) 3:49–241.

Griffiths, J. F. and R. Ranaivoson. 1972. Madagascar. In J. F. Griffiths, ed., *World Survey of Climatology* 10:87–144. The Hague: W. Junk.

Grubb, P. 1978. Patterns of speciation in African mammals. *Bull. Carnegie Mus. Nat. Hist.* 6:152–167.

Haffer, J. 1974. Avian speciation in tropical South America. *Nuttall Ornithol. Club Publ.* 14:1–390.

Hamilton, A. 1976. The significance of patterns of distribution shown by forest plants and animals in tropical Africa for the reconstruction of Upper Pleistocene palaeoenvironments: A review. *Palaeoecol. of Africa* 9:63–97.

Hamilton, A. E. and J. Buettner-Janusch. 1977. Chromosomes of Lemuriformes. III. The genus *Lemur*: Karyotypes of species, subspecies, and hybrids. *Ann. N.Y. Acad. Sci.* 293:125–159.

Hamilton, A. E., I. Tattersall, R. W. Sussman, and J. Buettner-Janusch. 1980. Chromosomes of Lemuriformes. VI. Comparative karyology of *Lemur fulvus*: A G-banded karyotype of *Lemur fulvus mayottensis* Schlegel, 1866. *Int. J. Primat.* 1:81–93.

Harrington, J. E. 1975. Field observations of social behavior of *Lemur fulvus fulvus* E. Geoffroy 1812. In I. Tattersall and R. W. Sussman, eds., *Lemur Biology,* pp. 259–279. New York: Plenum Press.

Harrington, J. E. 1977. Discrimination between males and females by scent in *Lemur fulvus*. *Anim. Behav.* 25:147–151.

Harrington, J. E. 1978. Diurnal behavior of *Lemur mongoz* at Ampijoroa, Madagascar. *Folia Primat.* 29:291–302.

Hayata, I., S.-I. Sonta, M. Hoh, and N. Kondo. 1971. Notes on the karyotypes of some prosimians, *Lemur mongoz*, *Lemur catta*, *Nycticebus coucang*, and *Galago crassicaudatus*. *Japanese J. Genet.* 46:61–64.

Heffner, H. and B. Masterton. 1970. Hearing in primitive primates: slow loris (*Nycticebus coucang*) and potto (*Perodicticus potto*). *J. Comp. Physiol. Psychol.* 71:175–182.

Heirtzler, J. R. and R. H. Burroughs. 1971. Madagascar's paleoposition: New data from the Mozambique Channel. *Science* 174: 488–490.

Hershkovitz, P. 1977. *Living New World Monkeys (Platyrrhini)*, vol. 1. Chicago: University of Chicago Press.

Hill, R. L. and J. Buettner-Janusch. 1964. Evolution of hemoglobin. *Federation Proc.* 23:1236–1242.

Hill, W. C. O. 1948. Rhinoglyphics: Epithelial sculpture of the mammalian rhinarium. *Proc. Zool. Soc. Lond.* 118:28–35.

Hill, W. C. O. 1953. *Primates: Comparative Anatomy and Taxonomy*. I. *Strepsirhini*. Edinburgh: The University Press.

Hladik, C. M. and P. Charles-Dominique. 1974. The behavior and ecology of the sportive lemur (*Lepilemur mustelinus*) in relation to its dietary pecularities. In R. D. Martin, G. A. Doyle, and A. C. Walker, eds., *Prosimian Biology*, pp. 23–37. London: Duckworth.

Hladik, C. M., P. Charles-Dominique, P. Valdebouze, J. Delort-Laval, and J. Flanzy. 1971. La caecotrophie chez un primate phyllophage du genre *Lepilemur* et les corrélations avec les pecularités de son appareil digestif. *C. R. Acad. Sci.* (Paris) 272:3191–3194.

Hofer, H. 1953. Über Gehirn und Schädel von *Megaladapis edwardsi* G. Grandidier (Lemuroidea). *Zeitschr. Wiss. Zool.* 157:220–284.

Hofer, H. 1969. The evolution of the brain of primates: Its influence on the form of the skull. *Ann. N.Y. Acad. Sci.* 167:341–356.

Humbert, H. 1927. Principaux aspects de la végétation à Madagascar: Déstruction d'une flore insulaire par le feu. *Mém. Acad. Malgache* 5:1–80.

Humbert, H. 1955. Les térritoires phytogéographiques de Madagascar, leur cartographie. *Ann. Biol.* (3 ser.) 31:195–204.

Humbert, H. 1959. Origines présumées et affinités de la flore de Madagascar. *Mém. Inst. Sci. Madagascar* (sér. B) 9:149–187.

Humbert, H. and G. Cours Darne. 1965. Notice de la carte Madagascar, Carte Internationale de Tapis Végétal. *Trav. Sect. Sci.
Tech. Inst. Français Pondichery.* Hors série 6:1–162.

Jolly, A. 1966a. *Lemur Behavior.* Chicago: University of Chicago
Press.

Jolly, A. 1966b. Lemur social behavior and primate intelligence. *Science* 153:501–507.

Jolly, A. 1972. Troop continuity and troop spacing in *Propithecus verreauxi* and *Lemur catta* at Berenty (Madagascar). *Folia Primat.*
17:335–362.

Jolly, A. 1980. *A World Like Our Own: Man and Nature in Madagascar.*
New Haven: Yale University Press.

Jolly, A., H. Gustafson, A. Mertl, and G. Ramanantsoa. In press.
Population, espace vital et composition des groupes chez le maki
(Lemur catta) et le sifaka *(Propithecus verreauxi verreauxi)* à Ber-
enty, République Malagasy. *Bull. Acad. Malgache.*

Jolly, C. J. 1970a. The seed-eaters: A new model of hominid differentiation based on a baboon analogy. *Man* (n.s.) 5:5–26.

Jolly, C. J. 1970b. *Hadropithecus*, a lemuroid small-object feeder. *Man*
(n.s.) 5:525–529.

Jolly, C. J. 1972. The classification and natural history of *Theropithecus (Simopithecus)* (Andrews, 1916), baboons of the African Plio-
Pleistocene. *Bull. Brit. Mus. (Nat. Hist.)* Geol. 22:1–123.

Jouffroy, F.-K. 1962. La musculature des membres chez les lémuriens
de Madagascar. Étude descriptive et comparative. *Mammalia* 26
(Suppl. 2):1–326.

Jouffroy, F.-K. 1963. Contribution à la connaisance du genre *Archaeolemur* Filhol, 1895. *Ann. Paléont.* 49:129–155.

Jouffroy, F.-K. 1975. Osteology and myology of the lemuriform
postcranial skeleton. In I. Tattersall and R. W. Sussman, eds.,
Lemur Biology, pp. 149–192. New York, Plenum Press.

Jouffroy, F.-K. and J. Lessertisseur. 1959. La main des lémuriens
malgaches comparée à celle des autres primates. *Mém. Inst. Sci.
Madagascar* (sér A) 13:195–219.

Jouffroy, F.-K. and J. Lessertisseur. 1979. Relationships between
limb morphology and locomotor adaptations among prosimians: An osteometric study. In M. E. Morbeck, H. Preuschoft,
and N. Gomberg, eds., *Environment, Behavior, and Morphology:
Dynamic Interactions in Primates*, pp. 143–181. New York:
G. Fischer.

Jungers, W. L. 1977. Hindlimb and pelvic adaptations to vertical
climbing and clinging in *Megaladapis,* a giant subfossil prosimian

from Madagascar. *Yrbk. Phys. Anthrop.* 20 (for 1976):508–524.

Katz, M. B. and C. Premoli. 1979. India and Madagascar in Gondwanaland based on matching Precambrian lineaments. *Nature* 279:312–315.

Kaudern, W. 1914. Einige Beobachtungen über die Zeit der Fortpflanzung der madagassischen Säugetiere. *Ark. Zool.* 9 (1):1–22.

Kaudern, W. 1915. Säugetiere aus Madagaskar. *Ark. Zool.* 9(18): 1–101.

Kauffman, E. G. 1975. Overview: Western Interior Cretaceous Basin. *Mountain Geologist* 14:85.

Kay, R. F. 1975. The functional adaptations of primate molar teeth. *Amer. J. Phys. Anthrop.* 43:195–215.

Kent, P. E. 1972. Mesozoic history of the east coast of Africa. *Nature* 238:147–148.

Kent, P. E. 1974. Leg 25 results in relation to East African coastal stratigraphy. In E. S. W. Simpson, R. Schlich, et al., *Initial Reports of the Deep Sea Drilling Project*, 25:679–684. Washington, D.C.: Govt. Printing Office.

Klopfer, P. H. 1974. Mother-young relations in lemurs. In R. D. Martin, G. A. Doyle, and A. C. Walker, eds., *Prosimian Biology*, pp. 273–292. London: Duckworth.

Klopfer, P. H. and A. Jolly. 1970. The stability of territorial boundaries in a lemur troop. *Folia Primat.* 12:199–208.

Klopfer, P. H. and M. S. Klopfer. 1970. Patterns of maternal care in three species of *Lemur*. I. Normative description. *Zeitschr. Tierpsychol.* 27:984–996.

Koechlin, J. 1972. Flora and vegetation of Madagascar. In R. Battistini and G. Richard-Vindard, eds., *Biogeography and Ecology of Madagascar*, pp. 145–226. The Hague: W. Junk.

Kollman, M. and L. Papin. 1925. Études sur les lémuriens: anatomie comparée des fosses nasales et de leurs annexes. *Arch. Morph. Gén. et Exp.* 22:1–61.

Kolmer, W. 1930. Zur Kenntnis des Auges der Primaten. *Z. Anat. Entwickl.* 93:679–722.

Kress, J. H., J. M. Conley, R. H. Eaglen, and A. E. Ibanez. 1978. The behavior of *Lemur variegatus* Kerr 1792. *Zeitschr. Tierpsychol.* 48:87–99.

Lamberton, C. 1929. Sur les *Archaeoindris* de Madagascar. *C. R. Acad. Sci.* (Paris) 188:1572–1574.

Lamberton, C. 1934. Contribution à la connaissance de la faune subfossile de Madagascar: Lémuriens et ratites. *Mém. Acad. Malgache* 17:1–168.

Lamberton, C. 1937. Fouilles faites en 1936. *Bull. Acad. Malgache* 19 (for 1936) :1–19.

Lamberton, C. 1938a. Contribution à la connaissance de la faune subfossile de Madagascar. Note 3. *Bull. Acad. Malgache* 20 (for 1937):1–44.

Lamberton, C. 1938b. Dentition de lait de quelques lémuriens subfossiles. *Mammalia* 2:57–80.

Lamberton, C. 1939. Contribution à la connaissance de la faune subfossile de Madagascar: Lémuriens et cryptoproctes. *Mém. Acad. Malgache* 27:1–203.

Lamberton, C. 1941. Contribution à la connaissance de la faune subfossile de Madagascar. Note 9. *Mém. Acad. Malgache* 35: 7–132.

Lamberton, C. 1947. Contribution à la connaissance de la faune subfossile de Madagascar. Note 16. *Bull. Acad. Malgache* 26 (for 1944):1–52.

Lamberton, C. 1948. Contribution à la connaissance de la faune de Madagascar. Note 20. *Bull. Acad. Malgache* 27 (for 1946):24–28.

Lamberton, C. 1952. L'hypophyse et le gigantisme chez certains lémuriens fossiles de Madagascar. *Bull. Acad. Malgache* 29 (for 1949):26–42.

Livingstone, D. A. 1975. Late Quaternary climatic change in Africa. *Ann. Rev. Ecol. Syst.* 6:249–280.

Lorenz, L. R. von Liburnau. 1899. Einen fossilen Anthropoiden von Madagaskar. *Anz. Kais. Akad. Wiss. Wien.* 19:255–257.

Lorenz, L. R. von Liburnau. 1900a. Über einige reste ausgestorbener primaten von Madagaskar. *Denkschr. Kais Akad. Wiss. Wien.* (math-nat.) 70:1–15.

Lorenz, L. R. von Liburnau. 1900b. *Palaeolemur destructus. Anz. Akad. Wiss. Wien* 20:8.

Lorenz, L. R. von Liburnau. 1902. Über *Hadropithecus stenognathus* Lz. *Denkschr. Kais. Akad. Wiss. Wien* 72:243–254.

Lorenz, L. R. von Liburnau. 1905. *Megaladapis edwardsi* G. Grandidier. *Denkschr. Kais. Akad. Wiss. Wien.* (math.-nat.) 77:451–490.

Luckett, W. P. 1974. Comparative development and evolution of the placenta in primates. *Contrib. Primatol.* 3:142–234.

Luckett, W. P. 1975. Ontogeny of the fetal membranes and the placenta: Their bearing on primate phylogeny. In W. P. Luckett and F. S. Szalay, eds., *Phylogeny of the Primates: A Multidisciplinary Approach*, pp. 157–182. New York: Plenum Press.

Luyendyk, B. P. 1974. Gondwanaland dispersal and the early formation of the Indian Ocean. In T. A. Davies, B. P. Luyendyk, et al., *Initial Reports of the Deep Sea Drilling Project,* 26:923–945. Washington, D.C.: Govt. Printing Office.

McElhinny, M. W. and B. J. J. Embleton. 1976. The palaeoposition of Madagascar: Remanence and magnetic properties of late Palaeozoic sediments. *Earth Planet. Sci. Lett.* 31:101–112.

Mahé, J. 1965. *Les Subfossiles malgaches.* Tananarive, Imprimerie Nationale:1–11.

Mahé, J. 1968. Conséquence biologique tirée de l'orientation vestibulaire du crâne de *Palaeopropithecus. Bull. Mus. Natl. Hist. Nat.* (Paris), (sér. 2), 40:634–639.

Mahé, J. 1976. Craniométrie des lémuriens. Analyses multivariables—phylogénie. *Mém. Mus. Natl. Hist. Nat.* (Paris), (sér. C), 32:1–342.

Mahé, J. and M. Sourdat. 1972. Sur l'extinction des vertébrés subfossiles et l'aridification du climat dans le sud-ouest de Madagascar. *Mém. Soc. Geol. France* 17(14):295–309.

Major, C. I. Forsyth. 1893. Exhibition of a subfossil lemuroid skull from Madagascar. *Proc. Zool. Soc. Lond.* 36:532–535.

Major, C. I. Forsyth. 1894. On *Megaladapis madagascariensis,* an extinct gigantic lemuroid from Madagascar, with remarks on the associated fauna and on its geological age. *Phil. Trans. Roy Soc. Lond.* (series B) 185:15–38.

Major, C. I. Forsyth. 1896. Preliminary notice on fossil monkeys from Madagascar. *Geol. Mag.* (decade 4) 3:433–436.

Major, C. I. Forsyth. 1897. On the brains of two sub-fossil Malagasy lemuroids. *Proc. Roy. Soc. Lond.* 62:47–50.

Major, C. I. Forsyth. 1900a. A summary of our present knowledge of extinct primates from Madagascar. *Geol. Mag.* (decade 4) 7:492–499.

Major, C. I. Forsyth. 1900b. Extinct mammalia from Madagascar: *Megaladapis insignis,* sp. nov. *Phil. Trans. Roy. Soc. Lond.* (series B) 193:47–50.

Martin, R. D. 1972a. A preliminary field-study of the Lesser Mouse Lemur (*Microcebus murinus* J. F. Miller, 1777). *Z. Comp. Ethol.,* Suppl. 9:43–89.

Martin, R. D. 1972b. Adaptive radiation and behavior of the Malagasy lemurs. *Phil. Trans. Roy. Soc. Lond.* (series B) 264:295–352.

Martin, R. D. 1972c. A laboratory breeding colony of the Lesser Mouse Lemur. In W. I. B. Beveridge, ed., *Breeding Primates,* pp. 161–171. Basel: Karger.

Martin, R. D. 1973. A review of the behavior and ecology of the Lesser Mouse Lemur (*Microcebus murinus* J. F. Miller, 1777). In

R. P. Michael and J. H. Crook, eds., *Comparative Ecology and Behavior of Primates*, pp. 1–68. London: Academic Press.

Martin, R. D., G. A. Doyle, and A. C. Walker, eds., 1974. *Prosimian Biology*. London: Duckworth.

Mason, G. A. and J. Buettner-Janusch. 1977. Codominant autosomal inheritance of polymorphic red cell acid phosphatases of lemurs and some properties of the enzymes. *Biochem. Genet.* 15:487–507.

Masterton, B., H. Heffner, and R. Ravizza. 1969. The evolution of human hearing. *J. Acoust. Soc. Amer.* 45:966–985.

Matthews, R. K. and R. Z. Poore. 1980. The Tertiary $\delta^{18}O$ record: an alternative view concerning glacio-eustatic sea level fluctuations. *Geology* 8:501–504.

Mayr, E. 1969. *Principles of Systematic Zoology*. New York: McGraw Hill.

Megiser, H. 1609. Beschriebung der uberaussreichen Insul Madagascar. Altenburg.

Mertl, A. S. 1975. Discrimination of individuals by scent in a primate. *Behav. Biol.* 14:505–509.

Mertl, A. S. 1976. Olfactory and visual cues in social interactions of *Lemur catta*. *Folia Primat.* 26:151–161.

Mertl, A. S. 1977. Habitation to territorial scent marks in the field by *Lemur catta*. *Behav. Biol.* 21:500–507.

Mertl-Milhollen, A. S. 1979. Olfactory demarcation of territorial boundaries by a primate—*Propithecus verreauxi*. *Folia Primat.* 32:35–42.

Mertl-Milhollen, A., H. Gustafson, N. Budnitz, K. Dainis, and A. Jolly. 1979. Population and territory stability of the *Lemur catta* at Berenty, Madagascar. *Folia Primat.* 31:106–122.

Milne-Edwards, A. and A. Grandidier. 1875. *Histoire physique, naturelle, et politique de Madagascar*. Vol. 1, 1, text. Paris: Imprimerie Nationale for A. Grandidier.

Milton, K. 1978. Role of the upper canine and P^2 in increasing the harvesting efficiency of *Hapalemur griseus* Link, 1795. *J. Mammal.* 59:188–190.

Mitchell, C., J. Vernon, and P. Herman. 1971. What does the lemur really hear? *J. Acoust. Soc. Amer.* 50(2):710–711.

Mitchell, C., R. Gillette, J. Vernon, and P. Herman. 1970. Pure-tone auditory thresholds in two species of lemurs. *J. Acoust. Soc. Amer.* 48(2):531–535.

Mivart, St. George. 1864. Notes on the crania and dentition of the Lemuridae. *Proc. Zool. Soc. Lond.* (for 1864), pp. 611–648.

Morat, P. 1969. Note sur l'application à Madagascar du quotient pluviothermique d'Emberger. *Cah. O.R.ST.O.M.* (Sér. Biol.), 10:117–132.

Moreau, R. E. 1966. *The Bird Faunas of Africa and Its Islands.* New York: Academic Press.

Mundy, Peter. 1907–36. The travels of Peter Mundy in Europe and Asia, 1608–1667. R. C. Temple, ed. Cambridge: The Hakluyt Society. Vol. 1, 1907; vol. 2, 1914; vol. 3(1 and 2), 1919; vol. 4, 1924; vol. 5, 1936.

Myers, R. D. 1971. Primates. In G. C. Whittow, ed., *Comparative Physiology of Thermoregulation*, 2:283–326. New York: Academic Press.

Niaussat, M.-M and D. Molin. 1978. Hearing and vocalization in a Malagasy lemur: *Phaner furcifer.* In D. Chivers and J. Herbert, eds., *Recent Advances in Primatology*, 1:821–825. London: Academic Press.

Nute, P. E. and J. Buettner-Janusch. 1969. Genetics of polymorphic transferrins in the genus *Lemur. Folia Primat.* 10:181–194.

Pagès, E. 1978. Home range, behaviour and tactile communication in a nocturnal Malagasy lemur *Microcebus coquereli.* In D. Chivers and K. Joysey, eds., *Recent Advances in Primatology*, 3: 171–177. London: Academic Press.

Pariente, G. 1974. Influence of light on the activity rhythms of two Malagasy lemurs: *Phaner furcifer* and *Lepilemur mustelinus leucopus.* In R. D. Martin, G. A. Doyle, and A. C. Walker, eds., *Prosimian Biology*, pp. 183–198. London: Duckworth.

Pariente, G. F. 1975. Observation ophthalmologique de zones fovéales vraies chez *Lemur catta* et *Hapalemur griseus*, primates de Madagascar. *Mammalia* 39:487–497.

Pariente, G. F. 1976. Les differents aspects de la limite du tapétum lucidum chez les prosimiens. *Vision Res.* 16:387–391.

Pariente, G. F. 1979. The role of vision in prosimian behavior. In G. A. Doyle and R. D. Martin, eds., *The Study of Prosimian Behavior*, pp. 411–459. New York: Academic Press.

Perrier de la Bathie, H. 1921. La végétation malgache. *Ann. Mus. Colon,* Marseille (3 sér.) 9:1–268.

Perrier de la Bathie, H. 1927. Fruits et graines du gisement de subfossiles d'Ampasambazimba. *Bull. Acad. Malgache* 10 (for 1927):24–25.

Perrier de la Bathie, H. 1936. *Biogéographie des plantes de Madagascar.* Paris: Soc. Edit. Géogr. Marit. et Colon.

Petit, M. and F. Bourgeat. 1965. Les "lavaka" malgaches: un agent naturel d'évolution des versants. *Bull. Assoc. Geogr. France* 332:29–33.

Petter, J.-J. 1962. Recherches sur l'écologie et l'éthologie des lémuriens malgaches. *Mém. Mus. Natl. Hist. Nat.* (Paris) 27: 1–146.

Petter, J.-J. 1977. The aye-aye. In Prince Rainier and G. H. Bourne, eds., *Primate Conservation*, pp. 37–57. New York: Academic Press.

Petter, J.-J. 1978. Ecological and physiological adaptations of five sympatric nocturnal lemurs to seasonal variations in food production. In D. Chivers and J. Herbert, eds., *Recent Advances in Primatology*, 1:211–223. London: Academic Press.

Petter, J.-J. and P. Charles-Dominique. 1979. Vocal communication in prosimians. In G. A. Doyle and R. D. Martin, eds., *The Study of Prosimian Behaviour*, pp. 247–304. London: Academic Press.

Petter, J.-J. and A. Petter-Rousseaux. 1960. Remarques sur la systématique du genre *Lepilemur. Mammalia* 24:76–86.

Petter, J.-J. and A. Petter-Rousseaux. 1964. Première tentative d'estimation des densités de peuplement des lémuriens malgaches. *Terre et Vie* 18:427–435.

Petter, J.-J. and A. Petter-Rousseaux. 1967. Contribution à la systématique des *Cheirogaleinae* [sic] (lémuriens malgaches). *Allocebus*, gen. nov. pour *Cheirogaleus trichotis* Gunther 1875. *Mammalia* 31:574–582.

Petter, J.-J. and A. Peyrieras. 1970a. Nouvelle contribution à l'étude d'un lémurien malgache, le aye-aye (*Daubentonia madagascariensis* E. Geoffroy) [sic]. *Mammalia* 34:167–193.

Petter, J.-J. and A. Peyrieras. 1970b. Observations éco-éthologiques sur les lémuriens malgaches du genre *Hapalemur. Terre et Vie* 24:356–382.

Petter, J.-J., R. Albignac, and Y. Rumpler. 1977. Faune de Madagascar 44: Mammifères Lémuriens (Primates Prosimiens). Paris, ORSTOM/CNRS.

Petter, J.-J., A. Schilling, and G. Pariente. 1971. Observations éco-éthologiques sur deux lémuriens malgaches nocturnes: *Phaner furcifer* et *Microcebus coquereli. Terre et Vie* 25:287–327.

Petter-Rousseaux, A. 1962. Recherches sur la biologie de la réproduction des primates inférieurs. *Mammalia* 26 (Suppl. 1): 1–88.

Petter-Rousseaux, A. 1964. Reproductive physiology and behavior of the Lemuroidea. In J. Buettner-Janusch, ed., *Evolution and Genetic Biology of the Primates*, 2:91–132. New York: Academic Press.

Petter-Rousseaux, A. 1968. Cycles génitaux saisonniers des lémuriens malgaches. *Entretiens de Chizé* 38:11–22.

Petter-Rousseaux, A. 1970. Observation sur l'influence de la photopériode sur l'activité sexuelle chez *Microcebus murinus* (Miller, 1777). *Ann. Biol. Anim. Biochim. Biophys.* 12:367–375.

Pilbeam, D. R. and A. C. Walker. 1968. Fossil monkeys from the Miocene of Napak, north-east Uganda. *Nature* 220:657–660.

Piveteau, J. 1950. Recherches sur l'encéphale de lémuriens disparus. *Ann. Paleont.* 36:87–103.

Pollen, F. P. L. 1863. Énumération des animaux vertébrés de l'île de Madagascar. *Ned. Tijdschr. Dierk.* 1:277.

Pollen, F. P. L. and J. C. Van Dam. 1877. Recherches sur la faune de Madagascar et de ses dépendences. 1: *Relation de Voyage*. Leiden: E. J. Brill.

Pollock, J. I. 1975a. The social behaviour and ecology of *Indri indri*. Ph.D. Thesis, London University.

Pollock, J. I. 1975b. Field observations on *Indri indri*: a preliminary report. In I. Tattersall and R. W. Sussman, eds., *Lemur Biology*, pp. 287–311. New York: Plenum Press.

Pollock, J. I. 1977. The ecology and sociology of feeding in *Indri indri*. In T. Clutton-Brock, ed., *Primate Ecology: Studies of Feeding and Ranging Behaviour in Lemurs, Monkeys, and Apes*, pp. 37–69. London: Academic Press.

Pollock, J. I. 1979a. Spatial distribution and ranging behaviour in lemurs. In G. A. Doyle and R. D. Martin, eds., *The Study of Prosimian Behavior*, pp. 359–409. New York: Academic Press.

Pollock, J. I. 1979b. Female dominance in *Indri indri*. *Folia Primat.* 31:143–164.

Prince, J. H. 1956. *Comparative Anatomy of the Eye*. Springfield, Ill.: Charles C. Thomas.

Purchas, Samuel. 1625. *Hakluytus Posthumus or Purchas His Pilgrimes*. 5 vols. London: Henry Fetherston.

Radinsky, L. 1968. A new approach to mammalian cranial analysis, illustrated by examples of prosimian primates. *J. Morph.* 124:167–180.

Radinsky, L. 1970. The fossil evidence of prosimian brain evolution. In C. R. Noback and W. Montagna, eds., *The Primate Brain*, pp. 209–224. New York: Appleton-Century-Crofts.

Radinsky, L. 1974. Prosimian brain morphology: Functional and phylogenetic implications. In R. D. Martin, G. A. Doyle, and A. C. Walker, eds., *Prosimian Biology*, pp. 781–798. London: Duckworth.

Raison, J.-P. and P. Vérin. 1968. Le site de subfossiles de Taolambiby (sud-ouest de Madagascar). *Ann. Univ. Tananarive* (Lettres et Sci. Humaines) 7:132–142.

Rand, A. L. 1935. On the habits of some Madagascar mammals. *J. Mammal.* 16:89–104.

Raybaud, M. 1902. Les gisements fossilifères d'Ampasambazimba. *Bull. Acad. Malgache* (anc. sér.) 1:64–66.

Reng, R. 1977. Die Placenta von *Microcebus murinus* Miller. *Z. Säugetierekunde* 42:201–214.

Richard, A. F. 1973. Social organization and ecology of *Propithecus verreauxi* Grandidier 1867. Ph.D. Thesis, University of London.

Richard, A. F. 1974a. Patterns of mating in *Propithecus* verreauxi. In R. D. Martin, G. A. Doyle, and A. C. Walker, eds., *Prosimian Biology*, pp. 49–74. London: Duckworth.

Richard, A. 1974b. Intra-specific variation in the social organization and ecology of *Propithecus verreauxi. Folia Primat.* 22:178–207.

Richard, A. 1976. Preliminary observations on the birth and development of *Propithecus verreauxi* to the age of six months. *Primates* 17:357–366.

Richard, A. 1977. The feeding behavior of *Propithecus verreauxi.* In T. Clutton-Brock, ed., *Primate Ecology: Studies of Feeding and Ranging Behaviour in Lemurs, Monkeys, and Apes*, pp. 71–96. London: Academic Press.

Richard, A. 1978a. *Behavioral Variation: Case Study of a Malagasy Lemur.* Lewisburg: Bucknell University Press.

Richard, A. 1978b. Variability in the feeding behavior of a Malagasy prosimian, *Propithecus verreauxi*: Lemuriformes. In G. G. Montgomery, ed., *The Ecology of Arboreal Folivores*, pp. 519–533. Washington, D.C.: Smithsonian Institution Press.

Richard, A. F. and R. Heimbuch. 1975. An analysis of the social behavior of three groups of *Propithecus verreauxi.* In I. Tattersall and R. W. Sussman eds., *Lemur Biology*, pp. 313–333. New York: Plenum Press.

Richard, A. F. and R. W. Sussman. 1975. Future of the Malagasy lemurs: Conservation or extinction? In I. Tattersall and R. W. Sussman, eds., *Lemur Biology*, pp. 335–350. New York: Plenum Press.

Roberts, D. and I. Davidson. 1975. The lemur scapula. In I. Tattersall and R. W. Sussman, eds., *Lemur Biology*, pp. 125–147. New York: Plenum Press.

Roberts, D. and I. Tattersall. 1974. Skull form and the mechanics of mandibular elevation in mammals. *Amer. Mus. Novitates* 2536: 1–9.

Rode, P. 1939. Catalogue des types de mammifères du Muséum National d'Histoire Naturelle. I. Ordre des Primates. B. Sous-ordre des Lémuriens. *Bull. Mus. Natl. Hist. Nat.* (Paris) (2 ser.) 11:434–449.

Rohen, J. W. 1962. Sehorgan. *Primatologia* 11:1–210.

Rohen, J. W. and A. Castenholz. 1967. Über die Zentralisation der Retina bei Primaten. *Folia Primat.* 5:92–147.

Rose, M. D. 1973. Quadrupedalism in primates. *Primates* 14: 337–357.

Rumbaugh, D. and R. Arnold. 1971. Learning: A comparative study of *Lemur* and *Cercopithecus*. *Folia Primat.* 14:154–160.

Rumbaugh, D. and T. Gill. 1973. The learning skills of great apes. *J. Hum. Evol.* 2:171–179.

Rumpler, Y. 1975. The significance of chromosomal studies in the systematics of the Malagasy lemurs. In I. Tattersall and R. W. Sussman, eds., *Lemur Biology*, pp. 25–40. New York: Plenum Press.

Rumpler, Y. and R. Albignac. 1969a. Existence d'un variabilité chromosomique intraspécifique chez certains lémuriens. *C. R. Soc. Biol.* 163:1989–1992.

Rumpler, Y. and R. Albignac. 1969b. Etude cytogénétique de quelques hybrides intraspécifiques et interspécifiques de lémuriens. *Ann. Sci. Univ. Besancon* (sér. 3) Med., 6:15–18.

Rumpler, Y. and R. Albignac. 1971. Étude cytogénétique de *Varecia variegata* et de *Lemur rubriventer*. *C. R. Soc. Biol.* 165:741–747.

Rumpler, Y. and R. Albignac. 1973a. Cytogenetic study of the endemic Malagasy lemur: *Hapalemur* I. Geoffroy, 1851. *J. Hum. Evol.* 2:267–270.

Rumpler, Y. and R. Albignac. 1973b. Cytogenetic study of the endemic Malagasy lemur subfamily Cheirogaleinae Gregory, 1915. *Amer. J. Phys. Anthrop.* 38:261–264.

Rumpler, Y. and R. Albignac. 1975. Intraspecific chromosome variability in a lemur from the North of Madagascar: *Lepilemur septentrionalis,* species nova. *Amer. J. Phys. Anthrop.* 42:425–429.

Rumpler, Y. and R. Andriamiandra. 1971. Etude histologique de glandes de marquage de la face antérieure du cou des lémuriens malgaches. *C. R. Soc. Biol.* 165:436–440.

Rumpler, Y. and J. H. Oddou. 1970. Comportement de marquage et structure histologique des glandes de marquage chez *Lemur macaco. C. R. Soc. Biol.* 164:2686–2689.

Rumpler, Y., R. Albignac, and A. Rumpler-Randriamonta. 1972. Étude cytogénétique de *Lepilemur ruficadatus. C. R. Soc. Biol.* 166:1208–1211.

Russell, R. J. 1975. Body temperatures and behavior of captive cheirogaleids. In I. Tattersall and R. W. Sussman, eds., *Lemur Biology*, pp. 193–206. New York: Plenum Press.

Russell, R. J. 1977. The behavior, ecology, and environmental physiology of a nocturnal primate, *Lepilemur mustelinus* (Strepsirhini, Lemuriformes, Lepilemuridae). Ph.D. Thesis, Duke University.

Russell, R. J. and L. McGeorge. 1977. Distribution of *Phaner* (Primates, Lemuriformes, Cheirogaleidae, Phanerinae) in southern Madagascar. *J. Biogeogr.* 4:169–170.

Saban, R. 1956. L'os temporal et ses rapports chez les lémuriens subfossiles de Madagascar. I. Type à molaires quadrituberculées, formes archaïques. *Mém. Inst. Sci. Madagascar* (sér. A) 10: 251–297.

Saban, R. 1963. Contribution à l'étude de l'os temporal des primates. *Mém. Mus. Natl. Hist. Nat.* (Paris) (n.s., A) 29:1–377.

Saban, R. 1975. Structure of the ear region in living and subfossil lemurs. In I. Tattersall and R. W. Sussman, eds., *Lemur Biology*, pp. 83–109. New York: Plenum Press.

Sarich, V. M. and J. E. Cronin. 1976. Molecular systematics of the primates. In M. Goodman and R. Tashian, eds., *Molecular Anthropology*, pp. 141–170. New York: Plenum Press.

Schaeffer, B., M. Hecht, and N. Eldredge. 1972. Phylogeny and paleontology. *Evol. Biol.* 6:31–46.

Schilling, A. 1970. L'organe de Jacobson du lémurien malgache *Microcebus murinus* (Miller, 1777). *Mém. Mus. Natl. Hist. Nat.* (Paris) (n.s., A) 61 (4):203–280.

Schilling, A. 1974. A study of marking behaviour in *Lemur catta*. In R. D. Martin, G. A. Doyle, and A. C. Walker, eds., *Prosimian Biology* pp. 347–362. London: Duckworth.

Schilling, A. 1979. Olfactory communication in prosimians. In G. A. Doyle and R. D. Martin, eds. *The Study of Prosimian Behavior*, pp. 461–542. New York: Academic Press.

Schlegel, A. 1866. Contributions à la faune de Madagascar et des îles avoisinantes, d'après les découvertes et observations de Mm François Pollen et D. C. Van Dam. *Ned. Tijdschr. Dierk.* 3: 73–89.

Schultz, A. H. 1930. The skeleton of the trunk and limbs of higher primates. *Hum. Biol.* 2:303–435.

Schwartz, J. H. 1974. Dental development and eruption in the prosimians and its bearing on their evolution. Ph.D. dissertation, Columbia University.

Schwartz, J. H. 1975. Development and eruption of the premolar region of prosimians and its bearing on their evolution. In I. Tattersall and R. Sussman, eds., *Lemur Biology,* pp. 41–63. New York: Plenum Press.

Schwartz, J. and I. Tattersall. 1979. The phylogenetic relationships of Adapidae (Primates, Lemuriformes). *Anthrop Papers Amer. Mus. Nat. Hist.* 55:271–283.

Schwartz, J., I. Tattersall, and N. Eldredge. 1978. Phylogeny and classification of the primates revisited. *Yrbk. Phys. Anthrop.* 21:95–133.

Schwarz, Ernst. 1931. A revision of the genera and species of Madagascar Lemuridae. *Proc. Zool. Soc. Lond.* pp. 399–428.

Scrutton, R. A. 1973. Structure and evolution of the sea floor south of South Africa. *Earth Planet. Sci. Lett.* 19:250–256.

Scrutton, R. A. 1978. Davie fracture zone and the movement of Madagascar. *Earth Planet. Sci. Lett.* 39:84–88.

Ségoufin, J. 1978. Anomalies magnétiques mésozoïques dans le bassin de Mozambique. *C. R. Acad. Sci.* (Paris) 287:109–112.

Sera, G. 1935. I caratteri morfologici di *Palaeopropithecus* e l'addatamento aquatico primitivo dei mammiferi e dei Primati in particulare. *Arch. Ital. Anat. Embriol.,* pp. 229–270.

Sheine, W. 1979. The effect of variations in molar morphology on masticatory effectiveness and digestion of cellulose in prosimian primates. Ph.D. Thesis, Duke University.

Simons, E. L. 1962. A new Eocene primate genus, *Cantius,* and a revision of some allied European lemuroids. *Bull. Brit. Mus. (Nat. Hist.)* 7:1–36.

Simons, E. L. 1972. Primate evolution. New York: Macmillan.

Simpson, G. G. 1945. The principles of classification and a classification of mammals. *Bull. Amer. Mus. Nat. Hist.* 85:1–350.

Simpson, E. S. W., R. Schlich et al. 1974. *Initial reports of the Deep Sea Drilling Project,* vol. 25. Washington, D.C.: Govt. Printing Office.

Smith, A. G. and A. Hallam. 1970. The fit of the southern continents. *Nature* 225:139–144.

Smith, A. G. and J. C. Briden. 1977. *Mesozoic and Cenozoic Palaeocontinental Maps.* Cambridge: Cambridge University Press.

Smith, G. E. 1903. On the morphology of the brain in the Mammalia, with special reference to that of lemurs, recent and extinct. *Trans. Linn. Soc. Lond.* Zool. ser., 8:319–432.

Smith, G. E. 1908. On the form of the brain in the extinct lemurs of Madagascar, with some remarks on the affinities of the Indrisinae. *Trans. Zool. Soc. Lond.* 18:163–177.

Sonnerat, M. 1782. *Voyage aux Indes orientales et à la Chine.* Paris: Froulé.

Spuhler, O. 1935. Genitalzyklus und Spermiogenese der Mausmaki. *Z. Zellforsch. u. mikroskop. Anat.* 23:442–463.

Standing, H. 1903. Rapport sur des ossements sub-fossiles provenant d'Ampasambazimba. *Bull. Acad. Malgache* (anc. sér.) 2: 227–235.

Standing, H. 1905. Rapport sur des ossements sub-fossiles provenant d'Ampasambazimba. *Bull. Acad. Malgache* (anc. sér.) 4: 305–310.

Standing, H. 1908a. On recently discovered subfossil primates from Madagascar. *Trans. Zool. Soc. Lond.* 18:69–162.

Standing, H. 1908b. Subfossiles provenant des fouilles d'Ampasambazimba. *Bull. Acad. Malgache* (anc. sér.) 6:9–11.

Standing, H. 1910. Note sur les ossements subfossiles provenant des fouilles d'Ampasambazimba. *Bull. Acad. Malgache* (anc. sér.) 7(for 1909):61–64.

Starmühlner, F. 1960. Beobachtungen am Mausmaki (*Microcebus murinus*). *Natur u. Volk.* 90:194–204.

Stephan, H. and O. J. Andy. 1969. Quantitative comparative neuroanatomy of primates: an attempt at a phylogenetic interpretation. *Ann. N.Y. Acad. Sci.* 167(1):370–387.

Stephan, H., R. Bauchot, and O. J. Andy. 1970. Data on size of the brain and various brain parts in insectivores and primates. In C. R. Noback and W. Montagna, eds., *The Primate Brain,* pp. 289–297. New York: Meredith.

Strauss, R. W. 1978. The ovoimplantation of *Microcebus murinus* Miller (Primates, Lemuroidea, Strepsirhini). *Amer. J. Anat.* 152:99–109.

Sussman, R. W. 1972. An ecological study of two Madagascan primates: *Lemur fulvus rufus* Audebert and *Lemur catta* Linnaeus. Ph.D. Dissertation, Duke University.

Sussman, R. W. 1974. Ecological distinctions in sympatric species of *Lemur.* In R. D. Martin, G. A. Doyle, and A. C. Walker, eds.,

Prosimian Biology, pp. 75–108. London: Duckworth.

Sussman, R. W. 1975. A preliminary study of the behavior and ecology of *Lemur fulvus rufus* Audebert, 1800. In I. Tattersall and R. W. Sussman, eds., *Lemur Biology,* pp. 237–258. New York: Plenum Press.

Sussman R. W. 1977a. Distribution of the Malagasy lemurs. Part 2: *Lemur catta* and *Lemur fulvus* in southern and western Madagascar. *Ann. N.Y. Acad. Sci.* 293:170–184.

Sussman, R. W. 1977b. Socialization, social structure, and ecology of two sympatric species of *Lemur.* In S. Chevalier-Skolnikoff and F. Poirier, eds., *Primate Bio-Social Development,* pp. 515–528. New York: Garland.

Sussman, R. W. 1977c. Feeding behaviour of *Lemur catta* and *Lemur fulvus.* In T. Clutton-Brock, ed., *Primate Ecology: Studies of Feeding and Ranging Behaviour in Lemurs, Monkeys, And Apes,* pp. 1–36. London: Academic Press.

Sussman, R. W. and A. Richard. 1974. The role of aggression among diurnal prosimians. In R. Holloway, ed., *Primate Aggression, Territoriality, and Xenophobia,* pp. 49–76. New York: Academic Press.

Sussman, R. W. and I. Tattersall. 1976. Cycles of activity, group composition, and diet of *Lemur mongoz mongoz* Linnaeus, 1766 in Madagascar. *Folia Primat.* 26:270–283.

Szalay, F. 1972. Cranial morphology of the early Tertiary *Phenacolemur* and its bearing on primate phylogeny. *Amer. J. Phys. Anthrop.* 36:59–76.

Szalay, F. S. and C. C. Katz. 1973. Phylogeny of lemurs, galagos, and lorises. *Folia Primat.* 19:88–103.

Tattersall, I. 1971. Revision of the subfossil Indriinae. *Folia Primat.* 16:257–269.

Tattersall, I. 1972. The functional significance of airorhynchy in *Megaladapis. Folia Primat.* 18:20–26.

Tattersall, I. 1973a. A note on the age of the subfossil site of Ampasambazimba, Miarinarivo Province, Malagasy Republic. *Amer. Mus. Novitates* 2520:1–6.

Tattersall, I. 1973b. Cranial anatomy of the Archaeolemurinae (Lemuroidea, Primates). *Anthrop. Papers Amer. Mus. Nat. Hist.* 52: 1–110.

Tattersall, I. 1973c. Subfossil lemuroids and the "adaptive radiation" of the Malagasy lemurs. *Trans. N.Y. Acad. Sci.* 35:314–324.

Tattersall, I. 1974. Facial structure and mandibular mechanics in

Archaeolemur. In R. D. Martin, G. A. Doyle, and A. C. Walker, eds., *Prosimian Biology,* pp. 563–577. London: Duckworth.

Tattersall, I. 1975. Notes on the cranial anatomy of the subfossil Malagasy lemurs. In I. Tattersall and R. W. Sussman, eds., *Lemur Biology,* pp. 111–124. New York: Plenum Press.

Tattersall, I. 1976a. Note sur la distribution et sur la situation actuelle des lémuriens des Comores. *Mammalia* 40:519–521.

Tattersall, I. 1976b. Notes on the status of *Lemur macaco* and *Lemur fulvus* (Primates, Lemuriformes). *Anthrop. Papers Amer. Mus. Nat. Hist.* 53:255–261.

Tattersall, I. 1976c. Group structure and activity rhythm in *Lemur mongoz* (Primates, Lemuriformes) on Anjouan and Mohéli islands, Comoro Archipelago. *Anthrop. Papers Amer. Mus. Nat. Hist.* 53:367–380.

Tattersall, I. 1977a. The lemurs of the Comoro Islands. *Oryx* 13:445–448.

Tattersall, I. 1977b. Distribution of the Malagasy lemurs. Part 1: The lemurs of northern Madagascar. *Ann. N.Y. Acad. Sci.* 293:160–169.

Tattersall, I. 1977c. Ecology and behavior of *Lemur fulvus mayottensis* (Primates, Lemuriformes). *Anthrop. Papers Amer. Mus. Nat. Hist.* 54:421–482.

Tattersall, I. 1978. Behavioural variation in *Lemur mongoz* (=*L. m. mongoz*). In D. J. Chivers and K. A. Joysey, eds., *Recent Advances in Primatology,* 3:127–132. London: Academic Press.

Tattersall, I. 1979. Patterns of activity in the Mayotte lemur, *Lemur fulvus mayottensis. J. Mammal.* 60:314–323.

Tattersall, I., and J. H. Schwartz. 1974. Craniodental morphology and the systematics of the Malagasy lemurs (Primates, Prosimii). *Anthrop. Papers Amer. Mus. Nat. Hist.* 52:139–192.

Tattersall, I. and J. H. Schwartz. 1975. Relationships among the Malagasy lemurs: The craniodental evidence. In W. P. Luckett and F. S. Szalay, eds., *Phylogeny of the Primates: A Multidisciplinary Approach,* pp. 299–312. New York: Plenum Press.

Tattersall, I. and R. W. Sussman. 1975. Observations on the ecology and behavior of the mongoose lemur *Lemur mongoz mongoz* Linnaeus (Primates, Lemuriformes) at Ampijoroa, Madagascar. *Anthrop. Papers Amer. Mus. Nat. Hist.* 52:193–216.

Tattersall, I. and R. W. Sussman, eds. 1975. *Lemur Biology.* New York: Plenum Press.

Uzzell, T. and D. Pilbeam. 1971. Phyletic divergence dates of hominoid primates: a comparison of fossil and molecular data. *Evolution* 25:615–635.

Vail, P. R. 1977. Seismic recognition of depositional facies on slopes and rises. *Am. Assoc. Petroleum Geologists Short Course Notes* 5: F1–F9.

Van Couvering, J. A. and J. A. H. Van Couvering. 1975. African isolation and the Tethys seaway. In J. Seneš, ed., *Proc. of the 6th Congress, R. C. M. N. S.* (Bratislava), pp. 363–370.

Van Horn, R. N. 1972. Structural adaptations to climbing in the gibbon hand. *Amer. Anthrop.* 74:326–334.

Van Horn, R. N. 1975. Primate breeding season: photoperiodic regulation in captive *Lemur catta*. *Folia Primat.* 24:203–220.

Van Horn, R. N. and G. G. Eaton. 1979. Reproductive physiology and behavior in prosimians. In G. A. Doyle, ed., *The Study of Prosimian Behavior*, pp. 79–122. New York: Academic Press.

Van Valen, L. 1965. Treeshrews, primates, and fossils. *Evolution* 19:137–151.

Vick, L. G. and J. M. Conley. 1976. An ethogram for *Lemur fulvus*. *Primates* 17:125–144.

Walker, A. C. 1967a. Locomotor adaptation in recent and subfossil Madagascan lemurs. Ph.D. Thesis, University of London.

Walker, A. C. 1967b. Patterns of extinction among the subfossil Madagascan lemuroids. In P. S. Martin and H. E. Wright, Jr., eds., *Pleistocene Extinctions*, pp. 425–432. New Haven: Yale University Press.

Walker, A. C. 1972. The dissemination and segregation of early primates in relation to continental configuration. In W. W. Bishop and J. A. Miller, eds., *Calibration of Hominoid Evolution*, pp. 195–218. Edinburgh: Scottish Academic Press.

Walker, A. C. 1974. Locomotor adaptations in past and present prosimian primates. In F. A. Jenkins, ed., *Primate Locomotion*, pp. 349–381. New York: Academic Press.

Walker, A. C. 1979. Prosimian locomotor behavior. In G. A. Doyle and R. D. Martin, eds., *The Study of Prosimian Behavior*, pp. 543–565. New York: Academic Press.

Ward, S. C. and R. W. Sussman. 1979. Correlates between locomotor anatomy and behavior in two sympatric species of *Lemur*. *Amer. J. Phys. Anthrop.* 50:575–590.

Webb, C. S. 1953. *A Wanderer in the Wind: The Odyssey of an Animal Collector*. London: Hutchinson.

Wilkerson, B. and D. Rumbaugh. 1979. Learning and intelligence in prosimians. In G. A. Doyle and R. D. Martin, eds., *The Study of Prosimian Behavior*, pp. 207–246. New York: Academic Press.

Wolin, L. R. and L. C. Massopust. 1970. Morphology of the primate retina. In C. R. Noback and W. Montagna eds., *The Primate Brain*, pp. 1–27. New York: Meredith.

Wright, J. B. and P. McCurry. 1970. Letter to the Editor on : The significance of sandstone inclusions in lavas of the Comores archipelago. *Earth Planet. Sci. Lett.* 8:267.

Zapfe, H. 1963. Lebensbild von *Megaladapis edwardsi* (Grandidier): ein Rekonstruktionversuch. *Folia Primat.* 1:178–187.

Zilles, K. and A. Schleicher. 1980. Similarities and differences in the cortical areal patterns of *Galago demidovii* (E. Geoffroy, 1796: Lorisidae, Primates) and *Microcebus murinus* (E. Geoffroy, 1828; Lemuridae, Primates). *Folia Primat.* 33:161–171.

Zilles, K., G. Rehkämper, and A. Schleicher. 1979. A quantitative approach to cytoarchitectonics. V. The areal pattern of the cortex of *Microcebus murinus* (E. Geoffroy 1828), Lemuridae, Primates). *Anat. Embryol.* 157:269–289.

INDEX

Note: Since references to particular topics are in most cases closely grouped in the text, they are generally listed separately, rather than under the taxa to which they apply.

no damage noted 5/20/91-NR